2ND EDITION

DESIGNING AND CONDUCTING
MIXED METHODS
RESEARCH

This book is dedicated to all of my students and audience members who have participated in my classes and in my workshops on mixed methods. Thanks for your advice.

—John

This book is dedicated to Mark for all of his support, encouragement, friendship, and love. I thank him to the moon and back.

—Vicki

2ND EDITION

DESIGNING AND CONDUCTING
MIXED METHODS
RESEARCH

JOHN W. CRESWELL
University of Nebraska-Lincoln

VICKI L. PLANO CLARK
University of Nebraska-Lincoln

Los Angeles | London | New Delhi
Singapore | Washington DC

For information:

SAGE Publications, Inc.
2455 Teller Road
Thousand Oaks, California 91320
E-mail: order@sagepub.com

SAGE Publications Ltd.
1 Oliver's Yard
55 City Road
London EC1Y 1SP
United Kingdom

SAGE Publications India Pvt. Ltd.
B 1/I 1 Mohan Cooperative Industrial Area
Mathura Road, New Delhi 110 044
India

SAGE Publications Asia-Pacific Pte. Ltd.
33 Pekin Street #02-01
Far East Square
Singapore 048763

Printed in the United States of America

Library of Congress Cataloging-in-Publication Data

Creswell, John W.
Designing and conducting mixed methods research/John W. Creswell, Vicki L. Plano Clark—2nd ed.
   p. cm.
Includes bibliographical references and index.
ISBN 978-1-4129-7517-9 (pbk.)
  1. Social sciences—Research–Methodology. 2. Research—Evaluation. I. Plano Clark, Vicki L. II. Title.

H62.C6962 2011
001.4'2—dc22   2010010924
This book is printed on acid-free paper.

10 11 12 13 14 10 9 8 7 6 5 4 3 2 1

Acquisitions Editor:	Vicki Knight
Associate Editor:	Lauren Habib
Editorial Assistant:	Ashley Dodd
Managing Editor:	Claudia A. Hoffman
Production Editor:	Brittany Bauhaus
Copy Editor:	Megan Markanich
Typesetter:	C&M Digitals (P) Ltd.
Proofreader:	Jenifer Kooiman
Indexer:	Molly Hall
Cover Designer:	Glenn Vogel
Marketing Manager:	Stephanie Adams

BRIEF CONTENTS

DETAILED CONTENTS

LIST OF FIGURES

LIST OF TABLES

PREFACE

The basic idea of this book is to provide an introduction to the design and conduct of mixed methods research. In the past 5 to 10 years, we have seen tremendous interest in this approach to research. Although mixed methods has had its roots over the last 20 years in several disciplines and fields of study, the interest in it has expanded rapidly to many social and human sciences fields, many arenas for research, and many countries. This is in distinct contrast to the time period when we wrote the first edition. During that time, researchers were mostly curious about this developing approach called "mixed methods." Today, from our workshops, presentations, and classes, we now know that people no longer wonder what this approach is and whether it is a legitimate model of inquiry. Their interests now have gravitated toward the procedures of research, actually how to conduct a mixed methods study. To this end, we have maintained our original premise that those reading about mixed methods not only need to know the steps in the process of designing and conducting a study but they also are curious about the actual procedures involved and the many new techniques and strategies that have unfolded in the mixed method field.

This book is an introduction as well as a detailed assessment about how to conduct a mixed methods study. We fold into our discussion many examples of recently published mixed methods empirical articles as well as methodological discussions. We attempt to highlight the most important steps through the ample use of bullet points, and we introduce the reader to some of the latest writings in the field. Since the inception of the *Journal of Mixed Methods Research* (JMMR), which we helped to cofound and coedit, we have reviewed hundreds of manuscripts for publication from diverse disciplines, different parts of the world, and from varied perspectives about this form of inquiry. From these articles and from our personal experiences in mixed methods research teams, classes, and presentations, we draw together a detailed rendering of how to design and conduct a mixed methods study. We hope that the beginning mixed methods researcher will find useful techniques for designing their own study and that the experienced researcher will see useful summaries of the latest thinking about mixed methods.

● AUDIENCE FOR THE BOOK

The primary audience for this book is individuals learning about mixed methods research for the first time. Graduate students who have some experience with both qualitative and quantitative research will find this book useful for designing their first mixed methods project. Writers in the field of mixed methods will hopefully see this book as including state-of-the-art ideas. Policy makers and practitioners will find this book a useful introduction to mixed methods as they review published studies or establish their own mixed methods projects. With the discipline expansion of mixed methods, this book should be applicable across many social and human science fields. We have attempted to incorporate examples from fields as diverse as sociology, psychology, education, management, marketing, social work, family studies, communication studies, and leadership. We have also brought in examples from the health sciences, such as nursing, family medicine, and mental health, to name a few. Finally, we see this book as core reading in a mixed methods research course—a type of course that is increasingly being found among the lists of research courses on college and university campuses. For new fields being introduced to mixed methods, we will use many of the tables and figures in this book in our future workshops both here in the United States and abroad.

● BOOK FEATURES

We have maintained many of the book features found in the first edition. The general layout of the book follows the process of conducting a study, from the initial assessment as to whether mixed methods is the best approach to study a research problem; to the philosophical assumptions and theoretical stances that guide research; and on to developing an introduction, collecting and analyzing data, and writing the proposal and final report for a study. To augment this process approach, we highlight six popular designs in mixed methods research and provide examples of good published journal article studies that illustrate each of the designs. At the end of each chapter, we provide a summary of the chapter's content, as well as suggest practical activities to make concrete the major points of the chapter. One activity in particular threads throughout the book: We ask the reader to incorporate the ideas from the chapter into the development of a mixed methods study that is being actively designed. Each step in this process of design unfolds with each chapter. We also do not favor either quantitative or qualitative research and switch the emphasis from one chapter to another so as to balance the importance of both approaches to research. This balance is central to understanding mixed methods research. At the end of each chapter, we provide further readings so that the ideas presented in the chapter might be expanded. We have attempted to define key terms used throughout the book and provide a glossary of these terms at the end to help understand the unique language that is emerging in mixed methods. We have also maintained and expanded in this second edition the references to Web sites and resources that readers should find helpful. A new Instructor's Resource, which includes sample activities, links to example articles and Web sites, and PowerPoint slides, is also being developed to accompany the second edition.

NEW FEATURES ADDED TO THE SECOND EDITION ●

In this second edition, we have considerably expanded the information and ideas that were presented in our first edition. Frankly speaking, we are overwhelmed by the new knowledge that has emerged in recent years about mixed methods, and we realize that our discussions here have been selective but representative of current discussions in the field of mixed methods.

We can summarize, however, the new material that is found in this second edition:

• In examining how we introduced the topic of mixed methods in the first edition, we felt that this second edition should start simply with the key questions that we often receive about mixed methods: what it is, why it is used, and what are its advantages and challenges. Today we are in a better position to answer these questions than when we drafted the first edition of this book. Chapter 1 reflects our answers to these questions.

• We also know today much more about the history of mixed methods and its roots so that scholars can convey to others what the field is all about. We also have had an ongoing development and assessment about the philosophy that underlies mixed methods. While the general outline of this philosophy was apparent in our first edition, we are able now to expand on it considerably. We also know much more today about the theoretical orientations—whether social science or emancipatory—that mixed methods researchers use, a topic not addressed in the first edition. And perhaps more importantly, we know more about "how" they are using these theoretical lenses. Chapter 2 in this edition expands on these topics.

● One comment that readers and workshop participants have remarked to us since the publication of the first edition is that the designs we advanced were not inclusive enough for the many types of designs being used today. Accordingly, in Chapter 3, we have added to the four designs discussed in the first edition two additional designs—the transformative and the multiphase designs—that are appearing with greater frequency in the field of mixed methods. We have also expanded how we talk about all of the six designs in this book, such as describing in more detail their basic purpose, the procedures, the philosophy, and the advantages and challenges. We have added flowcharts for each design that detail the steps involved in conducting each design.

● We have updated the examples of mixed methods studies throughout the book. We cite many articles published in the JMMR, and in Chapter 4, we include several newly published studies that illustrate each of our six major mixed methods designs. As a guide to understanding these studies, we provide the latest notation system available and diagrams of the procedures cited in the illustrative studies. These diagrams incorporate new ways of conceptualizing the designs and features that have grown in importance, such as the "point of interface" of the qualitative and quantitative approaches in studies.

● We now have many more examples of titles for mixed methods studies, better scripts for designing purpose statements, and a clearer understanding of the various forms of mixed methods research questions. Chapter 5 includes a more detailed discussion of each of these topics than what was found in the first edition.

● For this second edition, we rewrote our procedures for data collection in mixed methods, moving away from a more generic discussion to a detailed approach to data collection decisions necessary for each of the six major designs. Now a reader should have a clearer sense of some of the types of decisions made about collecting data and how they might be resolved using practices that we feel work well. Chapter 6 reflects this new discussion about collecting mixed methods data.

● Chapter 7 on data analysis has been reworked as well to focus on the specific procedures of data analysis for each of the major designs. We include new topics in this chapter that reflect the growing understanding of mixed methods data analysis: the use of joint displays and the numerous examples that now populate the literature, the emerging discussion about validity, and the newest thinking about the use of software in the process of mixed methods analysis, including the use of computers to generate joint displays for analysis across the various types of mixed methods designs.

● As with the first edition, we still feel that having an understanding of the structure of the mixed methods report—whether it is a dissertation or thesis proposal, a dissertation, a journal article, or a proposal for extramural funding—is an important conceptual step in the solid design and conduct of research. In this edition, we have added into our structural outlines an example of a mixed method dissertation outline so that we now have a more complete set of guidelines. We have also updated our discussion about evaluating mixed methods research to include some of the latest thinking on this topic. These ideas are found in Chapter 8.

● We conclude this book, in Chapter 9, by tying together many ideas flowing throughout the book and advancing recommendations for the practice of designing and conducting a study. Our recommendations are based on an assessment of the current state of the field and our practical experiences working on and advising many mixed methods projects.

ACKNOWLEDGMENTS ●

Our work and this book have benefited greatly from the contributions of many. We begin by thanking our acquisitions editor at SAGE Publications, Vicki Knight, for her encouragement, coordination, and support throughout the project. We also thank the entire staff at SAGE for their encouragement of the field of mixed methods research. As the reader can see in our many references, both of us have collaborated extensively with staff and colleagues working with our Office of Qualitative and Mixed Methods Research (OQMMR) at the University of Nebraska–Lincoln. We want to highlight the importance of collaboration with Ron Shope, Manijeh Badiee, Amanda Garrett, Sherry Wang, Dr. Michael Fetters, Nataliya Ivankova, and the many individuals whom we have collaborated with in family medicine at the University of Michigan; at the Health Services Research Center of the

Department of Veterans Affairs, Ann Arbor, Michigan; and at the University of Nebraska Medical Center. We are also indebted to the many workshop participants over the years who have provided useful ideas and questions about mixed methods. These individuals are located in many fields, in many parts of the United States, the United Kingdom, South Africa, Australia, and Canada, and other countries across the globe.

This is an exciting time in the evolution of the field that has become mixed methods research. We hope that this book is a useful tool for researchers to use in learning about this approach to research and in conducting their own mixed methods studies.

ABOUT THE AUTHORS

John W. Creswell, PhD, is a professor of educational psychology and teaches courses and writes about mixed methods research, qualitative methodology, and general research design. He has been at the University of Nebraska–Lincoln (UNL) for over 30 years and has authored 12 books and multiple editions of these books, many of which focus on alternative types of research designs, comparisons of different qualitative methodologies, and the nature of and use of mixed methods research. His books are read by audiences in the social and health sciences around the world. In addition, he founded and codirected the Office of Qualitative and Mixed Methods Research (OQMMR) at UNL, which provides support for scholars incorporating qualitative and mixed methods research into projects for extramural funding. He served as the founding coeditor for the SAGE journal *Journal of Mixed Methods Research*, and he has been an adjunct professor of family medicine at the University of Michigan and assisted investigators in the health sciences and education on the research methodology for National Institutes of Health (NIH) and National Science Foundation (NSF) projects. He was a Senior Fulbright Scholar to South Africa in 2008 and lectured at five universities to faculty in education and the health sciences. His hobbies include exercising, playing the piano, and writing short stories.

Vicki L. Plano Clark, PhD, is director of the Office of Qualitative and Mixed Methods Research (OQMMR) and is a Research Assistant Professor in the Quantitative, Qualitative, and Psychometric Methods program at the University of Nebraska–Lincoln (UNL). She teaches research methods courses, including foundations of educational research, qualitative research, and mixed methods research. She has authored and coauthored over 30 articles, chapters, and student manuals, including the books *The Mixed Methods Reader* (SAGE, 2008) and *Understanding Research: A Consumer's Guide* (Pearson, 2010) with John W. Creswell, and currently serves as an associate editor with the *Journal of Mixed Methods Research*. Her methodological writings focus on the procedural issues that arise when implementing different mixed methods designs and how mixed methods research is adopted and adapted within disciplines. Through her work in the research service center, she engages in research in the areas of education, family research, counseling psychology, family medicine, and nursing. Prior to focusing on applied research methods, she served 12 years as laboratory manager in UNL's Department of Physics and Astronomy, working with the Research in Physics Education Group to develop and evaluate innovative curricular materials to help students understand introductory physics concepts. In her spare time, she is currently working with her husband, Mark, to develop their own understanding of quilt making, the game of golf, and home remodeling.

CHAPTER 1

THE NATURE OF MIXED
METHODS RESEARCH

What is it about the nature of mixed methods that draws researchers to its use? Its popularity can be easily documented through journal articles, conference proceedings, books, and the formation of special interest groups (Creswell, in press-b; Plano Clark, 2010). It has been called the "third methodological movement" following the developments of first quantitative and then qualitative research (Tashakkori & Teddlie, 2003a, p. 5), the "third research paradigm" (Johnson & Onwuegbuzie, 2004, p. 15), and "a new star in the social science sky" (Mayring, 2007, p. 1). Why does it merit such superlatives? One answer is that it is an intuitive way of doing research that is constantly being displayed through our everyday lives.

Consider for a moment *An Inconvenient Truth*, the award-winning documentary on global warming featuring the former U.S. vice president and Nobel Prize winner Al Gore (http://www.climatecrisis.net/an-inconvenient-truth.php). In the documentary, Gore narrated both the statistical trends and the stories of his personal journey related to the changing climate and global warming. This documentary brings together both quantitative and qualitative data to tell the story. Also, listen closely to CNN's broadcast reports about hurricanes or about the votes cast in elections. The trends are again supported by the individual stories. Or listen to commentators at sporting events. There is often a play-by-play commentator who describes the somewhat linear unfolding of the game (a quantitative perspective) and then the additional commentary by the "color" announcer who tells us about the individual stories and highlights of the personnel on the playing field. Again, both quantitative and qualitative data come together in these broadcasts.

In these instances, we see mixed methods thinking in ways that Greene (2007) called the "multiple ways of seeing and hearing" (p. 20). Multiple ways are visible in everyday life, and mixed methods becomes a natural outlet for research. But other factors also contribute to this interest in mixed methods. Researchers recognize it as an accessible approach to inquiry. They have research questions (or problems) that can best be answered using mixed methods, and they see the value of using it (as well as the challenges it poses).

Understanding the nature of mixed methods research is an important first step to using it in research. This chapter reviews several preliminary considerations necessary before a researcher designs a mixed methods study. This chapter addresses the following considerations:

- Understanding what mixed methods research means
- Viewing examples of mixed methods studies
- Recognizing what types of research problems merit a mixed methods study
- Knowing the advantages of using mixed methods
- Realizing the challenges of using mixed methods

● DEFINING MIXED METHODS RESEARCH

Several definitions for mixed methods have emerged over the years that incorporate various elements of methods, research processes, philosophy, and research design. These different stances are summarized in Table 1.1.

An early definition of mixed methods came from writers in the field of evaluation. Greene, Caracelli, and Graham (1989) emphasized the mixing of methods and the disentanglement of methods and philosophy (i.e., paradigms) when they said,

> In this study, we defined mixed-method designs as those that include at least one quantitative method (designed to collect numbers) and one qualitative method (designed to collect words), where neither type of method is inherently linked to any particular inquiry paradigm. (p. 256)

Ten years later, the definition shifted from mixing two methods to mixing in all phases of the research process—a methodological orientation (Tashakkori & Teddlie, 1998). Included within this orientation would be mixing philosophical (i.e., worldview) positions, inferences, and the interpretations of

Table 1.1	Authors and the Focus or Orientation of Their Definition of Mixed Methods

Author(s) and Year	Focus of the Definition
Greene, Caracelli, and Graham (1989)	Methods Philosophy
Tashakkori and Teddlie (1998)	Methodology
Johnson, Onwuegbuzie, and Turner (2007)	Qualitative and quantitative research Purpose
Journal of Mixed Methods Research (JMMR) (call for submissions)	Qualitative and quantitative research Methods
Greene (2007)	Multiple ways of seeing, hearing, and making sense of the social world
Creswell and Plano Clark (2007)	Methods Philosophy
Core characteristics (presented and used in this book)	Methods Philosophy Research design

results. Thus, Tashakkori and Teddlie (1998) defined mixed methods as the combination of "qualitative and quantitative approaches in the methodology of a study" (p. ix). These authors reinforced this methodological orientation in their preface to the *SAGE Handbook of Mixed Methods in Social & Behavioral Research* by writing "mixed methods research has evolved to the point where it is a separate methodological orientation with its own worldview, vocabulary, and techniques" (Tashakkori & Teddlie, 2003a, p. x).

In a highly cited *Journal of Mixed Methods Research* (JMMR) article, Johnson, Onwuegbuzie, and Turner (2007) sought a consensus about a definition by suggesting a composite understanding based on 19 different definitions provided by 21 highly published mixed methods researchers. The authors commented about the definitions, citing the variations in them, from what was being mixed (e.g., methods, methodologies, or types of research), the place in the research process in which mixing occurred (e.g., data collection, data analysis), the scope of the mixing (e.g., from data to worldviews), the purpose or rationale for mixing (e.g., breadth, corroboration), and the elements driving the research (e.g., bottom-up, top-down, a core component). Incorporating these diverse perspectives, Johnson et al. (2007) ended with their composite definition:

> Mixed methods research is the type of research in which a researcher or team of researchers combines elements of qualitative and quantitative research approaches (e.g., use of qualitative and quantitative viewpoints, data collection, analysis, inference techniques) for the purposes of breadth and depth of understanding and corroboration. (p. 123)

In this definition, the authors did not view mixed methods simply as methods but more as a methodology that spanned viewpoints to inferences and that included the combination of qualitative and quantitative research. They incorporated diverse viewpoints but did not specifically mention paradigms (as in the Greene et al., 1989 definition). Their purposes for mixed methods—breadth and depth of understanding and corroboration—meant that they related the definition of mixed methods to a rationale for conducting it. Most importantly, perhaps, they suggested that there is a common definition that should be used.

When the call for paper submissions to the JMMR was issued for our first issue, we, as editors, felt that a general definition of mixed methods should be provided. Our approach incorporated both a general qualitative and quantitative research orientation as well as a methods orientation. Our intent was also to cast our definition within accepted approaches to mixed methods, to encourage submissions as broad as possible, and to "keep the discussion open about the definition of mixed methods" (Tashakkori & Creswell, 2007b, p. 3). Hence, the definition announced in the first issue of the journal was

> mixed methods research is defined as research in which the investigator collects and analyzes data, integrates the findings, and draws inferences using both qualitative and quantitative approaches or methods in a single study or a program of inquiry. (Tashakkori & Creswell, 2007b, p. 4)

Then, Greene (2007) provided a definition of mixed methods that conceptualized this form of inquiry differently as a way of looking at the social world

> . . . that actively invites us to participate in dialogue about multiple ways of seeing and hearing, multiple ways of making sense of the social world, and multiple standpoints on what is important and to be valued and cherished. (p. 20)

Defining mixed methods as "multiple ways of seeing" opens up broad applications beyond using it as only a research method. It can be used, for example, as an approach to think about designing documentaries (Creswell & McCoy, in press) or as a means for "seeing" participatory approaches to HIV-infected populations in the Eastern Cape of South Africa (Olivier, de Lange, Creswell, & Wood, 2010).

Also in 2007, in the first edition of this book, we provided a definition that had both a methods and a philosophical orientation. We said,

> Mixed methods research is a research design with philosophical assumptions as well as methods of inquiry. As a methodology, it involves philosophical assumptions that guide the direction of the collection and analysis and the mixture of qualitative and quantitative approaches in many phases of the research process. As a method, it focuses on collecting, analyzing, and mixing both quantitative and qualitative data in a single study or series of studies. Its central premise is that the use of quantitative and qualitative approaches, in combination, provides a better understanding of research problems than either approach alone. (Creswell & Plano Clark, 2007, p. 5)

This definition was patterned on describing an approach using multiple meanings, such as found in Stake's (1995) definition of a case study in which he talked about case study research as stemming from several distinct ideas.

At present, we feel that a definition for mixed methods should incorporate many diverse viewpoints. In this spirit, we rely on a **definition of core characteristics of mixed methods research**. It is a definition that we suggest in our workshops and in our presentations on mixed methods research. It combines methods, a philosophy, and a research design orientation. It also highlights the key components that go into designing and conducting a mixed methods study; thus, it will be the one emphasized in this book. In mixed methods, the researcher

- collects and analyzes persuasively and rigorously both qualitative and quantitative data (based on research questions);
- mixes (or integrates or links) the two forms of data concurrently by combining them (or merging them), sequentially by having one build on the other, or embedding one within the other;
- gives priority to one or to both forms of data (in terms of what the research emphasizes);
- uses these procedures in a single study or in multiple phases of a program of study;
- frames these procedures within philosophical worldviews and theoretical lenses; and
- combines the procedures into specific research designs that direct the plan for conducting the study.

[handwritten margin note: Definition is both descriptive and methodological]

These core characteristics, we believe, adequately describe mixed methods research. They evolved from many years of reviewing mixed methods articles and determining how researchers use both quantitative and qualitative methods in their studies.

● EXAMPLES OF MIXED METHODS STUDIES

One way to better understand the nature of mixed methods research beyond a definition is to examine published studies in journal articles. Although philosophical assumptions often lie in the background of published mixed methods studies, the core characteristics of our definition can be seen in the following examples:

- A researcher collects data on quantitative instruments and on qualitative data reports based on focus groups to see if the two types of data show similar results but from different perspectives (see the study of developing a health promotion perspective for older driver safety in the occupational science area by Classen et al., 2007).

- A researcher collects data using quantitative experimental procedures and follows up with interviews with a few individuals who participated in the experiment to help explain their scores on the experimental outcomes (see the study of college students' copy-and-paste note taking by Igo, Kiewra, & Bruning, 2008).

- A researcher explores how individuals describe a topic by starting with interviews, analyzing the information, and using the findings to develop a survey instrument. This instrument, in turn, is then administered to a sample from a population to see if the qualitative findings can be generalized to a population (see the study of lifestyle behaviors of Japanese college women by Tashiro, 2002; also see the psychological study of the tendency to perceive the self as significant to others in young adults' romantic relationships by Mak & Marshall, 2004).

- A researcher conducts an experiment in which quantitative measures assess the impact of a treatment on an outcome. Before the experiment begins, the researcher collects qualitative data to help design the treatment or, alternatively, to better design strategies to recruit participants to the trial (see the study of physical activity and diet for families in one community by Brett, Heimendinger, Boender, Morin, & Marshall, 2002).

- A researcher seeks to bring about change in understanding the issues facing women. The researcher gathers data through instruments and focus groups to explore the meaning of the issues for women. The larger framework of change guides the researcher and informs all aspects of the study from the issues being studied, to the data collection, and to the call for reform at the end of the study (see the study exploring student–athlete culture and understanding specific rape myths by McMahon, 2007).

• A researcher seeks to evaluate a program that has been implemented in the community. The first step is to collect qualitative data in a needs assessment to determine what questions need to be addressed. This is followed by the design of an instrument to measure the impact of the program. This instrument is then used to compare certain outcomes both before and after the program has been implemented. From this comparison, follow-up interviews are conducted to determine why the program did or did not work. This multiphase mixed methods study is often found in long-term evaluation projects (see the study of the long-term impacts of interpretive programs at a historical site by Farmer & Knapp, 2008).

These examples all illustrate the collection of both quantitative and qualitative data, their integration or mix, and an underlying assumption that mixed methods research could be a useful approach to research.

WHAT RESEARCH PROBLEMS FIT MIXED METHODS? •

Authors of the example studies crafted their research as mixed methods projects based on their assumption that mixed methods could also best address their research problems. When preparing a research study employing mixed methods, the researcher needs to provide a justification for the use of this approach. Not all situations justify the use of mixed methods. There are times when qualitative research may be best, because the researcher aims to explore a problem, honor the voices of participants, map the complexity of the situation, and convey multiple perspectives of participants. At other times, quantitative research may be best, because the researcher seeks to understand the relationship among variables or determine if one group performs better on an outcome than another group. In our discussion of mixed methods, we do not want to minimize the importance of choosing either a quantitative or qualitative approach when it is merited by the situation. Further, we would not limit mixed methods to certain fields of study or topics. Mixed methods research seems applicable to a wide variety of disciplines in the social and health sciences. Certainly some disciplinary content specialists may select not to use mixed methods because of a lack of interest in qualitative research, but most content area problems can be addressed using mixed methods. Instead of thinking about fitting different methods to specific content topics, we suggest thinking about fitting methods to different types of research problems. For example, we find that a survey best fits a quantitative approach because of the need to understand the views of participants in an entire population. An experiment best fits a quantitative approach because of the need to determine whether a treatment works better than a control condition. Likewise, ethnography best fits a qualitative approach because of the need to understand how culture-sharing groups work. What situations, then, warrant an approach that combines quantitative and qualitative research—a mixed methods inquiry? Research problems suited for mixed methods are those in which one data source may be insufficient, results need to be explained, exploratory findings need to be generalized, a second method is needed to enhance a primary method, a theoretical stance needs to be employed, and an overall research objective can be best addressed with multiple phases, or projects.

A Need Exists Because One Data Source May Be Insufficient

We know that qualitative data provide a detailed understanding of a problem while quantitative data provide a more general understanding of a problem. This qualitative understanding arises out of studying a few individuals and exploring their perspectives in great depth whereas the quantitative understanding arises from examining a large number of people and assessing responses to a few variables. Qualitative research and quantitative research provide different pictures, or perspectives, and each has its limitations. When researchers study a few individuals qualitatively, the ability to generalize the results to many is lost. When researchers quantitatively examine many individuals, the understanding of any one individual is diminished. Hence, the limitations of one method can be offset by the strengths of the other method, and the combination of quantitative and qualitative data provide a more complete understanding of the research problem than either approach by itself.

There are several ways in which one data source may be inadequate. One type of evidence may not tell the complete story, or the researcher may lack confidence in the ability of one type of evidence to address the problem. The results from the quantitative and qualitative data may be contradictory, which could not be known by collecting only one type of data. Further, the type of evidence gathered from one level in an organization might differ from evidence looked at from other levels. These are all situations in which using only one approach to address the research problem would be deficient. A mixed methods design best fits this problem. For example, when Knodel and Saengtienchai (2005) studied the role that older-aged parents play in the care and support of adult sons and daughters with HIV and AIDS and AIDS orphans in Thailand, they collected both quantitative survey data and open-ended interviews. Reflecting on the use of both forms of data to understand the problem because quantitative data alone would be inadequate, they said,

The issues (in the interviews) covered were similar to the AIDS parents survey, but the conversational nature of the interview and the fact it allowed open-ended responses provided parents the opportunity to elaborate on the issues and the circumstances affecting them. (Knodel & Saengtienchai, 2005, p. 670)

A Need Exists to Explain Initial Results

Sometimes the results of a study may provide an incomplete understanding of a research problem and there is a need for further explanation. In this case, a mixed methods study is used with the second database helping to explain the first database. A typical situation is when quantitative results require an explanation as to what they mean. Quantitative results can net general explanations for the relationships among variables, but the more detailed understanding of what the statistical tests or effect sizes actually mean is lacking. Qualitative data and results can help build that understanding. For example, Weine et al. (2005) conducted a mixed methods study investigating family factors and processes involved in Bosnian refugees engaging in multiple-family support and education groups in Chicago. The first quantitative phase of the study addressed the factors that predicted engagement while the second qualitative phase consisted of interviews with family members to assess the family processes involved in engagement as multiple-family groups. The rationale for using mixed methods to study this situation was "quantitative analysis addressed the factors that predicted engagement. In order to better understand the processes by which families experience engagement, we conducted a qualitative content analysis to gain additional insight" (Weine et al., 2005, p. 560).

A Need Exists to Generalize Exploratory Findings

In some research projects, the investigators may not know the questions that need to be asked, the variables that need to be measured, and the theories that may guide the study. These unknowns may be due to the specific, remote population being studied (e.g., Native American in Alaska) or the newness of the research topic. In these situations, it is best to explore qualitatively to learn what questions, variables, theories, and so forth need to be studied and then follow up with a quantitative study to generalize and test what was learned from the exploration. A mixed methods project is ideal in these situations. The researcher begins with a qualitative phase to explore and then follows up with a quantitative phase to test whether the qualitative results generalize. For example, Kutner, Steiner, Corbett, Jahnigen, and Barton (1999) studied issues important to terminally ill patients. Their study began with qualitative interviews, and these were then used to develop an instrument that was administered to a second sample of terminally ill patients to test whether the identified issues varied by demographic characteristics. Kutner et al. (1999) said, "The use of initial open-ended interviews to explore the important issues allowed us to formulate relevant questions and discover what were truly concerns to this population" (p. 1350).

A Need Exists to Enhance a Study With a Second Method

In some situations, a second research method can be added to the study to provide an enhanced understanding of some phase of the research. For example, researchers can enhance a quantitative design (e.g., experiment or correlational study) by adding qualitative data or by adding quantitative data to a qualitative design (e.g., grounded theory or case study). In both of these cases, a second method is embedded, or nested, within a primary research method. The embedding of qualitative data within a quantitative study is a typical approach. For example, Donovan et al. (2002) conducted an experimental trial comparing the outcomes for three groups of men with prostate cancer receiving different treatment procedures. They began their study, however, with a qualitative component in which they interviewed the men to determine how best to recruit them into the trial (e.g., how best to organize and present the information) because all the men had received abnormal results and sought the best treatment. Toward the end of their article, Donovan et al. (2002) reflected on the value of this preliminary, smaller, qualitative component used to design procedures for recruiting individuals to the trial:

> We showed that the integration of qualitative research methods allowed us to understand the recruitment process and elucidate the changes necessary to the content and delivery of information to maximize recruitment and ensure effective and efficient conduct of the trial. (p. 768)

A Need Exists to Best Employ a Theoretical Stance

A situation may exist in which a theoretical perspective provides a framework for the need to gather both quantitative and qualitative data in a mixed methods study. The data to be collected might be all gathered at the same time or in a sequence with one form of data building on the other. The theoretical perspective could seek to bring about change or simply provide a lens through which the entire study might be viewed. For example, Fries (2009) conducted a study using Bourdieu's reflexive sociology ("the interplay of objective social structure with subjective agency in social behavior," p. 327) as a theoretical lens for gathering both quantitative and qualitative data in the use of complementary and alternative medicine. He gathered survey and interview data in the first strand, analyzed statistical population health data in the second strand, and analyzed interviews in the third strand. Fries (2009) concluded that "this study has presented a case study from the sociology of alternative medicine to show how reflexive sociology might provide a theoretical basis for mixed methods research oriented toward understanding the interplay of structure and agency in social behavior" (p. 345).

A Need Exists to Understand a Research Objective Through Multiple Research Phases

In projects that span several years and have many components, such as evaluation studies and multiyear health investigations, the researchers may need to connect several studies to reach an overall objective. These studies may involve projects that gather both quantitative and qualitative data simultaneously or gather the information sequentially. We can consider them multiphase or multiproject mixed methods studies. These projects often involve teams of researchers working together over many phases of the project. For example, Ames, Duke, Moore, and Cunradi (2009) conducted a multiphase study of the drinking patterns of young U.S. Navy-enlisted recruits during their first 3 years of military service. To understand the drinking patterns, they conducted a study over a 5-year period, gathered data to develop an instrument in one phase, to modify their model in another phase, and to analyze their data through a final phase. Ames et al. (2009) presented a figure of the phases of their research over 5 years and introduced the implementation sequence this way:

> The complexity of the resulting research design, consisting of both longitudinal survey data collection with a highly mobile population coupled with qualitative interviewing in diverse settings, required the formation of a methodologically diverse research team and a clear delineation of the temporal sequence by which qualitative and quantitative findings would be used to inform and enrich one another. (p. 130)

These scenarios serve to illustrate situations in which mixed methods research fits the problems under study. They also begin to lay the groundwork for understanding the designs of mixed methods that will be discussed later and the reasons authors cite for undertaking a mixed methods study. Although we cite a single reason for mixed methods in each illustration, many authors cite multiple reasons, and we recommend that aspiring (and experienced) researchers begin to take note of the rationales in published studies cited by authors for using mixed methods approaches.

● WHAT ARE THE ADVANTAGES OF USING MIXED METHODS?

Understanding the nature of mixed methods involves more than knowing its definition and when it should be used. In addition, at the outset of selecting a mixed methods approach, researchers need to know the advantages that accrue from using it so that they can convince others of the value of mixed methods. Next we enumerate some of the advantages.

Mixed methods research provides strengths that offset the weaknesses of both quantitative and qualitative research. This has been the historical argument for mixed methods research for more than 30 years (e.g., see Jick, 1979). One might argue that quantitative research is weak in understanding the context or setting in which people talk. Also, the voices of participants are not directly heard in quantitative research. Further, quantitative researchers are in the background, and their own personal biases and interpretations are seldom discussed. Qualitative research makes up for these weaknesses. On the other hand, qualitative research is seen as deficient because of the personal interpretations made by the researcher, the ensuing bias created by this, and the difficulty in generalizing findings to a large group because of the limited number of participants studied. Quantitative research, it is argued, does not have these weaknesses. Thus, the combination of strengths of one approach makes up for the weaknesses of the other approach.

Mixed methods research provides more evidence for studying a research problem than either quantitative or qualitative research alone. Researchers are enabled to use all of the tools of data collection available rather than being restricted to the types of data collection typically associated with quantitative research or qualitative research.

Mixed methods research helps answer questions that cannot be answered by quantitative or qualitative approaches alone. For example, "Do participant views from interviews and from standardized instruments converge or diverge?" is a mixed methods question. Others would be, "In what ways do qualitative interviews explain the quantitative results of a study?" (using qualitative data to explain the quantitative results) and "How can a treatment be adapted to work with a particular sample in an experiment?" (exploring qualitatively before an experiment begins). To answer these questions, quantitative *or* qualitative approaches would not provide a satisfactory answer. The array of possibilities of mixed methods questions will be explored further in the discussion in Chapter 5.

Mixed methods provides a bridge across the sometimes adversarial divide between quantitative and qualitative researchers. We are social, behavioral, and human sciences researchers first, and divisions between quantitative and qualitative research only serve to narrow the approaches and the opportunities for collaboration.

Mixed methods research encourages the use of multiple worldviews, or paradigms (i.e., beliefs and values), rather than the typical association of certain paradigms with quantitative research and others for qualitative research. It also encourages us to think about a paradigm that might encompass all of quantitative and qualitative research, such as pragmatism. These paradigm stances will be discussed further in the next chapter.

Mixed methods research is "practical" in the sense that the researcher is free to use all methods possible to address a research problem. It is also "practical" because individuals tend to solve problems using both numbers and words, combine inductive and deductive thinking, and employ skills in observing people as well as recording behavior. It is natural, then, for individuals to employ mixed methods research as a preferred mode for understanding the world.

WHAT ARE THE CHALLENGES IN USING MIXED METHODS? ●

We must admit that mixed methods is not the answer for every researcher or every research problem. Its use does not diminish the value of conducting a study that is exclusively either quantitative or qualitative. It does, however, require having certain skills, time, and resources for extensive data collection and analysis, and perhaps, most importantly, educating and convincing others of the need to employ a mixed methods design so that a researcher's mixed methods study will be accepted by the scholarly community.

The Question of Skills

We believe that mixed methods is a realistic approach if the researcher has the requisite skills. We strongly recommend that researchers first gain experience with both quantitative research and qualitative research separately before undertaking a mixed methods study. At a minimum, researchers should be acquainted with both quantitative and qualitative data collection and analysis techniques. This point was emphasized in our definition of mixed methods. Mixed methods researchers should be familiar with common methods of collecting quantitative data, such as using measurement instruments and closed-ended attitudinal scales. Researchers need an awareness of the logic of hypothesis testing and the ability to use and interpret statistical analyses, including common descriptive and inferential procedures available in statistical software packages. Finally, researchers need to understand essential issues of rigor in quantitative research, including reliability, validity, experimental control, and generalizability. In later chapters, we will delve into what constitutes a rigorous quantitative approach.

A similar set of qualitative research skills is necessary. Researchers should be able to identify the central phenomenon of their study; to pose qualitative, meaning-oriented research questions; and to consider participants as the experts. Researchers should be familiar with common methods of collecting qualitative data, such as semistructured interviews using open-ended questions and qualitative observations. Researchers need basic skills in analyzing qualitative text data, including coding text and developing themes and descriptions based on these codes, and should be acquainted with a qualitative data analysis software package. Finally, it is important that researchers understand essential issues of persuasiveness in qualitative research, including credibility, trustworthiness, and common validation strategies.

Finally, those undertaking this approach to research should have a solid grounding in mixed methods research. This requires reading the literature on mixed methods that has accumulated since the late 1980s and noting the best procedures and the latest techniques for conducting a good inquiry. It may also mean taking courses in mixed methods research that are beginning to appear both online and in residence on many campuses. It may mean apprenticing with someone familiar with mixed methods who can provide an understanding of the skills involved in conducting this form of research.

The Question of Time and Resources

Even when researchers have basic quantitative and qualitative research skills, they should ask themselves if a mixed methods approach is feasible, given time and resources. These are important issues to consider early in the planning stage. Mixed methods studies may require extensive time, resources, and effort on the part of the researchers. Researchers should consider the following questions:

- Is there sufficient time to collect and analyze two different types of data?
- Are there sufficient resources from which to collect and analyze both quantitative and qualitative data?
- Are the skills and personnel available to complete this study?

In answering these questions, researchers must consider how long it will take to gain approval for the study, to gain access to participants, and to complete the data collection and analysis. Researchers should keep in mind that qualitative data collection and analysis often require more time than that needed for quantitative data. The length of time required for a mixed methods study is also dependent on whether the study will be using a one-phase, two-phase, or multiphase design. Researchers need to think about the expenses that will be part of the study. These expenses may include, for example, printing costs for quantitative instruments, recording and transcription costs for qualitative interviews, and the cost of quantitative and qualitative software programs.

Because of the increased demands associated with mixed methods designs, mixed methods researchers should consider working in teams. We realize that this is impractical for graduate students who are expected to work independently. If a team can be formed, however, it has the advantage of bringing together individuals with diverse methodological and content expertise and of involving more personnel in the mixed methods project. Working with a team can be a challenge. It can increase the costs associated with the research. In addition, individuals with the necessary skills need to be located, and team leaders need to create and maintain a successful collaboration among team members. However, the diversity of a team may be a strength because of enhanced communications among members representing different specialties and content areas.

The Question of Convincing Others

Mixed methods research is relatively new in terms of methodologies available to researchers. As such, others may not be convinced of or understand the value of mixed methods. Some may see it as a "new" approach. Others may feel that they do not have time to learn a new approach to research, and some may object to mixed methods on philosophical grounds regarding the mixing of different philosophical positions, as we will see in the next chapter. Still others might be so ensconced in their own methods and approaches to research that they might not be open to the possibility of mixed methods research.

One way to help convince others of the utility of mixed methods is to locate exemplary mixed methods studies in the literature on a topic or in a content area and share these studies to educate others. These studies can be selected from prestigious journals with a national and international reputation. How does a researcher find these mixed methods studies?

Mixed methods studies can be difficult to locate in the literature, because only recently have researchers begun to use the term *mixed methods* in their titles or in their methods' discussions. Also, some disciplines may use different terms for naming this research approach. Based on our extensive work with the literature, we have developed a short list of terms that we use to search for mixed methods studies within electronic databases and journal archives. These terms include

- mixed method* (where * is a wildcard that will allow hits for "mixed method," "mixed methods," and "mixed methodology"),

- quantitative AND qualitative,
- multimethod, and
- survey AND interview.

Note that the second search term uses the logic operator AND (i.e., quantitative AND qualitative). This requires that both words appear in the document so it will satisfy the search criteria. If too many articles are found, try limiting the search so that the terms must appear within the abstract or restricting it to recent years. If not enough articles result, try searching for combinations of common data collection techniques, such as "survey AND interview." By using these strategies, researchers may locate a few good examples of mixed methods research that illustrate the core characteristics introduced in this chapter. Sharing these examples with stakeholders can be helpful when convincing them of the utility and feasibility of a mixed methods approach.

SUMMARY

Before deciding on a mixed methods study, the researcher needs to consider several preliminary considerations about the nature of mixed methods research. First, the researcher needs some understanding as to what constitutes a mixed methods study to determine if this approach is the best to use for their particular study. Several core characteristics have been recommended: the collection and analysis of both quantitative and qualitative data; the mixing of the two types of data either by merging them, having one build on the other, or embedding one within the other; the emphasis or priority of one or both forms of data; the use of the two forms of data in a single study or a sustained line of research inquiry; the use of a philosophical or theoretical orientation that informs all aspects of the study; and the use of a specific type of mixed methods design for procedures. Most important in this list of characteristics would be the availability of two sets of data, one quantitative and one qualitative. Second, some assessment needs to occur as to whether the research problem best fits mixed methods. Many topics and problems are suitable for mixed methods (e.g., violence has escalated in our schools or children have poor nutrition in their families). Consider if the research problem can be best addressed using mixed methods procedures. Some problems are best studied by using two data sources and collecting only one may provide an incomplete understanding. Another study may need a second database to help explain the first database. Another type of problem may require that the researcher first explore qualitatively before undertaking a quantitative study, use a theoretical lens to study the problem, or conduct multiple phases of studies to build an overall understanding of the problem.

Not only are multiple data sources helpful in understanding research problems but there are other advantages of using mixed methods. The strength of one method may offset the weaknesses of the other. Using multiple sources of data simply provides more evidence for studying a problem than a single method of data. Oftentimes research questions are posed that require both an exploration as well as an explanation drawing from different data sources. Mixed methods also is well suited for interdisciplinary research that brings scholars together from different fields of study, and it enables researchers to employ multiple philosophical perspectives that guide their research. Finally, mixed methods is both practical and intuitive in that it helps offer multiple ways of viewing problems—something found in everyday living.

This does not mean that using mixed methods will be easy. It requires that the researchers have skills in several areas: quantitative research, qualitative research, and mixed methods research. Because of the extensive data collected, it takes time to gather data from both quantitative and qualitative sources, and it takes resources to fund these data collection (and data analysis) efforts. Perhaps most importantly, individuals planning a mixed methods study need to convince others of the value of mixed methods. It is a relatively new approach to inquiry, and it requires an openness to using multiple perspectives in research. A search through the literature will yield good examples of mixed methods studies today, and these can be shared with important stakeholders to help educate them about mixed methods studies.

ACTIVITIES

1. Locate a mixed methods study in your field or discipline. Engage in these steps:

   a) Suspend your interest in the content of the articles, and focus instead on the research methods used.

   b) Review the core characteristics of a mixed methods study, and identify how the study represents a good mixed methods study because it addresses the core characteristics.

2. Consider the value of mixed methods research for different audiences, such as policy makers, graduate advisors, individuals in jobs or the workplace, and graduate students. Discuss the value for each audience.

3. Consider whether a mixed methods approach is feasible for your study. List out the skills, resources, and time that you have available for the project.

4. Consider designing a mixed methods project. State in your own words how you will define mixed methods research, mention why mixed methods is well suited to address your research problem, and cite both the advantages and challenges of using it as an approach to research.

ADDITIONAL RESOURCES TO EXAMINE

For definitions of mixed methods, consult the following resources:

Creswell, J. W. (2009). *Research design: Qualitative, quantitative, and mixed methods approaches* (3rd ed.). Thousand Oaks, CA: Sage.

Greene, J. C. (2007). *Mixed methods in social inquiry*. San Francisco: Jossey-Bass.

Greene, J. C., Caracelli, V. J., & Graham, W. F. (1989). Toward a conceptual framework for mixed-method evaluation designs. *Educational Evaluation and Policy Analysis, 11*(3), 255–274.

Johnson, R. B., Onwuegbuzie, A. J., & Turner, L. A. (2007). Toward a definition of mixed methods research. *Journal of Mixed Methods Research, 1*(2), 112–133.

For the rationale or purpose for using mixed methods to address problems, see the following resources:

Bryman, A. (2006). Integrating quantitative and qualitative research: How is it done? *Qualitative Research, 6*(1), 97–113.

Mayring, P. (2007). Introduction: Arguments for mixed methodology. In P. Mayring, G. L. Huber, L. Gurtler, & M. Kiegelmann (Eds.), *Mixed methodology in psychological research* (pp. 1–4). Rotterdam/Taipei: Sense Publishers.

For the advantages of mixed methods research, see the following resources:

Creswell, J. W., & McCoy, B. R. (in press). The use of mixed methods thinking in documentary development. In S. N. Hesse-Biber (Ed.), *The handbook of emergent technologies in social research*. Oxford, UK: Oxford University Press.

Plano Clark, V. L. (2005). Cross-disciplinary analysis of the use of mixed methods in physics education research, counseling psychology, and primary care (Doctoral dissertation, University of Nebraska–Lincoln, 2005). *Dissertation Abstracts International, 66,* 02A.

For the skills needed to conduct mixed methods research, see the following resource:

Creswell, J. W., Tashakkori, A., Jensen, K. D., & Shapley, K. L. (2003). Teaching mixed methods research: Practices, dilemmas, and challenges. In A. Tashakkori & C. Teddlie (Eds.), *Handbook of mixed methods in social & behavioral research* (pp. 619–637). Thousand Oaks, CA: Sage.

CHAPTER 2

THE FOUNDATIONS
OF MIXED METHODS
RESEARCH

P rior to designing a mixed methods study, researchers need to consider more than whether their research problems or questions are best suited for mixed methods. They also should develop a deep understanding of mixed methods so that they can not only define and justify mixed methods and recognize its essential characteristics but also so that they can reference important works that have established this approach. This means understanding some of the history of mixed methods and key writings that have informed its development. Another step prior to designing a study is to understand what assumptions about knowledge and the acquisition of knowledge a researcher makes when selecting mixed methods. This understanding requires knowing philosophical assumptions. Finally, mixed methods researchers today often select a theory as a lens in their study, which threads through an entire study. Thus, an initial step in planning a mixed methods study is to give some consideration to whether a theory will be used in a study and how the theory is incorporated into a project.

This chapter reviews historical, philosophical, and theoretical foundations for planning and conducting a mixed methods study. In this chapter, we will address

- the historical foundations of mixed methods,
- the philosophical assumptions made when choosing a mixed methods study, and
- theoretical lenses that may be used in mixed methods research.

● HISTORICAL FOUNDATIONS

In planning a mixed methods project, researchers need to know something about its history, how it has evolved, and the current interest in mixed methods. Besides providing a definition for mixed methods, a mixed methods plan or study includes references to the literature, a justification for its use, and documentation about its previous use in a particular field of study. This all requires some knowledge of the historical foundations of mixed methods research, such as knowing when it began, who has been writing about it, and recent applications of its use.

When Did Mixed Methods Begin?

We often date the beginnings of mixed methods back to the late 1980s with the coming together of several publications all focused on describing and defining what is now known as mixed methods. Several writers working in different disciplines and countries all came to the same idea at roughly the same time. Writers from sociology in the United States (Brewer & Hunter, 1989) and in the United Kingdom (Fielding & Fielding, 1986), from evaluation (Greene, Caracelli, & Graham, 1989) in the United States, from management in the United Kingdom (Bryman, 1988), from nursing in Canada (Morse, 1991), and from education in the United States (Creswell, 1994) were sketching out the concept of mixed methods during the late 1980s to the early 1990s. All of these individuals were writing books, book chapters, and articles on an approach to research that moved beyond simply using quantitative and qualitative methods as distinct, separate strands in a study. They were giving serious thought to ways to link or combine these methods. The authors began a discussion about how to integrate, or "mix," the data and their reasons for it; Bryman (2006) would pull these integrative approaches together several years later. The authors also discussed the possible research designs and the names for designs; Creswell and Plano Clark (2007) would later assemble a list of the classifications of types of design. A shorthand notation system was developed to convey these designs; Morse (1991) gave specific attention to the notation. Debates emerged about the philosophy behind this form of inquiry; Reichardt and Rallis (1994) would make explicit the debate forming in the United States.

It is true that antecedents to these procedural and philosophical developments in mixed methods had taken form much earlier than the late 1980s (Creswell, in press-a). As early as 1959, Campbell and Fiske had discussed the inclusion of multiple sources of quantitative information in the validation of psychological traits. Others had advocated the use of multiple data sources—both quantitative and qualitative this time—to conduct scholarly studies (Denzin, 1978), and several well-known figures in quantitative research, such as Campbell (1974) and Cronbach (1975), advocated for the inclusion of qualitative data in quantitative experimental studies. The combination and interplay of survey research and fieldwork was a central feature in the writings of Sieber in 1973. In the field of evaluation, Patton, in 1980, suggested "methodological mixes" for experimental and naturalistic designs, and he advanced several diagrams to illustrate different combinations of these mixes. In short, these developments signaled key antecedents to what would later be more systematic attempts to forge mixed methods into a complete research design and to create a distinct approach to research (Creswell, in press-a).

Why Mixed Methods Emerged

A number of factors have contributed to the evolution of mixed methods research, as we know it today following the early 1990s period of research. The complexity of our research problems calls for answers beyond simple numbers in a quantitative sense or words in a qualitative sense. A combination of both forms of data provides the most complete analysis of problems. Researchers situate numbers in the contexts and words of participants, and they frame the words of participants with numbers, trends, and statistical results. Both forms of data are necessary today. In addition, qualitative research has evolved to a point where writers consider it a legitimate form of inquiry in the social and human sciences (see Denzin & Lincoln, 2005). On the other hand, quantitative researchers, we believe, recognize that qualitative data can play an important role in quantitative research. Qualitative researchers, in turn, realize that reporting only qualitative participant views of a few individuals may not permit generalizing the findings to many individuals. Audiences such as policy makers, practitioners, and others in applied areas need multiple forms of evidence to document and inform the research problems. A call for increased sophistication of evidence leads to a collection of both quantitative and qualitative data.

The Development of the Name

There has been much discussion about the name for this form of inquiry. During the past 50 years, writers have used different names, making it difficult to locate specific research studies that we would call "mixed methods" research. It has been called "integrated" or "combined" research, advancing the notion that two forms of data are blended together (Steckler, McLeroy, Goodman, Bird, & McCormick, 1992), and it is sometimes called "quantitative and qualitative methods" (Fielding & Fielding, 1986), which acknowledges that the approach is actually a combination of methods. It has been called "hybrid" research (Ragin, Nagel, & White, 2004) or "methodological triangulation" (Morse, 1991), which recognizes the convergence of quantitative and qualitative data, "combined research" (Creswell, 1994), and "mixed methodology," which acknowledges that it is both a method and a philosophical worldview (Tashakkori & Teddlie, 1998). Along the same line, it has recently been called "mixed research" to reinforce the idea that this approach is more than simply methods and ties into other facets of research, such as philosophical assumptions (Onwuegbuzie & Leech, 2009). Today, we believe that the most frequently used name is "mixed methods research," a name associated with the *Handbook of Mixed Methods in Social & Behavioral Research* (Tashakkori & Teddlie, 2003a, in press) as well as with the SAGE journal, the *Journal of Mixed Methods Research* (JMMR). Although the term *mixed methods* is becoming increasingly used by a large number of social, behavioral, and human science scholars, its continued use will encourage researchers to see this approach as a distinct model of inquiry.

Stages in the Evolution of Mixed Methods

Our approach to mixed methods research has grown out of the work of others as well as the historical and philosophical discussions of the last several decades. For those designing and conducting mixed methods studies, a historical overview is not an idle exercise in recapping the past. Knowing this history helps researchers defend their use of this approach, justify their use of it as a research approach, and cite leading proponents of the approach in their "methods" discussions.

There have been several stages in the history of mixed methods (e.g., Tashakkori & Teddlie, 1998). Here we will review this history and organize it into five, often overlapping, time periods, as shown in Table 2.1.

Table 2.1	Selected Writers and Their Contributions to the Development of Mixed Methods Research

Stage of Development	Author(s) and Year	Contribution to Mixed Methods Research
Formative period	Campbell and Fiske (1959)	Introduced the use of multiple quantitative methods
	Sieber (1973)	Combined surveys and interviews
	Denzin (1978)	Discussed using both quantitative and qualitative data in a study
	Jick (1979)	Discussed triangulating quantitative and qualitative data
	Cook and Reichardt (1979)	Presented 10 ways to combine quantitative and qualitative data
Paradigm debate period	Rossman and Wilson (1985)	Discussed stances toward combining methods—purists, situationalists, and pragmatists
	Bryman (1988)	Reviewed the debate and established connections within the two traditions
	Reichardt and Rallis (1994)	Discussed the paradigm debate and reconciled two traditions
	Greene and Caracelli (1997)	Suggested that we move past the paradigm debate
Procedural development period	Greene, Caracelli, and Graham (1989)	Identified a classification system of types of mixed methods designs
	Brewer and Hunter (1989)	Focused on the multimethod approach as used in the process of research
	Bryman (1988)	Addressed reasons for combining quantitative and qualitative research
	Morse (1991)	Developed a notation system
	Creswell (1994)	Identified three types of mixed methods designs
	Morgan (1998)	Developed a typology for determining type of design to use
	Newman and Benz (1998)	Provided an overview of procedures

(Continued)

Table 2.1 (Continued)

Stage of Development	Author(s) and Year	Contribution to Mixed Methods Research
	Tashakkori and Teddlie (1998)	Presented topical overview of mixed methods research
	Bamberger (2000)	Provided an international policy focus to mixed methods research
Advocacy and expansion period	Tashakkori and Teddlie (2003a)	Provided a comprehensive treatment of many aspects of mixed methods research
	Johnson and Onwuegbuzie (2004)	Positioned mixed methods research as a natural complement to traditional quantitative and qualitative research
	Creswell (2009c)	Compared quantitative, qualitative, and mixed methods approaches in the process of research
	Greene (2007)	Emphasized the rationales, purposes, and potential for mixing methods in social research and evaluation
	Plano Clark and Creswell (2008)	Compiled published methodological and empirical studies in mixed methods
	Teddlie & Tashakkori (2009)	Chronicled changes that have occurred over the past 5 to 10 years in mixed methods research
	Morse & Niehaus (2009)	Argued for mixed methods designs that had a core component and a supplemental component
Reflective period	Tashakkori & Teddlie (2003b)	Presented issues and priorities in the mixed methods field
	Greene (2008)	Identified four methodological domains and discussed what we know and what we need to know to consider mixed methods a distinctive methodology
	Creswell (2008a, 2009b, in press-b)	Developed a map of the mixed methods literature
	Howe (2004)	Critiqued mixed methods as constraining qualitative methods to a largely auxiliary role and failing to use qualitative research in an interpretive way

Stage of Development	Author(s) and Year	Contribution to Mixed Methods Research
	Giddings (2006)	Critiqued mixed methods as marginalizing nonpositivist research methodologies and privileging the positivist tradition
	Holmes (2006)	Critiqued the ways in which mixed methods research were described by mixed methods writers
	Freshwater (2007)	Interrogated the assumptions underpinning mixed methodology and its discourse using a postmodern perspective
	Creswell (in press-a)	Identified and gave voice to controversies in mixed methods research

Formative period. The formative period in the history of mixed methods began in the 1950s and continued up until the 1980s. This period saw the initial interest in using more than one method in a study. It found momentum in psychology in the 1950s through the combination of multiple quantitative methods in a study (Campbell & Fiske, 1959), the use of surveys and fieldwork in sociology (Sieber, 1973) and multiple methods in general (Denzin, 1978), the initiatives in triangulating both quantitative and qualitative approaches (Jick, 1979; Patton, 1980), and discussions in psychology about combining quantitative and qualitative data when they arose from different perspectives (see Cook & Reichardt, 1979). These were the early antecedents of mixed methods as it is known today (Creswell, in press-a)

Paradigm debate period. The paradigm debate period in the history of mixed methods developed during the 1970s and 1980s when qualitative researchers were adamant that different assumptions provided the foundations for quantitative and qualitative research (see Bryman, 1988; Guba & Lincoln, 1988; Smith, 1983). The paradigm debate involved scholars arguing whether or not qualitative and quantitative data could be combined, because qualitative data were linked with certain philosophical assumptions and quantitative data were connected to other philosophical assumptions. If this was true, then, as some commented, mixed methods research was untenable (or incommensurable), because it asked for paradigms to be combined (Smith, 1983). Rossman and Wilson (1985) called these individuals who could not mix paradigms, "purists." The discussion came to a head by 1994 with vocal advocates on both sides arguing their points at the American Evaluation Association meeting (Reichardt & Rallis, 1994). Today, the links between the methods of data collection and the larger philosophical assumptions are not as tightly drawn as envisioned in the 1990s. Denzin and Lincoln (2005), for example, have advanced the idea that different types of methods can be associated with different types of worldviews or philosophies. Other perspectives have also developed, such as the situationalists, who adapted their methods to the situation, and pragmatists, who believed that multiple paradigms can be used to address research problems (Rossman & Wilson, 1985). Although the issue of reconciling paradigms is still apparent (see the writings of Giddings, 2006; Holmes, 2006), calls have been made to embrace pragmatism as the best philosophical foundation for mixed methods research (see Tashakkori & Teddlie, 2003a) and to use different paradigms in mixed methods research but to honor each and be explicit about when each is used (Greene & Caracelli, 1997).

Procedural development period. Although the debate about which paradigms provide a foundation for mixed methods research has not disappeared, attention during the 1980s began to shift toward the **procedural development period** in the history of mixed methods in which writers focused on methods of data collection, data analysis, research designs, and the purposes for conducting a mixed methods study. In 1989, Greene et al. authored a classic article in the field of evaluation that laid the groundwork for mixed methods research design. In their article, they analyzed 57 evaluation studies, developed a classification system of five types, and talked about the design decisions that go into each of the types. Following this article, many authors have identified types of mixed methods designs with distinct names and procedures. At roughly the same time, two sociologists, Brewer and Hunter (1989), contributed to the discussion by linking multimethod research to the steps in the process of research (e.g., formulating problems, sampling, and collecting data). Bryman (1988) also discussed the reasons for combining quantitative and qualitative data. By 1991, Morse, a nursing researcher, had designed a notation system to convey how the quantitative and qualitative components of a study were implemented. Building on these classifications and notations, writers began discussing specific types of mixed methods designs. For example, Creswell (1994) created a parsimonious set of three types of designs and found studies that illustrated each type. Morgan (1998) provided a decision matrix for determining the type of design to use, and books, such as those of Bamberger (2000), Newman and Benz (1998), and Tashakkori and Teddlie (1998), began to map the contours of mixed methods procedures in policy research and in attending to issues such as validity and inferences.

Advocacy and expansion period. In recent years, we have moved into an **advocacy and expansion period** in the history of mixed methods in which many authors have advocated for mixed methods research as a separate methodology, method, or approach to research, and interest in mixed methods has extended to many disciplines and many countries.

We have become advocates for mixed methods as well, providing workshops on the topic to disciplines and fields seeking to learn more about this approach, and noting unfolding developments that span from conferences, to journals, to fields of study, and on to international countries. Much growth in the field of mixed methods research has occurred since the publication of the 2003 768-page *Handbook of Mixed Methods in Social & Behavioral Research* (Tashakkori & Teddlie, 2003a), a compendium of writings including 26 chapters devoted to controversies, methodological issues, applications in different discipline fields, and future directions. As this handbook suggested back in 2003, we have seen much evidence for growth of interest in mixed methods through funding initiatives, publications, conferences, and applications in different disciplines and countries. In the second edition of the handbook (Tashakkori & Teddlie, in press), the range of topics now has expanded to include 31 chapters and new writers to the field.

In funding initiatives, the National Institutes of Health (NIH) took the lead several years ago in discussing guidelines (National Institutes of Health, 1999) for "combined' quantitative and qualitative research, although these guidelines, as seen from the present-day perspective, are in need of a revision and update. In 2004, NIH held a workshop titled "Design and Conduct of Qualitative and Mixed-Methods Research in Social Work and Other Health Professions," which was sponsored by seven NIH Institutes and two research offices. In 2003, the U.S. National Science Foundation (NSF) held a workshop on the scientific foundations of qualitative research with several papers devoted to the topic of combining quantitative and qualitative methods (Ragin et al., 2004). The National Research Council (2002) discussed scientific research in education and concluded that three questions need to guide inquiries: "Description—What is happening? Cause—Is there a systematic effect? And the process or mechanism—Why or how is it happening?" (p. 99). These questions, in combination, suggest both a quantitative and a qualitative approach to scientific inquiry. Private U.S. foundations, such as the Robert Wood Johnson Foundation and the W. T. Grant Foundation, have had workshops on mixed methods research. In the United Kingdom, the Economic and Social Research Council (ESRC) has funded through its Research Methods Programme inquiries into the use of mixed methods research (Bryman, 2007).

Plano Clark (2010) examined funded projects by the NIH and their use of the mixed methods terms in the proposal abstracts. Examining only the new funding awards (identified in the first year of funding) and using the search terms of mixed methods or multimethod, Plano Clark obtained 272 hits from RePORTER (known as the National Institutes of Health Expenditures and Results query tool, http://projectreporter.nih.gov/reporter.cfm) during the period of 1997 to 2008. Her review of these projects showed a steady increase in the use of these terms in abstracts for funded projects during this time period. Funding for the projects came from 25 different NIH agencies (with the National Institute of Mental Health funding the largest percentage of identified projects at 24%) as one indicator of the widespread interest in this approach. As might be expected from the health sciences, 27% of the projects included an experimental or control trial component, and many projects revealed complex designs and design names, such as a "mixed methods prospective randomized controlled study," a "longitudinal mixed methods descriptive study," or an "equivalent, sequential, transformative, mixed-methods study" (Plano Clark, 2010). The names alone present the immense variation that exists in undertaking health science mixed methods projects. In other work with the NIH database, we have explored the K-awards given to new scholars who present both a plan for career development as well as a substantive project. Looking solely at the funded projects for 2007, a number of these projects funded included a training component related to qualitative research and mixed methods.

In journals and in disciplines, the number of published mixed methods studies continues to increase. We found more than 60 articles in the social and human sciences that employed mixed methods research between 1995 and 2005 (Plano Clark, 2005). Mixed methods research is being published in special journal issues, such as in the *Annals of Family Medicine* (e.g., see Creswell, Fetters, & Ivankova, 2004) and the *Journal of Counseling Psychology* (e.g., see Hanson, Creswell, Plano Clark, Petska, & Creswell, 2005). Calls for increased use of qualitative data in traditional experimental trials in the health sciences have been reported in prestigious journals such as the *Journal of the American Medical Association* (Flory & Emanuel, 2004), in *Lancet* (Malterud, 2001), in *Circulation* (e.g., see Curry, Nembhard, & Bradley, 2009), in the *Journal of Traumatic Stress* (e.g., see Creswell & Zhang, 2009), and in *Psychology in the Schools* (e.g., Powell, Mihalas, Onwuegbuzie, Suldo, & Daley, 2008). Several journals are now devoted to publishing both empirical mixed methods studies as well as methodological discussions, such as the JMMR, *Quality and Quantity, Field Methods*, and the online journal *International Journal of Multiple Research Approaches* (IJMRA). Mixed methods is appearing with increased frequency in titles to empirical mixed methods journal articles (e.g., see Slonim-Nevo & Nevo, 2009). In addition, cross-disciplinary reviews of mixed methods research are available in the field of evaluation (Greene et al., 1989), in higher education studies (Creswell, Goodchild, & Turner, 1996), in educational research (Johnson & Onwuegbuzie, 2004), in family medicine, physics education, and counseling psychology (Plano Clark, 2005), in four social science disciplines (Bryman, 2006), in marketing research (Harrison, 2010), in family research (Plano Clark, Huddleston-Casas, Churchill, Green, & Garrett, 2008), and in multicultural counseling research (Plano Clark & Wang, 2010).

There is increased use of mixed methods, as these journals indicate, in different discipline fields. Intervention researchers are incorporating qualitative data into their clinical trials in evidence-based medicine (see the discussion about mixed methods intervention trials, Creswell, Fetters, Plano Clark, & Morales, 2009). Although such experimental trials have raised questions about the subversion of qualitative research to the dominant, quantitative methodology in the health sciences (see Howe, 2004), they do serve to bring qualitative research into the health sciences—where it has not gained much entry—in an acceptable manner to many investigators. Also, discipline-based approaches, such as geographic information systems (GIS) are being seen as applications of mixed methods procedures in fields, such as sociology (Fielding & Cisneros-Puebla, 2009). Books on mixed methods up until now have been general in scope, aimed broadly at the social or health sciences (e.g., Creswell, 2009c; Creswell & Plano Clark, 2007; Greene, 2007; Morse & Niehaus, 2009; Plano Clark & Creswell, 2008; Teddlie & Tashakkori, 2009). More recently, discipline-based books on research methods and mixed methods have emerged with a chapter on mixed methods or with the entire book focused on mixed methods, such as in media and communication (Berger, 2000), education and psychology (Mertens, 2005), social work (Engel & Schutt, 2009), family research (Greenstein, 2006), and nursing and the health sciences (Andrew & Halcomb, 2009).

On the international scene, interest has grown in mixed methods in many countries around the world. Recent publications in JMMR attest to strong international participation from such countries as Sri Lanka (Nastasi et al., 2007), Germany (Bernardi, Keim, & von der Lippe, 2007), Japan (Fetters, Yoshioka, Greenberg, Gorenflo, & Yeo, 2007), and the United Kingdom (O'Cathain, Murphy, & Nicholl, 2007). The Mixed Methods Conference, now hosted by Leeds University in the United Kingdom and based in England, has completed five successful conferences. Over the years, American scholars have been involved in this conference, thus lessening the "Atlantic gap" that often occurs between the U.S. academics and those from other countries. An international community is forming around mixed methods, with discussions about the quantitative and qualitative skills needed to undertake this form of inquiry and the need, especially in countries such as South Africa (Olivier, de Lange, Creswell, & Wood, 2010), for involvement of individuals with quantitative skills amidst the preponderance of qualitative talent. Such an international community is also assembling through conference groups, such as the Special Interest Group on Mixed Methods Research formed in the American Educational Research Association. Its initial meeting was held in April 2005 in Montreal, Canada. In addition, SAGE Publications has started an online network, Methodspace, to link researchers, including mixed methods scholars, worldwide (see http://www.methodspace.com/group/mixedmethodsresearchers).

For teaching about mixed methods, courses have developed on college and university campuses encouraged by commentary about the content and instructional approaches of the courses (Creswell, Tashakkori, Jensen, & Shapley, 2003), teaching graduate students to learn, use, and appreciate both quantitative and qualitative research within a mixed methods framework (Onwuegbuzie & Leech, 2009) and identifying the strengths, challenges, and lessons learned from teaching such courses (see Christ, 2009). Several international online mixed methods courses are now available, offered in the United States at the University of Nebraska–Lincoln (UNL), the University of Arkansas, and the University of Alabama at Birmingham. Articles such as Christ's (2009) highlight the importance of examining pedagogical issues.

Reflective period. We feel that in the last 5 to 7 years, mixed methods has entered into a new historical period. This **reflective period** in the history of mixed methods is characterized by two intersecting themes: (1) a current assessment of the field and a look into the future and (2) constructive criticisms challenging the emergence of mixed methods and what it has become.

Three discussions have appeared in recent years that help to map the current state of the field of mixed methods: Creswell (2008a, 2009b), Greene (2008), and Tashakkori and Teddlie (2003b). The issues and topics in these three discussions are summarized in Table 2.2. The first discussion was presented by Tashakkori and Teddlie (2003b) in the beginning and ending chapters of the first edition of their handbook. It detailed five major unresolved issues and controversies in the use of mixed methods in social and behavioral research. A few years later, Greene (2008) published an analysis of key domains in mixed methods in the JMMR based on a keynote address presented to the Mixed Methods Special Interest Group at the American Educational Research Association in 2007. In setting forth her domains, Greene (2008) asked, "What important questions remain to be engaged?" and she raised questions about "priorities for a mixed methods research agenda" (p. 8). Creswell's (2008a) mapping of topics in the field of mixed methods was first presented as a keynote address to the 2008 Mixed Methods Conference

Table 2.2 Issues, Priorities, and Topics About Mixed Methods Currently Being Addressed

General Domain	Areas and Domains for Tashakkori & Teddlie (2003b)	Specific Issues and Questions	Areas and Domains for Greene (2008)	Specific Priorities	Areas and Domains for Creswell (2008a, 2009b)	Specific Topics
Essence of mixed methods domain	• The nomenclature and the basic definitions used in mixed methods research • The utility of mixed methods research (Why do we do it?)	• Should we use QUAN and QUAL terms or develop new mixed methods terms? • What are the reasons for conducting mixed methods research?			• Nature of mixed methods	• Definition of bilingual language for incorporation of mixed methods into existing designs
Philosophical domain	• The paradigmatic foundations for mixed methods research	• What are the paradigm perspectives in mixed methods research (dialectical, single paradigm, multiple paradigm)?	• Philosophical assumptions and stances	• What actually does influence inquirers' methodological decisions in practice? • How do the assumptions and stances of pragmatism influence inquiry decisions?	• Philosophical and theoretical issues	• Combining philosophical positions, worldviews, and paradigms • Philosophical foundation of mixed methods • Use of qualitative theoretical lens in mixed methods

(Continued)

Table 2.2 (Continued)

General Domain	Areas and Domains for Tashakkori & Teddlie (2003b)	Specific Issues and Questions	Areas and Domains for Greene (2008)	Specific Priorities	Areas and Domains for Creswell (2008a, 2009b)	Specific Topics
						• False distinction between qualitative and quantitative research • Thinking in a mixed methods way—mental models
Procedures domain	• Design issues in mixed methods research • Issues in drawing inferences in mixed methods research	• How can mixed methods design be conceptualized (stages of research as conceptualization, method, and inference)? • How do monostrand designs differ from multistrand? • What are the types of multistrand designs?	• Inquiry logics	• What are the particular strengths and limitations of various methods of data collection? • How do we choose particular methods for a given inquiry purpose and design?	• Techniques of mixed methods	• Unusual blends of methods • Joint displays of quantitative and qualitative data • Longitudinal, evaluation studies • Transforming qualitative data into counts • Process steps of research (theory, questions, sampling, interpretation)

General Domain	Areas and Domains for Tashakkori & Teddlie (2003b)	Specific Issues and Questions	Areas and Domains for Greene (2008)	Specific Priorities	Areas and Domains for Creswell (2008a, 2009b)	Specific Topics
		• What are the rules and procedures for forming inferences? • What are the standards for evaluating and improving the quality of inferences?		• Around what does mixing occur? (Constructs? Questions? Purposes?) • What does a methodology of mixed methods look like?		• New thinking about research designs • Methodological issues in using designs • Notations for designs • Visual diagrams for designs • Software applications • Integration and mixing issues • Rationale for mixed methods • Validity • Ethics
Adoption and use domain	• The logistics of conducting mixed methods research	• What is involved in collaborating on a mixed methods project?	• Guidelines for practice	• What are the unique aspects of mixed methods practice that deal specifically with mixing?	• Adoption and use of mixed methods	• Fields and disciplines using it • Team approaches

(Continued)

Table 2.2 (Continued)

General Domain	Areas and Domains for Tashakkori & Teddlie (2003b)	Specific Issues and Questions	Areas and Domains for Greene (2008)	Specific Priorities	Areas and Domains for Creswell (2008a, 2009b)	Specific Topics
		• What are some of the unresolved pedagogical issues in teaching mixed methods research?		• What can be learned from conversations across disciplines and fields of applied inquiry practice?		• Linking mixed methods to discipline techniques • Teaching mixed methods to students • Writing up and reporting
Political domain			• Sociopolitical commitments	• Who is the audience? What perspective is represented? Whose voice is heard? And who is being advocated for?	• Politicization of mixed methods	• Funding of mixed methods research • Deconstructing mixed methods • Justifying mixed methods

SOURCE: Creswell (in press-b) reprinted with permission of SAGE Publications, Inc.

at Cambridge University in England. He compared papers being presented at the conference with his developing understanding of the field culled from over 300 submissions during three years as coeditor and cofounder of JMMR. From this conference presentation, he then drafted a shorter version as an editorial for JMMR focusing on a few specific issues (Creswell, 2009b).

As shown in Table 2.2, there are some common themes appearing across all three writings. These themes are philosophical issues, the procedures in conducting a mixed methods study, and the adoption and use of mixed methods. As for the philosophical issues, all three discussions point to understanding the philosophical foundations of mixed methods, with the last recent writings (Creswell, 2008a, 2009b; Greene, 2008) focusing much more on the *practice* of using philosophical perspectives in mixed methods studies (e.g., how to combine them, how they influence inquiry decisions).

In terms of procedures, Tashakkori and Teddlie (2003b) focused on the broader design issues, while Greene (2008) and Creswell (2008a, 2009b) went into the detailed areas of methods. This analysis suggests that discussions are becoming more analytic about how to conduct a study. This reinforces the assumption that many of us hold that the techniques of conducting mixed methods research have received considerable attention in the field. On the adoption and use of mixed methods, while the earlier discussions by Tashakkori and Teddlie (2003b) focused on collaboration and teaching mixed methods, the more recent writings by Greene (2008) and Creswell (2008a, 2009b) have examined increased use of mixed methods by new disciplines and across fields of inquiry practice. This analysis does suggest the trend of mixed methods spreading to many fields and being adapted to suit unique discipline approaches to research methodology.

As these writings indicate, the growth and interest in mixed methods has accelerated in recent years. It is not surprising, then, that it has drawn attention from individuals willing to challenge and critique its approaches. In the field of education, Howe (2004) addressed whether mixed methods privileged postpositivist thinking and marginalized qualitative interpretative approaches. His concern was mainly directed toward the National Research Council (2002), mentioned earlier, and how their report assigned a prominent role to quantitative experimental research and a lesser role to qualitative, interpretive research. Within this schema—which he called "mixed methods experimentalism" (p. 53)—not only was qualitative research limited to an auxiliary role but it also minimized the use of qualitative research in an interpretive role that included voices of stakeholders and dialogue.

From the field of nursing research have come several critiques. Giddings (2006), from New Zealand, challenged the claims made by mixed methods writers about inclusiveness and about how qualitative and quantitative methods would produce the "best of both worlds" (p. 195). She also challenged the use of binary terms in mixed methods, such as *qualitative* and *quantitative*, which reduced methodological diversity, the use of mixed methods as a "cover" for the continuing hegemony of positivism, and the use of mixed methods as a "quick fix" in response to economic and administrative pressures (p. 195). An Australian, Holmes (2006), also in nursing, critiqued the way in which mixed methods was being described. Like the others, he was concerned about the marginalization of qualitative interpretive frameworks in mixed methods and recommended that the mixed methods community provide a clearer concept of their terms and include a qualitative interpretive framework.

Another voice from nursing, Freshwater (2007), provided a postmodern critique of mixed methods. She was concerned about how mixed methods was being "read" and the discourse that followed. Discourse was defined as a set of rules or assumptions for organizing and interpreting the subject matter of an academic discipline or field of study in mixed methods. The uncritical acceptance of mixed methods as an emerging dominant discourse ("is nearing becoming a metanarrative," Freshwater, 2007, p. 139) impacts how it is located, positioned, presented, and perpetuated. Freshwater (2007) called on mixed methods writers to make explicit the internal power struggle between the mixed methods text as created by the researcher and the text as seen by the reader or audience. Mixed methods, she felt, was too "focused on fixing meaning" (p. 137). Expanding on this, she stated that mixed methods was mainly about doing away with "indeterminancy and moving toward incontestability" (p. 137), citing as key examples the objective third-person style of writing, the flatness, and the disallowance for competing interpretations to coexist. She requested that mixed methods researchers adopt a "sense of incompleteness" (p. 138) and recommended that reforms required a

> need to explore the possibility of hybridization in which a radical intertextuality of mixing forms, genres, conventions, and media . . . where there are no clear rules of representation and where the researcher, who is in reality working with radical undecidability and circumscribed indeterminacy, is able to make this experience freely available to readers and writers. (p. 144)

Creswell (in press-a) gave voice and focus to several of these critiques in a summary of controversies in mixed methods research. He discussed 11 controversies, examined multiple sides to the issues, and posed lingering questions. As shown in Table 2.3, these controversies related to definition, use of terms, philosophical issues, the discourse of mixed methods, the design possibilities, and the value of mixed methods research. Several of

Table 2.3	Eleven Key Controversies and Questions Being Raised in Mixed Methods Research

Controversies	Questions Being Raised
1. The changing and expanding definitions of mixed methods research	What is mixed methods research? How should it be defined? What shifts are being seen in its definition?
2. The questionable use of qualitative and quantitative descriptors	Are the terms *quantitative* and *qualitative* useful descriptors? What inferences are made when these terms are used? Is there a binary distinction being made that does not hold in practice?
3. Is mixed methods a "new" approach to research?	When did the conceptualization of mixed methods begin? Does mixed methods predate the period often associated with its beginning (1988–1989)? What initiatives began prior to 1988 and 1989?
4. What drives the interest in mixed methods?	How has mixed methods grown in interest? What is the role of funding agencies in its development?
5. Is the paradigm debate still being discussed?	Can paradigms be mixed? What stances on paradigm use in mixed methods have developed? Should the paradigm for mixed methods be based on scholarly communities?
6. Does mixed methods privilege postpositivism?	In the privileging of postpositivism in mixed methods, does it marginalize qualitative, interpretive approaches and relegate it to secondary status?
7. Is there a fixed discourse in mixed methods?	Who controls the discourse about mixed methods? Is mixed methods nearing a "metanarrative?"
8. Should mixed methods adopt a bilingual language for its terms?	What is the language of mixed methods research? Should the language be bilingual or new, or reflect quantitative and qualitative terms?
9. Are there too many confusing design possibilities for mixed methods procedures?	What designs should mixed methods researchers use? Are the present designs complex enough to reflect practice? Should entirely new ways of thinking about designs be adopted?
10. Is mixed methods research misappropriating designs and procedures from other approaches to research?	Are the claims of mixed methods overstated (because of misappropriation of other approaches to research)? Can mixed methods be seen as an approach lodged within a larger framework (e.g., ethnography)?
11. What value is added by mixed methods beyond the value gained through quantitative or qualitative research?	Does mixed methods provide a better understanding of a research problem than either quantitative or qualitative research alone? How could the value of mixed methods research be substantiated through scholarly inquiry?

SOURCE: Creswell (in press-a) reprinted with permission of SAGE Publications, Inc.

these controversies are addressed later in Chapter 9 in which we offer final recommendations for the design and conduct of mixed methods research.

● PHILOSOPHICAL FOUNDATIONS

Just as mixed methods has a history that can be chronicled, it also has a philosophy or perhaps philosophies that provide a foundation for conducting research. In fact, all research has a philosophical foundation, and inquirers should be aware of assumptions they make about gaining knowledge during their study. These assumptions shape the processes of research and the conduct of inquiry. Knowledge of these assumptions is especially important for graduate students who need to be able to identify and articulate the assumptions that they bring to research. Granted, philosophical assumptions often do not become explicit statements in published journal articles or books, but they do frequently arise at conference presentations or in graduate student committee meetings. As a general rule, we suggest that mixed methods researchers not only be aware of their philosophical assumptions but also clearly articulate their assumptions in their mixed methods projects.

What is involved in articulating philosophical assumptions in a mixed methods study? We believe that it includes acknowledging the worldview(s) providing a foundation for the study, describing the elements of the worldview, and relating these elements to specific procedures in a mixed methods project.

Philosophy and Worldviews

A framework is needed for thinking about how philosophy fits into the design of a mixed methods study. We like to use Crotty's (1998) conceptualization (as adapted) to position philosophy within a mixed methods study. As shown in Figure 2.1, Crotty contends that there are four major elements in developing a proposal or designing a study. At the broadest level are the issues of philosophical assumptions, such as the epistemology behind the study or how researchers gain knowledge about what they know. These philosophical assumptions, in turn, inform the use of a theoretical "stance" that the researcher might use (later we will refer to these stances as lenses drawn from social science theory or emancipatory theory). This stance then informs the methodology used, which is a strategy, a plan of action, or a research design. Finally, the methodology incorporates the methods, which are techniques or procedures used to gather, analyze, and interpret the data.

Figure 2.1 Four Levels for Developing a Research Study

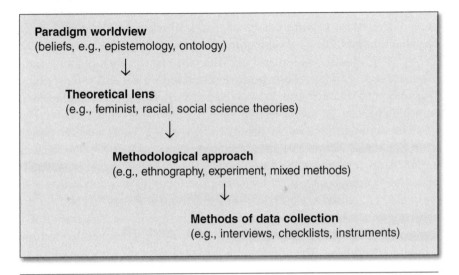

Paradigm worldview
(beliefs, e.g., epistemology, ontology)

↓

Theoretical lens
(e.g., feminist, racial, social science theories)

↓

Methodological approach
(e.g., ethnography, experiment, mixed methods)

↓

Methods of data collection
(e.g., interviews, checklists, instruments)

SOURCE: Adapted from Crotty (1998).

As we discussed in Chapter 1, mixed methods is largely a method, but it also involves a strategy for conducting research, and it could therefore be assigned in Crotty's classification at the level of a methodology.

Surrounding a mixed methods project, then, are philosophical assumptions that operate at a broad, abstract level. **Philosophical assumptions** in mixed methods research consist of a basic set of beliefs or assumptions that guide inquiries (see Guba & Lincoln, 2005). A term that we would use to describe these assumptions is **worldview**, and we say that mixed methods researchers bring to their inquiry a worldview composed of beliefs and assumptions about knowledge that informs their study. A term that is often used synonymously with worldview would be *paradigm*. Going back to the original use of the term by Thomas Kuhn (1970), a paradigm is a set of generalizations, beliefs, and values of a community of specialists. Although Kuhn himself pointed out the many uses of paradigm, the term that we favor is worldview, which may or may not be associated with a specific discipline or community of scholars but which suggests the shared beliefs and values of researchers. The most noted work on worldviews is available in qualitative research (Guba & Lincoln, 2005), but philosophical discussions are available for quantitative approaches as well (Phillips & Burbules, 2000). Most of these writings are by authors from the fields of social foundations of research or the philosophy of education (for overviews of many different worldviews in research, see Guba & Lincoln, 2005; Paul, 2005; Slife & Williams, 1995).

What worldviews might inform the practices of mixed methods researchers? Various writers have offered worldview possibilities, but we feel that four possible worldviews can inform mixed methods research. Like Crotty (1998), who holds that these different stances are not "watertight compartments" (p. 9), these worldviews provide a general philosophical orientation to research and, as we see later, they can be combined or used individually.

The four worldviews in Table 2.4 provide a good starting point. Postpositivism is often associated with quantitative approaches. Researchers make claims for knowledge based on (1) determinism or cause-and-effect thinking; (2) reductionism, by narrowing and focusing on select variables to interrelate; (3) detailed observations and measures of variables; and (4) the testing of theories that are continually refined (Slife & Williams, 1995). Constructivism, typically associated with qualitative approaches, works from a different worldview. The understanding or meaning of phenomena, formed through participants and their subjective views, make up this worldview. When participants provide their understandings, they speak from meanings shaped by social interaction with others and from their own personal histories. In this form of inquiry, research is shaped "from the bottom up"— from individual perspectives to broad patterns and, ultimately, to broad understandings.

Table 2.4 Basic Characteristics of Four Worldviews Used in Research

Postpositivist Worldview	Constructivist Worldview	Participatory Worldview	Pragmatist Worldview
Determination	Understanding	Political	Consequences of actions
Reductionism	Multiple participant meanings	Empowerment and issue oriented	Problem centered
Empirical observation and measurement	Social and historical construction	Collaborative	Pluralistic
Theory verification	Theory generation	Change oriented	Real-world practice oriented

SOURCE: Creswell (2009c) reprinted with permission of SAGE Publications, Inc.

Participatory worldviews are influenced by political concerns, and this perspective is more often associated with qualitative approaches than quantitative approaches, although it does not always have this association. The need to improve our society and those in it characterizes these views. Issues such as empowerment, marginalization, hegemony, patriarchy, and other issues affecting marginalized groups need to be addressed, and researchers collaborate with individuals experiencing these injustices. In the end, the participatory researcher plans for the social world to be changed for the better, so that individuals will feel less marginalized. A final worldview, **pragmatism**, is typically associated with mixed methods research. The focus is on the consequences of research, on the primary importance of the question asked rather than the methods, and on the use of multiple methods of data collection to inform the problems under study. Thus, it is pluralistic and oriented toward "what works" and practice.

All four worldviews have common elements but take different stances on these elements. Worldviews differ in the nature of reality (ontology), how we gain knowledge of what we know (epistemology), the role values play in research (axiology), the process of research (methodology), and the language of research (rhetoric) (Creswell, 2009c; Lincoln & Guba, 2000). These different stances influence how researchers conduct and report their inquiries. Examples of these common elements, the different worldviews, and how the elements and worldviews are translated into practice are shown in Table 2.5. Ontology refers to the nature of reality (and what is real) when researchers conduct their inquiries. The postpositivist tends to view reality as singular. An example would be a theory that hovers above the research study and helps to explain (in a single reality) the findings in the study. Another illustration would be the postpositivist tendency to reject or fail to reject a hypothesis. On the other hand, the constructivist views reality as multiple and actively looks for multiple perspectives from participants, such as perspectives developed through multiple interviews. The participatory researcher finds reality always negotiated and cast within a political context, while the pragmatist views reality as both singular (e.g., there may be a theory that operates to explain the phenomenon of study) as well as multiple (e.g., it is important to assess varied individual input into the nature of the phenomenon as well).

As another example of differences among the worldviews, consider the methodological differences (i.e., the process of research). In postpositivist research, the investigator works from the "top" down, from a theory to hypotheses to data to add to or contradict the theory. In constructivist approaches, the inquirer works more from the "bottom" up, using the participants' views to build broader themes and generate a theory interconnecting

e 2.5   Elements of Worldviews and Implications for Practice

Worldview Element	Postpositivism	Constructivism	Participatory	Pragmatism
Ontology (What is the nature of reality?)	Singular reality (e.g., researchers reject or fail to reject hypotheses)	Multiple realities (e.g., researchers provide quotes to illustrate different perspectives)	Political reality (e.g., findings are negotiated with participants)	Singular and multiple realities (e.g., researchers test hypotheses and provide multiple perspectives)
Epistemology (What is the relationship between the researcher and that being researched?)	Distance and impartiality (e.g., researchers objectively collect data on instruments)	Closeness (e.g., researchers visit participants at their sites to collect data)	Collaboration (e.g., researchers actively involve participants as collaborators)	Practicality (e.g., researchers collect data by "what works" to address research question)
Axiology (What is the role of values?)	Unbiased (e.g., researchers use checks to eliminate bias)	Biased (e.g., researchers actively talk about their biases and interpretations)	Negotiated (e.g., researchers negotiate their biases with participants)	Multiple stances (e.g., researchers include both biased and unbiased perspectives)
Methodology (What is the process of research?)	Deductive (e.g., researchers test an a priori theory)	Inductive (e.g., researchers start with participants' views and build "up" to patterns, theories, and generalizations)	Participatory (e.g., researchers involve participants in all stages of the research and engage in cyclical reviews of results)	Combining (e.g., researchers collect both quantitative and qualitative data and mix them)
Rhetoric (What is the language of research?)	Formal style (e.g., researchers use agreed-on definitions of variables)	Informal style (e.g., researchers write in a literary, informal style)	Advocacy and change (e.g., researchers use language that will help bring about change and advocate for participants)	Formal or informal (e.g., researchers may employ both formal and informal styles of writing)

Handwritten margin notes:

How do we gain knowledge of what we know?

in research

the themes. In participatory research, the researcher collaborates with other participants serving as active members of the research team, helping to form questions, analyzing the data, and implementing the results in practice. In pragmatism, the approach may combine deductive and inductive thinking, as the researcher mixes both qualitative and quantitative data.

Worldviews Applied to Mixed Methods

Up until this point, we have reviewed four different worldviews and discussed how they might differ in terms of broad philosophical elements of ontology, epistemology, axiology, methodology, and rhetoric. Which worldview(s) best fits a mixed methods study? Answers to this question have occupied the attention of mixed methods researchers for some time (Tashakkori & Teddlie, 1998, 2003a), and their responses have varied. In designing and conducting mixed methods research, researchers need to know the alternative stances on worldviews and mixed methods research and to be able to articulate the stance they are using. They might convey their stance in a separate section of a project, titled "philosophical assumptions" or in the methods section of their plan or study. Mixed methods researchers might consider which of the following four stances best relate to their studies and convey the stance that they have embraced in the philosophical assumptions section of their study.

One "best" worldview for mixed methods. Although some individuals still seek to participate in the paradigm debate, many mixed methods writers have moved on to identify the "best" worldview that provides a foundation for mixed methods research. Tashakkori and Teddlie (2003a) suggested that at least 13 different authors embrace pragmatism as the worldview or paradigm for mixed methods research. Although we have already introduced pragmatism, because of its importance, it merits further discussion.

Pragmatism is a set of ideas articulated by many people, from historical figures, such as John Dewey, William James, and Charles Sanders Peirce, to contemporaries, such as Cherryholmes (1992) and Murphy (1990). It draws on many ideas, including employing "what works," using diverse approaches, and valuing both objective and subjective knowledge. Tashakkori and Teddlie (2003a) formally linked pragmatism and mixed methods research, arguing the following points:

- Both quantitative and qualitative research methods may be used in a single study.

- The research question should be of primary importance—more important than either the method or the philosophical worldview that underlies the method.
- The forced-choice dichotomy between postpositivism and constructivism should be abandoned.
- The use of metaphysical concepts such as "truth" and "reality" should also be abandoned.
- A practical and applied research philosophy should guide methodological choices.

Another "best" paradigm approach is found in the transformative–emancipatory paradigm of Mertens (2003; see also Sweetman, Badiee, & Creswell, 2010). Mertens (2003) provided an original, insightful contribution to the mixed methods literature by bridging the philosophy of inquiry (i.e., paradigms) with the practice of research. In discussing this perspective, she said,

> Transformative . . . scholars recommend the adoption of an explicit goal for research to serve the ends of creating a more just and democratic society that permeates the entire research process, from the problem formulation to the drawing of conclusions and the use of results. (Mertens, 2003, p. 159)

Indeed, Mertens (2003) has given us a framework that has immediate applicability for assessing the inclusion of an emancipatory perspective in mixed methods studies. She has suggested that the name for this framework is the "transformative" framework and that it includes a person's worldview and implicit value assumptions. These assumptions are that knowledge is not neutral and is influenced by human interests. Knowledge reflects the power and social relationships within society, and the purpose of knowledge construction is to aid people to improve society. Issues, such as oppression and domination—found in critical theory perspectives—become important to study. She cited several groups that have extended the thinking about the place of values in research, including feminists, members of diverse ethnic and racial groups, and people with disabilities (Mertens, 2003). By 2009, Mertens expanded her list of marginalized groups to also include lesbian, gay, bisexual, transgender, and queer communities and enlarged her theoretical perspectives to include positive psychology and resilience theory.

The critical realist perspective is also being discussed as a potential contribution to mixed methods research (Maxwell & Mittapalli, in press). It is a philosophical perspective that validates and supports key aspects of both quantitative and qualitative approaches. While identifying some specific limitations of each, realism, they contended, can constitute a productive stance for mixed methods research and facilitate collaboration between quantitative and qualitative researchers. They discussed **critical realism** as an integration of a realist ontology (there is a real world that exists independently of our perceptions, theories, and constructions) with a constructivist epistemology (our understanding of this world is inevitably a construction from our own perspectives and standpoint). The authors, however, acknowledged that explicit use of realist perspectives in mixed methods research was still relatively uncommon except in Europe (and cited examples in accounting, economics, psychiatry, and nursing). We would add that it confounds the use of theory and the use of paradigms since "critical" is often associated with a theoretical lens more than a worldview and that it offers a challenge to the status quo (see the next section on the use of theoretical lens).

Multiple worldviews in mixed methods. This position states that multiple paradigms may be used in mixed methods research; researchers must simply be explicit in their use. This "dialectical" perspective (Greene & Caracelli, 1997) recognizes that different paradigms give rise to contradictory ideas and contested arguments—features of research that are to be honored but cannot be reconciled. These contradictions, tensions, and oppositions reflect different ways of knowing about and valuing the social world. This stance emphasizes using multiple worldviews (e.g., constructivism and participatory) during the study instead of using a single worldview, such as pragmatism.

Worldviews relate to the type of mixed methods design. In this third stance, a stance we embrace, we suggest that more than one worldview might be used in a mixed methods study (in contrast to Stance 1), and that the selection of multiple worldviews relate to the type of mixed methods design used rather than a worldview based on how the researcher attempts to "know" the social world (as stated in Stance 2). We believe that multiple paradigms can be used in a mixed methods study and that they best relate to type of mixed methods designs. Although a worldview is not always "linked" to procedures in research, the guiding assumptions of worldviews often shape how mixed methods researchers construct their procedures. Quantitative methods (e.g., surveys, experiments) are typically used within a postpositivist worldview in which some guiding determining theory is advanced at the beginning, and the study is delimited to certain variables that are empirically measured and observed. Therefore, if a study begins with a survey, the researcher is implicitly using a postpositivist worldview to inform the study beginning with specific variables, empirical measures, and often framed within an a priori theory that is being tested in the survey project.

Then, if the researcher moves to qualitative focus groups in the second phase to follow up on and explain the survey results, it seems like the worldview shifts to more of a constructivist perspective. In the focus group, the attempt is to elicit multiple meanings from the participants, to build a deeper under-standing than the survey would yield, and to possibly generate a theory or pattern of responses that explain the survey results. In effect, the researcher has shifted from a postpositivist worldview in the first phase of the research into a constructivist worldview in the second phase. Thus, our view is that worldviews relate to types of designs, that the worldviews can change during a study, that the worldview may be tied to different phases in the project, and that researchers need to honor and to write about their worldviews in use. If, instead of implementing the different approaches in phases, a mixed meth-ods researcher collects both quantitative and qualitative data in the same phase of the project and merges the two databases, then an all-encompassing worldview might be best for the study. We would look to pragmatism (or a transformative perspective) as that worldview, because it enables researchers to adopt a pluralistic stance of gathering all types of data to best answer the research questions. We will make explicit this connection—between the worldview in use and the design in use—for each mixed methods design in Chapter 3.

Worldviews depend on the scholarly community. Recently mixed methods writers have turned to Kuhn's (1970) idea of a community of practitioners. Two key writings appeared in 2007 and 2008 in JMMR articles by the American author David Morgan and by the British author Martin Descombe. Morgan's (2007) article is a fascinating piece of scholarship, and it was first presented in 2005 as the keynote address at the Mixed Methods Conference in Cambridge, United Kingdom. Morgan (2007) saw paradigms as "shared belief systems that influence the kinds of knowledge researchers seek and how they interpret the evidence they collect" (p. 50). However, he saw four types of paradigms, and they differed in terms of generality. First, paradigms can be viewed as worldviews, an all-encompassing perspective on the world, or sec-ond, they can be seen as epistemologies incorporating ideas from the phi-losophy of science, such as ontology, methodology, and epistemology. Third, paradigms can be viewed as the "best" or "typical" solutions to problems, and fourth, paradigms may represent shared beliefs of a research field. It is this last perspective that Morgan strongly endorsed. Researchers, he said, shared a consensus in specialty areas about what questions were most meaningful and which procedures were most appropriate for answering the questions. In short, many practicing researchers looked to worldview perspectives from a "community of scholars" perspective (p. 53). According to Morgan, this was the version of paradigms that Kuhn (1970) most favored as he talked about a community of practitioners.

Denscombe (2008) reinforced Morgan's position and added more details about the nature of a community of practitioners. He outlined how communities work using such ideas as shared identity, common research problems, social networks, knowledge formation, and informal groupings. The mixed methods field is becoming fragmented by discipline orientation, and it will ultimately be shaped, we believe, by strong subject matter interests. For example, when colleagues in the health sciences at the Veterans Administration Health Services Research Center in Ann Arbor, Michigan, refer to mixed methods from an evaluation perspective of "formative" and "summative" procedures, they are embracing mixed methods from a field orientation that makes sense within the health services research area (Forman & Damschroder, 2007).

THEORETICAL FOUNDATIONS ●

Referring back to Crotty's (1998) model in Figure 2.1, we find theory operating at a narrower perspective than worldview. A theoretical foundation in mixed methods is a stance (or lens or standpoint) taken by the researcher that provides direction for many phases of a mixed methods project. How does the researcher incorporate it into a study? What type of theory might the researcher use? We see two types of theory that might inform a mixed methods study: a social sciences theory and an emancipatory theory.

A theoretical orientation for a mixed methods study would be the use of an explanatory framework from the social sciences that predicts and shapes the direction of a research study. A social science theory is positioned at the beginning of a mixed methods study, and it provides a framework, or theory, from the social sciences that guides the nature of the questions asked and answered in a study. The data collected may be either quantitative or qualitative or both. This theory may be a leadership theory, an economic theory, a marketing theory, a theory of behavioral change, a theory of adoption or diffusion, or any number of social science theories. It may be presented as a literature review, as a conceptual model, or as a theory that helps to explain what the researcher seeks to find in a study.

An example of a social science theory can be found in a mixed methods study about chronic pain and its management through learned resourcefulness by Kennett, O'Hagan, and Cezer (2008). These authors presented a mixed methods study to understand how learned resourcefulness empowers individuals. In this study, they gathered measures on Rosenbaum's

Self-Control Schedule (SCS) and through interviews with patients coping with chronic pain. In the opening paragraph of their study, they advanced the purpose of their study:

> Taking a critical realist perspective informed by Rosenbaum's (1990, 2000) model of self-control, we combine a quantitative measure of learned resourcefulness with a qualitative text-based analysis to characterize the processes that come into play in the self-management of pain for high- and low-resourceful clients following a multimodel treatment-based pain program. (Kennett et al., 2008, p. 318)

Rosenbaum's model was used because it challenged the status quo about health programs, as well as stimulated the transformation of practice. The authors first introduced the major components of Rosenbaum's model. This was followed by the research literature on resourcefulness as an important predictor of adopting healthy behavior and a discussion of one of Rosenbaum's experiments relating resourcefulness to coping with pain. The authors discussed the factors of the model leading to self-control, such as factors related to process-regulating cognitions (e.g., supporting family and friends), coping strategies (e.g., ability to cope with pain, such as diverting attention and reinterpreting pain), and staying in (or dropping out of) programs. The authors at this point might have drawn a diagram of these factors that influenced self-control as a guiding theoretical framework for their theory. They provided next, however, a series of questions drawn from Rosenbaum's model and the literature that guided their study examining the impact of a cognitive–behavioral chronic pain management program on self-management and how resourcefulness and a sense of self-directedness influence self-management skills for chronic pain. Toward the end of the article, they revisited the factors leading to self-management and propose a diagram of the most salient factors.

Stepping back from this discussion, we can see how a mixed methods researcher might incorporate a social science theoretical lens into a mixed methods study (see Creswell, 2009c):

- Place the discussion of the theory (model or conceptual framework) at the beginning of the article as an a priori framework to guide the questions in the study.
- Write about the theory by first advancing the name of the theory to be used followed by a description of the major variables of the theory. Discuss previous studies that have used the theory. End by specifically stating how the theory will inform the questions and procedures of the current mixed methods study.

- Include a diagram of the theory that indicates the direction of the causal links in the theory and the major concepts, or variables, in the theory.
- Have the theory provide a framework for both the quantitative and the qualitative data collection efforts in the study.

In contrast to a social science theory as a guiding explanation in a mixed methods study, an **emancipatory theory** in mixed methods involves taking a theoretical stance in favor of underrepresented or marginalized groups, such as a feminist theory, a racial or ethnic theory, a sexual orientation theory, or a disability theory (Mertens, 2009) and calling for change. With one goal of qualitative research to address issues of social justice and the human condition (Denzin & Lincoln, 2005), this emphasis has come to be expected from some scholars in mixed methods research. However, we noted a couple of years ago that few studies incorporated this theoretical emancipatory lens (Creswell & Plano Clark, 2007). Today, mixed methods studies with an empancipatory lens are becoming more frequently reported in the mixed methods literature. For example, recent mixed methods studies have addressed topics such as African American girls' interest in science (Buck, Cook, Quigley, Eastwood, & Lucas, 2009), women's social capital in Australia (Hodgkin, 2008), and women's understanding of community-specific rape myths (McMahon, 2007). Methodological writings about linking feminist standpoint epistemology to mixed methods have also been recently published (Hesse-Biber & Leavy, 2006)

From a review of articles incorporating an emancipatory lens, we can see numerous examples as to how to incorporate this lens into a mixed methods study. A recent study analyzed 13 mixed methods studies (Sweetman, Badiee, & Creswell, 2010) that incorporated an emancipatory theoretical lens. Results showed a wide variety of social science journals that published these studies (e.g., *Women and Health*, *Families in Society*, *Social Work Research*, *Urban Studies*), and six different theoretical lenses were used by the authors. Feminism was the most common (six studies), with socioeconomic status as the next (two studies), followed by disability, human ecology, and general gender. Some articles spanned multiple social categorizations, such as income, ethnicity, and gender. From reviewing these studies, the authors made recommendations for incorporating an emancipatory lens into a mixed methods study:

- Introduce the emancipatory lens at the beginning of the study.
- Apply it when discussing the literature.
- Make it explicit in discussing the research problem.

- Write it into the research questions using emancipatory, advocacy language.
- Discuss collecting data in a way that will not further marginalize the community.
- Position the researchers in the study.
- Suggest a plan of action or change to end the study.

Even with these suggestions, more work needs to be done to establish how the procedures of mixed methods might change depending on the type of emancipatory lens used (e.g., feminist, racial). As more mixed methods studies begin incorporating an emancipatory lens, we can learn more about how to include such a lens and the variety of studies that use this type of theoretical lens.

SUMMARY

In planning a mixed methods study, researchers need to cite references to the latest literature, justify its use, and recognize how their study fits into the evolving field of mixed methods research. Although some of the elements of mixed methods approaches were evident prior to the 1980s, several writers from different disciplines and different countries came to the idea of mixed methods at roughly the same time—the late 1980s. Thus, the field is a little over 20 years old, and it has evolved because of the complexity of research problems, the legitimatization of qualitative inquiry, and the need for more evidence in applied settings. Its evolution has gone through five phases: (1) the formative period of considering multiple forms of data; (2) the paradigm debate period in which heated discussions occurred about whether mixed methods inappropriately integrated different philosophical perspectives; (3) the procedural phase in which writers pushed for increased understanding about conducting a mixed methods study; (4) the advocacy and expansion phase in which writers suggested mixed methods was a distinct methodology and its popularity spread to diverse disciplines and different countries around the world; and (5) the current reflective phase in which writers discussed the priorities, issues, and controversies surrounding mixed methods research.

Further, researchers bring to their mixed methods study philosophical assumptions that need to be made explicit and discussed. Researchers need to acknowledge the philosophical worldview they bring to a project, identify the components of their worldview, and relate them to the specific elements of their mixed methods study. Worldviews are beliefs and values that researchers bring to a study, and they may be drawn from at least one or more perspectives, such as postpositivism, constructivism, participatory worldviews, and pragmatism. The elements for each worldview differ, and they are reflected in different philosophical assumptions, such as ontology, epistemology, axiology, methodology, and rhetoric. In response to these philosophical ideas, mixed methods researchers have taken different stances of the use of worldviews in their research. Some believe that there is a single worldview that informs mixed methods, such as pragmatism, transformative approaches, or critical realism. Others hold that multiple worldviews can inform a mixed methods study and that the choice of worldview is related to the type of mixed methods design chosen. A recent stance is that worldviews form within scholarly communities and that they may vary from community to community. Regardless of worldview, the assumptions behind a mixed methods study need to be identified and stated in mixed methods projects.

Mixed methods researchers may also use a theoretical lens in their study, one drawn from social sciences theories or from an emancipatory perspective, such as a feminist, disability, or ethnic viewpoint. Social science theories are often positioned at the beginning of a mixed methods study, and they inform the questions asked and the interpretation of the results. Emancipatory theories are often threaded throughout a project, and they inform the lens being used, the types of research questions asked, the procedures used in data collection, and the call for action advanced at the end of studies.

ACTIVITIES

1. Do a search of the literature using databases to find books and articles on mixed methods research. Note recent writers who describe the essential characteristics of mixed methods research. Compile a list of authors who you would cite as recent writers about mixed methods as you define mixed methods in your study.

2. What philosophical worldview(s) will inform your mixed methods study? Identify one or more worldviews, discuss the elements that comprise the worldviews, and state specifically how the worldview will inform the conduct of your mixed methods research study.

3. Select a mixed methods study with a feminist lens and analyze it. Look at the article by McMahon (2007) on understanding community-specific rape myths. Identify how the authors incorporated a feminist lens into the research problem, the research questions, the data collection, and in the call for change or action at the end of the article.

ADDITIONAL RESOURCES TO EXAMINE

For a historical analysis of mixed methods research, consult the following resources:

Greene, J. C. (2007). *Mixed methods in social inquiry*. San Francisco: Jossey-Bass.

Tashakkori, A., & Teddlie, C. (1998). *Mixed methodology: Combining qualitative and quantitative approaches*. Thousand Oaks, CA: Sage.

For a discussion of philosophical worldviews related to mixed methods research, see the following resources:

Denscombe, M. (2008). Communities of practice: A research paradigm for the mixed methods approach. *Journal of Mixed Methods Research*, *2*, 270–283.

Morgan, D. L. (2007). Paradigms lost and pragmatism regained: Methodological implications of combining qualitative and quantitative methods. *Journal of Mixed Methods Research*, *1*(1), 48–76.

For discussions of the use of a theoretical lens in mixed methods research, see the following resources:

Mertens, D. M. (2009). *Transformative research and evaluation*. New York: Guilford Press.

Sweetman, D., Badiee, M., & Creswell, J. W. (2010). Use of the transformative framework in mixed methods studies. *Qualitative Inquiry*. Prepublished April 15, 2010, DOI: 10.1177/1077800410364610

CHAPTER 3

CHOOSING A MIXED METHODS DESIGN

R esearch designs are procedures for collecting, analyzing, interpreting, and reporting data in research studies. They represent different models for doing research, and these models have distinct names and procedures associated with them. Research designs are useful, because they help guide the methods decisions that researchers must make during their studies and set the logic by which they make interpretations at the end of their studies. Once the researcher has identified that the research problem calls for a mixed methods approach and reflected on the philosophical and theoretical foundations of the study, the next step is to choose a specific design that best fits the problem and the research questions in the study. What designs are available, and how do researchers decide which one is appropriate for their studies? Mixed methods researchers need to be acquainted with the major types of mixed methods designs and the key decisions behind these designs to adequately consider available options. Each major design has its own history, purpose, considerations, philosophical assumptions, procedures, strengths, challenges, and variants. With an understanding of the basic designs in hand, researchers are equipped to choose and describe the mixed methods design best suited to address a stated problem.

This chapter introduces the basic designs available to the researcher planning to engage in mixed methods research. It will address

- principles for designing a mixed methods study;
- decisions necessary in choosing a mixed methods design;

- characteristics of major mixed methods designs;
- the history, purpose, philosophical assumptions, procedures, strengths, challenges, and variants for each of the major designs; and
- a model for writing about a design in a written report.

● PRINCIPLES FOR DESIGNING A MIXED METHODS STUDY

Designing research studies is a challenging process in both quantitative and qualitative research. This process can become even more of a challenge when the researcher has decided to use a mixed methods approach due to the inherent complexity in mixed methods designs. Although the design and conduct of any two mixed methods studies will never be exactly alike, there are several key principles that researchers consider to help navigate this process: using a fixed and/or emergent design; identifying a design approach to use; matching a design to the study's problem, purpose, and questions; and being explicit about the reason for mixing methods.

Recognize That Mixed Methods Designs Can Be Fixed and/or Emergent

Mixed methods designs may be fixed and/or emergent, and researchers need to be cognizant of the approach that they are using and open to considering the best alternative for their circumstances. **Fixed mixed methods designs** are mixed methods studies where the use of quantitative and qualitative methods is predetermined and planned at the start of the research process, and the procedures are implemented as planned. **Emergent mixed methods designs** are found in mixed methods studies where the use of mixed methods arises due to issues that develop during the process of conducting the research. Emergent mixed methods designs generally occur when a second approach (quantitative or qualitative) is added after the study is underway because one method is found to be inadequate (Morse & Niehaus, 2009). For example, Ras (2009) described how she found the need to add a quantitative component to her qualitative case study of self-imposed curricular change at one elementary school. She addressed emergent concerns with the trustworthiness of her interpretations of what she learned from her participants. In this way, her qualitative case study became a mixed methods study during her process of implementing the research study.

We view these two categories—fixed and emergent—not as a clear dichotomy but as end points along a continuum. Many mixed methods designs actually fall somewhere in the middle with both fixed and emergent aspects to the design. For example, the researcher may plan to conduct a study in two phases from the start, such as beginning with a quantitative phase and then following up with a qualitative phase. The details of the design of the second, qualitative phase, however, may emerge based on the researcher's interpretation of the results from the initial quantitative phase. Therefore, the study becomes an example of combining both fixed and emergent elements.

Due to our focus on planning mixed methods studies and the linear and fixed nature of printed text, our writing may appear to emphasize fixed designs. Keep in mind, however, that we recognize the importance and value of emergent mixed methods approaches. We believe that most of the design elements that we address in this book apply well whether the use of mixed methods is planned from the start and/or emerges due to the needs of a study.

Identify an Approach to Design

In addition to using fixed and emergent mixed methods designs, researchers also use different approaches for designing their mixed methods studies. There are several approaches to design that have been discussed in the literature, and researchers can benefit from considering their personal approach to designing mixed methods studies. These design approaches fall into two categories: typology-based and dynamic.

A **typology-based** approach to mixed methods design emphasizes the classification of useful mixed methods designs and the selection and adaptation of a particular design to a study's purpose and questions. Unquestionably, this design approach has been discussed most extensively in the mixed methods literature, as shown by the amount of effort that has been spent on classifying mixed methods designs. There is a wide range of available classifications of types of mixed methods designs that methodologists have advanced. Creswell, Plano Clark, Gutmann, and Hanson (2003) summarized the range of these classifications in 2003, and we have updated the summary with a list of 15 classifications in Table 3.1. These classifications represent diverse disciplines, including evaluation, health sciences, and education, and span scholarly writings about mixed methods approaches since the late 1980s. They also tend to use different terminology and emphasize different features of mixed methods designs (a topic we will turn our attention to later

Table 3.1 Mixed Methods Design Classifications

Author	Discipline	Mixed Methods Designs
Greene, Caracelli, and Graham (1989)	Evaluation	Initiation Expansion Development Complementarity Triangulation
Patton (1990)	Evaluation	Experimental design, qualitative data, and content analysis Experimental design, qualitative data, and statistical analysis Naturalistic inquiry, qualitative data, and statistical analysis Naturalistic inquiry, quantitative data, and statistical analysis
Morse (1991)	Nursing	Simultaneous triangulation Sequential triangulation
Steckler, McLeroy, Goodman, Bird, and McCormick (1992)	Public health education	Model 1: Qualitative methods to develop quantitative measures Model 2: Qualitative methods to explain quantitative findings Model 3: Quantitative methods to embellish qualitative findings Model 4: Qualitative and quantitative methods used equally and parallel

Greene and Caracelli (1997)	Evaluation	Component designs
		  Triangulation
		  Complementarity
		  Expansion
		Integrated designs
		  Iterative
		  Embedded or nested
		  Holistic
		  Transformative

Morgan (1998)	Health research	Complementary designs
		  Qualitative preliminary
		  Quantitative preliminary
		  Qualitative follow-up
		  Quantitative follow-up

Author	Discipline	Mixed Methods Designs

Tashakkori and Teddlie (1998)	Educational research	Mixed methods designs
		  Equivalent status (sequential or parallel)
		  Dominant–less dominant (sequential or parallel)
		  Multilevel use
		Mixed model designs
		  I. Confirmatory, qualitative data, statistical analysis, and inference
		  II. Confirmatory, qualitative data, qualitative analysis, and inference
		  III. Exploratory, quantitative data, statistical analysis, and inference
		  IV. Exploratory, qualitative data, statistical analysis, and inference
		  V. Confirmatory, quantitative data, qualitative analysis, and inference
		  VI. Exploratory, quantitative data, qualitative analysis, and inference
		  VII. Parallel mixed model
		  VIII. Sequential mixed model

Creswell (1999)	Educational policy	Convergence model Sequential model Instrument-building model
Sandelowski (2000)	Nursing	Sequential Concurrent Iterative Sandwich
Creswell, Plano Clark, Gutmann, and Hanson (2003)	Educational research	Sequential explanatory Sequential exploratory Sequential transformative Concurrent triangulation Concurrent nested Concurrent transformative

(Continued)

Table 3.1 (Continued)

Author	Discipline	Mixed Methods Designs
Creswell, Fetters, and Ivankova (2004)	Primary medical care	Instrument design model Triangulation design model Data transformation design model
Tashakkori and Teddlie (2003b)	Social and behavioral research	Multistrand designs Concurrent mixed designs 　Concurrent mixed methods design 　Concurrent mixed model design Sequential mixed designs 　Sequential mixed methods design 　Sequential mixed model design Multistrand conversion mixed designs 　Multistrand conversion mixed methods design 　Multistrand conversion mixed model design Fully integrated mixed model design
Greene (2007)	Evaluation	Component designs 　Convergence 　Extension Integrated designs 　Iteration 　Blending 　Nesting or embedding 　Mixing for reasons of substance or values
Teddlie & Tashakkori (2009)	Educational research	Mixed methods multistrand designs 　Parallel mixed designs 　Sequential mixed designs 　Conversion mixed designs 　Multilevel mixed designs 　Fully integrated mixed designs

Author	Discipline	Mixed Methods Designs
Morse and Neihaus (2009)	Nursing	Mixed method simultaneous designs Mixed method sequential designs Complex mixed method designs 　Qualitatively driven complex mixed method design 　Quantitatively driven complex mixed method design 　Multiple method research program

SOURCE: Adapted from Creswell, Plano Clark, et al. (2003, pp. 216–217, Table 8.1) with permission of SAGE Publications, Inc.

in this chapter). The different types and various classifications speak to the evolving nature of mixed methods research and the utility of considering designs as a framework for thinking about mixed methods.

There are also dynamic approaches for thinking about the process of designing a mixed methods study. Dynamic approaches to mixed methods design focus on a design process that considers and interrelates multiple components of research design rather than placing emphasis on selecting an appropriate design from an existing typology. Maxwell and Loomis (2003) introduced an interactive, systems-based approach to mixed methods design. They argued that the researcher should weigh five interconnected components when designing a mixed methods study: the study's purposes, conceptual framework, research questions, methods, and validity considerations. Although research questions are at the heart of this process, they discuss how the interrelationships among the components need to be considered throughout the design process.

Hall and Howard (2008) recently described another dynamic approach to mixed methods design, which they called the synergistic approach. They suggested that the synergistic approach provided a way to combine a typological approach with a systemic approach. In a synergistic approach, two or more options interacted so that their combined effect was greater than the sum of the individual parts. Translated into mixed methods, this meant that the sum of quantitative and qualitative research was greater than either approach alone. They defined this approach through a set of core principles: the concept of synergy, the position of equal value, the ideology of difference, and the relationship between the researcher(s) and the study design. They argued that this approach provided an effective combination of structure and flexibility that helped them consider how epistemology, theory, methods, and analysis could work together within a mixed methods design.

We suggest that researchers, particularly those new to designing and conducting mixed methods studies, consider starting with a typology-based approach to mixed methods design. Typologies provide the researcher with a range of available options to consider that are well defined, facilitate the researcher's use of a solid approach for addressing the research problem, and help the researcher anticipate and resolve challenging issues. That said, we do not advocate that researchers adopt a typology-based design like a cookbook recipe but instead use it as a guiding framework to help inform design choices. As researchers gain more expertise with mixing methods, they are more able to effectively design their studies using a dynamic approach.

Match the Design to the Research Problem, Purpose, and Questions

The different approaches for mixed methods design differ in their emphases but also share many commonalities. In particular, each emphasizes the overall problem, purpose, and research questions that are guiding the study. Research problems and questions that interest researchers arise in many ways, such as from the literature, the researcher's experiences or values, logistical constraints, results that cannot be explained, and stakeholder expectations (Plano Clark & Badiee, in press). No matter how the research questions are generated, scholars writing about mixed methods research uniformly agree that the questions of interest play a central role in the process of designing any mixed methods study. The importance of the research problem and questions is a key principle of mixed methods research design. This perspective stems from the pragmatic foundations for conducting mixed methods research where the notion of "what works" applies well to selecting the methods that "work" best to address a study's problem and questions.

Recall the general research problems related to mixed methods introduced in Chapter 1. These included one data source alone is insufficient, results need to be explained, exploratory results need to be further examined, a study needs to be enhanced through adding a second method, a theoretical stance needs to be advanced through the use of both types of methods, and a problem needs to be studied through multiple phases of research that include multiple types of methods. Research problems like these not only call for the use of mixed methods but also call for the researcher to use different designs that are able to address the different types of problems. Therefore, researchers should articulate their research problem and questions and consider them carefully so as to choose a design that matches the problem and research questions. As we consider research questions in Chapter 5, we will also discuss how some research questions may be stated or refined to reflect the selected design.

Be Explicit About the Reasons for Mixing Methods

Another key principle of mixed methods design is to identify the reason(s) for mixing quantitative and qualitative methods within the study. Combining methods is challenging and should only be undertaken when there is a specific reason to do so. There are many good discussions of reasons for mixing methods found in the literature to help researchers guide their work. Two prominent frameworks are listed in Table 3.2. The first is the list of five broad reasons for mixing methods identified by Greene, Caracelli, and Graham in their 1989 work. These reasons—triangulation, complementarity, development, initiation, and expansion—are defined in the table. Although they were quite broad and general, this typology of reasons is still frequently used and discussed in the literature. As mixed methods research has continued to evolve in the past 20 years, however, more detailed descriptions of researchers' reasons have emerged. Recently, Bryman (2006) provided a detailed list of reasons based on researchers' practices (see Table 3.2). His list of 16 reasons offered a useful, more detailed examination of researchers' reasons and practices that added to the more general description found in Greene et al.'s (1989) work.

Keep in mind that the reasons listed for mixing methods should be viewed as a general framework from which researchers can weigh alternative choices and use to justify their mixing decisions. In his work, Bryman (2006) noted that many mixed methods studies make use of multiple reasons for mixing methods and that new reasons for mixing may emerge as the study is underway. Being responsive to new insights is an essential aspect of conducting mixed methods research, but we feel is it also important for researchers to design their mixed methods studies with at least one clear reason as to why they are planning to combine methods.

Table 3.2 Two Typologies of Reasons for Mixing Methods

Greene, Caracelli, and Graham (1989) [1]	Bryman (2006) [2]
• **Triangulation** seeks convergence, corroboration, and correspondence of results from the different methods. • **Complementarity** seeks elaboration, enhancement, illustration, and clarification of the results from one method with the results from the other method. • **Development** seeks to use the results from one method to help develop or inform the other method, where development is broadly construed to include sampling and implementation, as well as measurement decisions. • **Initiation** seeks the discovery of paradox and contradiction, new perspectives of frameworks, the recasting of questions or results from one method with questions or results from the other method. • **Expansion** seeks to extend the breadth and range of inquiry by using different methods for different inquiry components.	• **Triangulation or greater validity** refers to the traditional view that quantitative and qualitative research might be combined to triangulate findings in order that they may be mutually corroborated. • **Offset** refers to the suggestion that the research methods associated with both quantitative and qualitative research have their own strengths and weaknesses so that combining them allows the researcher to offset their weaknesses to draw on the strengths of both. • **Completeness** refers to the notion that the researcher can bring together a more comprehensive account of the area of inquiry in which he or she is interested if both quantitative and qualitative research are employed. • **Process** refers to when quantitative research provides an account of structures in social life but qualitative research provides sense of process. • **Different research questions** refers to the argument that quantitative and qualitative research can each answer different research questions. • **Explanation** refers to when one is used to help explain findings generated by the other. • **Unexpected results** refers to the suggestion that quantitative and qualitative research can be fruitfully combined when one generates surprising results that can be understood by employing the other. • **Instrument development** refers to contexts in which qualitative research is employed to develop questionnaire and scale items—for example, so that better wording or more comprehensive closed answers can be generated. • **Sampling** refers to situations in which one approach is used to facilitate the sampling of respondents or cases. • **Credibility** refers to suggestions that employing both approaches enhances the integrity of findings. • **Context** refers to cases in which the combination is rationalized in terms of qualitative research providing contextual understanding coupled with either generalizable, externally valid findings or broad relationships among variables uncovered through a survey.

Greene, Caracelli, and Graham (1989) [1]	Bryman (2006) [2]
	• **Illustration** refers to the use of qualitative data to illustrate quantitative findings, often referred to as putting "meat on the bones" of "dry" quantitative findings.
	• **Utility or improving the usefulness of findings** refers to a suggestion, which is more likely to be prominent among articles with an applied focus, that combining the two approaches will be more useful to practitioners and others.
	• **Confirm and discover** refers to using qualitative data to generate hypotheses and using quantitative research to test them within a single project.
	• **Diversity of views** includes two slightly different rationales—namely, combining researchers' and participants' perspectives through quantitative and qualitative research respectively and uncovering relationships between variables through quantitative research while also revealing meanings among research participants through qualitative research.
	• **Enhancement or building upon quantitative and qualitative findings** entails a reference to making more of or augmenting either quantitative or qualitative findings by gathering data using a qualitative or quantitative research approach.

[1] Reprinted from *Educational Evaluation and Policy Analysis*, Vol. 11, Issue 3, p. 259, 1989, with permission of SAGE Publications, Inc.

[2] Reprinted from *Qualitative Research*, Vol. 6, Issue 1, pp. 105–107, 2006, with permission of SAGE Publications, Inc.

KEY DECISIONS IN CHOOSING ●
A MIXED METHODS DESIGN

Building on the four principles previously discussed, researchers are in a position to make important choices that define the mixed methods design used in a study. These decisions address the different ways that the quantitative and qualitative strands of the study relate to each other. A **strand** is a component of a study that encompasses the basic process of conducting quantitative or qualitative research: posing a question, collecting data, analyzing data, and interpreting results based on that data (Teddlie & Tashakkori, 2009). Mixed methods studies meeting our definition of mixed methods research include at least one quantitative strand and one qualitative strand. For example, Figure 3.1 depicts a

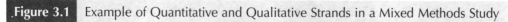

Figure 3.1   Example of Quantitative and Qualitative Strands in a Mixed Methods Study

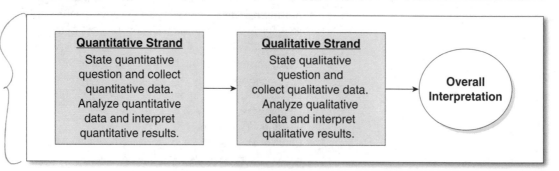

mixed methods study where the researcher starts with a quantitative strand and then conducts a qualitative strand. As shown in this figure, we will often portray strands as boxes in the figures of this text.

There are four key decisions involved in choosing an appropriate mixed methods design to use in a study. The decisions are (1) the level of interaction between the strands, (2) the relative priority of the strands, (3) the timing of the strands, and (4) the procedures for mixing the strands. We examine each of these decisions along with the available options.

Determine the Level of Interaction Between the Quantitative and Qualitative Strands

An important decision in mixed methods research is the level of interaction between the quantitative and qualitative strands in the study. The level of interaction is the extent to which the two strands are kept independent or interact with each other. Greene (2007) argued that this decision is the "most salient and critical" (p. 120) for designing a mixed methods study, and she noted two general options for a relationship: independent and interactive.

- An independent level of interaction occurs when the quantitative and qualitative strands are implemented so that they are independent from the other—that is, the two strands are distinct and the researcher keeps the quantitative and qualitative research questions, data collection, and data analysis separate. When the study is independent, the researcher only mixes the two strands when drawing conclusions during the overall interpretation at the end of the study.

- An interactive level of interaction occurs when a direct interaction exists between the quantitative and qualitative strands of the study. Through this direct interaction, the two methods are mixed before the final interpretation. This interaction can occur at different points in the research process and in many different ways. For example, the design and conduct of one strand may depend on the results from the other strand, the data from one strand may be converted into the other type and then the different data sets are analyzed together, or one strand may be implemented within a framework based on the other strand type.

Determine the Priority of the Quantitative and Qualitative Strands

Researchers also make decisions (implicitly or explicitly) about the relative importance of the quantitative and qualitative strands within the design. Priority refers to the relative importance or weighting of the quantitative and qualitative methods for answering the study's questions. There are three possible weighting options for a mixed methods design:

- The two methods may have an **equal priority** so that both play an equally important role in addressing the research problem.
- The study may utilize a **quantitative priority** where a greater emphasis is placed on the quantitative methods and the qualitative methods are used in a secondary role.
- The study may utilize a **qualitative priority** where a greater emphasis is placed on the qualitative methods and the quantitative methods are used in a secondary role.

Determine the Timing of the Quantitative and Qualitative Strands

Researchers also make decisions regarding the timing of the two strands. Timing (also referred to as pacing and implementation) refers to the temporal relationship between the quantitative and qualitative strands within a study. Timing is often discussed in relation to the time the data sets are collected, but most importantly, it describes the order in which the researchers use the results from the two sets of data within a study—that is, timing relates to the entire quantitative and qualitative strands, not just data collection. Timing within mixed methods designs can be classified in three ways: concurrent, sequential, or multiphase combination.

- **Concurrent timing** occurs when the researcher implements both the quantitative and qualitative strands during a single phase of the research study.
- **Sequential timing** occurs when the researcher implements the strands in two distinct phases, with the collection and analysis of one type of data occurring after the collection and analysis of the other type. A researcher using sequential timing may choose to start by either collecting and analyzing quantitative data first or collecting and analyzing qualitative data first.
- **Multiphase combination timing** occurs when the researcher implements multiple phases that include sequential and/or concurrent timing over a program of study. Examples of multiphase combination timing include studies conducted over three or more phases as well as those that combine both concurrent and sequential elements within one mixed methods program.

N=1 Phase 1 = Concurrent
Phase II = Sequential

④ Determine Where and How to Mix the Quantitative and Qualitative Strands

Finally, researchers need to decide the approach for mixing the two approaches within their mixed methods designs. **Mixing** is the explicit interrelating of the study's quantitative and qualitative strands and has been referred to as combining and integrating—that is, it is the process by which the researcher implements the independent or interactive relationship of a mixed methods study. Two concepts are useful for understanding when and how mixing occurs: the point of interface and mixing strategies. The **point of interface**, also known as the stage of integration, is a point within the process of research where the quantitative and qualitative strands are mixed (Morse & Niehaus, 2009). We conceptualize mixing occurring at four possible points during a study's research process: interpretation, data analysis, data collection, and design. Researchers employ mixing strategies that directly relate to these points of interface. These mixing strategies are (1) merging the two data sets, (2) connecting from the analysis of one set of data to the collection of a second set of data, (3) embedding of one form of data within a larger design or procedure, and (4) using a framework (theoretical or program) to bind together the data sets.

h-

- **Mixing during interpretation** occurs when the quantitative and qualitative strands are mixed during the final step of the research process after the researcher has collected and analyzed both sets of data. It involves the researcher drawing conclusions or inferences that reflect what was learned from the combination of results from the two strands of the study, such as by comparing or synthesizing the results in a discussion. All mixed methods designs should reflect on what was learned by the combination of methods in the final interpretation. For mixed methods designs that keep the two strands independent, this is the only point in the research process where mixing occurs.

- **Mixing during data analysis** occurs when the quantitative and quali- *b —* tative strands are mixed during the stage of the research process when the researcher is analyzing the two sets of data. First, the researcher quantita- tively analyzes the data from the quantitative strand and qualitatively analyzes the data from the qualitative strand. Then, using an interactive strategy of *N = 1* merging, the researcher explicitly brings the two sets of results together through a combined analysis. For example, the researcher further analyzes the quantitative and qualitative results by relating them to each other in a matrix that facilitates comparisons and interpretations. Another merging approach involves transforming one result type into the other type of data and merging through additional analyses of the transformed data.

- **Mixing during data collection** occurs when the quantitative and qual- *c —* itative strands are mixed during the stage of the research process when the researcher collects a second set of data. The researcher mixes by using a *e.g.,* strategy of **connecting** where the results of one strand build to the collection *N = 1 + Long-testing;)* of the other type of data. For example, the researcher may obtain quantita- *Interview;* tive results that lead to the subsequent collection of qualitative data in a sec- *FCE, etc.* ond strand. A researcher can also obtain qualitative results that build to the subsequent collection of quantitative data. The mixing occurs in the way that the two strands are connected. This connection occurs by using the results of the first strand to shape the collection of data in the second strand by spec- ifying research questions, selecting participants, and developing data collec- tion protocols or instruments.

- **Mixing at the level of design** occurs when the quantitative and quali- *d —* tative strands are mixed during the larger design stage of the research process. Mixing at the design level can involve mixing within a traditional quantitative or qualitative research design, an emancipatory theory, a substantive social science theory, or an overall program objective (Greene, 2007). Building from these ideas, we find researchers using three strategies for mixing at the design level: embedded mixing, theoretical framework-based mixing, and program objective framework-based mixing. When using an embedded mixing strategy, the researcher embeds quantitative and qualitative methods within a design associated with one of these two methods. For example, the researcher may embed a supplemental qualitative strand within a larger quantitative (e.g., experimental) design or embed a quantitative strand within a larger qualitative (e.g., narrative) design. The embedded nature occurs at the design level, in that the embedded method is conducted specifically to fit the context of the larger quantitative or qualitative design framework. When mixing within a theoretical framework, the researcher mixes quantitative and qualitative strands within a transformative framework (e.g., feminism) or a substantive framework (e.g., a social science theory) that guides the overall design. In this case, the two methods are mixed within a theoretical perspective. When mixing within a program-objective framework, the researcher mixes quantitative and qualitative strands within an overall program objective that guides the joining of multiple projects or studies in a multiphase project.

A persuasive and strong mixed methods design addresses the decisions of level of integration, priority, timing, and mixing. The many design typologies that were presented in Table 3.1, along with the wide array of decision options available to researchers presented in this section, illustrate the complexity and variety inherent in the conduct of mixed methods research. While there are potentially a limitless number of unique combinations, from our work with researchers across disciplines and based on reading hundreds of mixed methods studies, we have found that there is a relatively small set of combinations that are used most frequently in practice. Therefore, we next present a typology of major mixed methods designs that conveys the basic designs used as well as tries to encapsulate the richness available to mixed methods researchers.

● THE MAJOR MIXED METHODS DESIGNS

A mixed methods researcher thinks through these decision points and selects a design that reflects interaction, priority, timing, and mixing. As we will show, the various design options vary on these decision points. We include here the design options that are most commonly used in practice, and we advance a parsimonious and functional classification. Thus, we recommend six major mixed methods designs that provide a useful framework for researchers working to design their own studies. We urge researchers to carefully select a design that best matches the research problem and reasons for mixing in order to make the study manageable and simple to implement and describe. In addition, by selecting a typology-based design, the researcher is provided with a framework and logic to guide the implementation of the research methods to ensure that the resulting design is rigorous, persuasive, and of high quality.

The four basic mixed methods designs are the convergent parallel design, the explanatory sequential design, the exploratory sequential design, and the embedded design. In addition, our list of major designs includes two examples of designs that bring multiple design elements together: the transformative design and the multiphase design.

Prototypes of the Major Designs

Prototypical versions of these six designs are portrayed in Figure 3.2. We start with a brief introduction to these designs, including simple examples of studies

Figure 3.2 Prototypical Versions of the Six Major Mixed Methods Research Designs

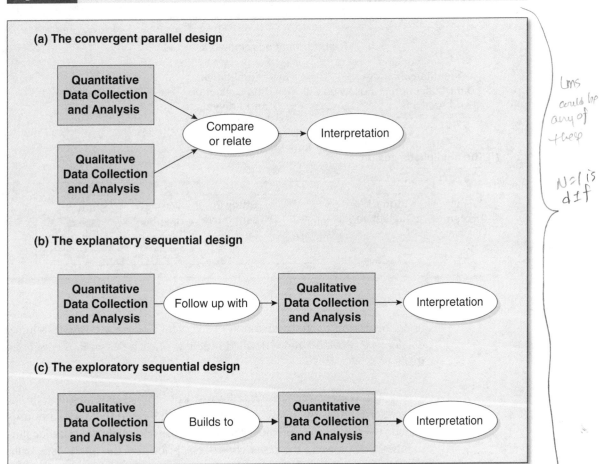

(Continued)

Figure 3.2 (Continued)

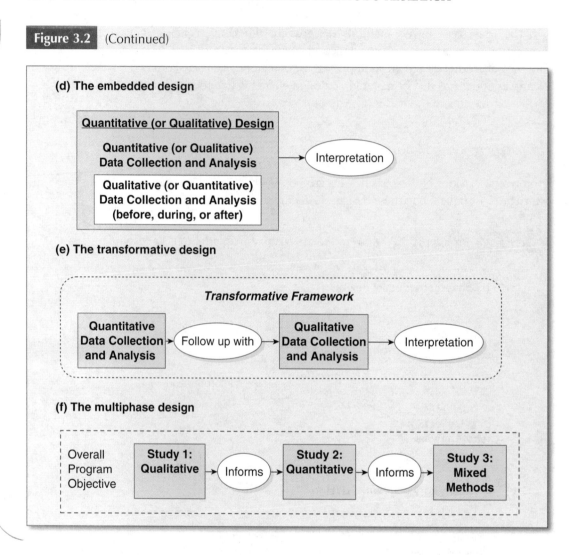

using the designs to study the topic of adolescent tobacco use. After this introduction, we provide a more detailed overview of each design in the sections that follow.

● The convergent parallel design. The convergent parallel design (also referred to as the convergent design) occurs when the researcher uses concurrent timing to implement the quantitative and qualitative strands during the same phase of the research process, prioritizes the methods equally, and keeps the strands independent during analysis and then mixes the results during the overall interpretation, as shown in Figure 3.1a. For example, the researcher might use a convergent design to develop a complete understanding of high school students' attitudes about tobacco use. During one semester, the researcher surveys high school students about their attitudes and also conducts focus group interviews on the topic with students. The researcher analyzes the survey data quantitatively and the focus group qualitatively and then merges the two sets of results to assess in what ways the results about adolescent attitudes converge and diverge.

● **The explanatory sequential design.** The explanatory sequential design (also referred to as the explanatory design) occurs in two distinct interactive phases (see Figure 3.1b). This design starts with the collection and analysis of quantitative data, which has the priority for addressing the study's questions. This first phase is followed by the subsequent collection and analysis of qualitative data. The second, qualitative phase of the study is designed so that it follows from the results of the first, quantitative phase. The researcher interprets how the qualitative results help to explain the initial quantitative results. For example, the researcher collects and analyzes quantitative data to identify significant predictors of adolescent tobacco use. Finding a surprising association between participation in extracurricular activities and tobacco use, the researcher conducts qualitative interviews with adolescents who are actively involved in extracurricular activities to attempt to explain the unexpected result.

● **The exploratory sequential design.** As shown in Figure 3.1c, the exploratory sequential design (also referred to as the exploratory design) also uses sequential timing. In contrast to the explanatory design, the exploratory design begins with and prioritizes the collection and analysis of qualitative data in the first phase. Building from the exploratory results, the researcher conducts a second, quantitative phase to test or generalize the initial findings. The researcher then interprets how the quantitative results build on the initial qualitative results. For example, the researcher collects qualitative stories about adolescents' attempts to quit smoking and analyzes the stories to identify the conditions, contexts, strategies, and consequences of adolescent quit attempts. Considering the resulting categories as variables, the researcher develops a quantitative instrument and uses it to assess the overall prevalence of these variables for a large number of adolescent smokers.

● **The embedded design.** The embedded design occurs when the researcher collects and analyzes both quantitative and qualitative data within a traditional quantitative or qualitative design, as depicted in Figure 3.1d. In an embedded design, the researcher may add a qualitative strand within a quantitative design, such as an experiment, or add a quantitative strand within a qualitative design, such as a case study. In the embedded design, the supplemental strand is added to enhance the overall design in some way. For example, the researcher may want to develop a peer intervention to help adolescents develop strategies for resisting pressure to smoke. The researcher begins by conducting a few focus groups with adolescents to learn when pressure is felt and how some adolescents resist. Using these results, the researcher develops a relevant intervention and tests it with a quantitative experimental design involving students at different schools.

- The transformative design. The transformative design is a mixed methods design that the researcher shapes within a transformative theoretical framework. All other decisions (interaction, priority, timing, and mixing) are made within the context of the transformative framework. The important role of the theoretical perspective is highlighted by the dotted line in Figure 3.1e, which depicts the possible methods that may have been selected within a transformative design. For example, the researcher using a feminist perspective may utilize a transformative design to quantitatively uncover and then qualitatively illuminate how the stereotypes of female smokers have served to marginalize them as "at risk" students within their school context.

- The multiphase design. As shown in Figure 3.1f, the multiphase design combines both sequential and concurrent strands over a period of time that the researcher implements within a program of study addressing an overall program objective. This approach is often used in program evaluation where quantitative and qualitative approaches are used over time to support the development, adaptation, and evaluation of specific programs. For example, a research team may want to help lower smoking rates for adolescents living in a particular Native American community. The researchers might first start by conducting a qualitative needs assessment study to understand the meaning of smoking and health from the perspective of adolescents in this community. Using these results, the researchers might develop an instrument and assess the prevalence of different attitudes across the community. In a third phase, the researchers might develop an intervention based on what they have learned and then examine both the process and outcomes of this intervention program.

With this brief introduction to six common mixed methods designs in hand, we now discuss each design in more detail. The detailed discussions address the history, purpose, reasons to use, philosophical assumptions, procedures, strengths, challenges, and variants of these mixed methods designs. We will examine examples of the major designs in depth in Chapter 4, but here we focus on the basic characteristics of the designs. These characteristics are also summarized in Table 3.3.

Table 3.3 Prototypical Characteristics of the Major Mixed Methods Types of Designs

Prototypical Characteristics	Convergent Design	Explanatory Design	Exploratory Design	Embedded Design	Transformative Design	Multiphase Design
Definition	• Concurrent quantitative and qualitative data collection, separate quantitative and qualitative analyses, and the merging of the two data sets	• Methods implemented sequentially, starting with quantitative data collection and analysis in Phase 1 followed by qualitative data collection and analysis in Phase 2, which builds on Phase 1	• Methods implemented sequentially, starting with qualitative data collection and analysis in Phase 1 followed by quantitative data collection and analysis in Phase 2, which builds on Phase 1	• Either the concurrent or sequential collection of supporting data with separate data analysis and the use of the supporting data before, during, or after the major data collection procedures	• Framing the concurrent or sequential collection and analysis of quantitative and qualitative data sets within a transformative, theoretical framework that guides the methods decisions	• Combining the concurrent and/or sequential collection of quantitative and qualitative data sets over multiple phases of a program of study
Design purpose	• Need a more complete understanding of a topic • Need to validate or corroborate quantitative scales	• Need to explain quantitative results	• Need to test or measure qualitative exploratory findings	• Need preliminary exploration before an experimental trial (sequential/before) • Need a more complete understanding of an experimental trial, such as the process and outcomes (concurrent/during) • Need follow-up explanations after an experimental trial (sequential/after)	• Need to conduct research that identifies and challenges social injustices	• Need to implement multiple phases to address a program objective, such as for program development and evaluation

(Continued)

Table 3.3 (Continued)

Prototypical Characteristics	Convergent Design	Explanatory Design	Exploratory Design	Embedded Design	Transformative Design	Multiphase Design
Typical paradigm foundation	• Pragmatism as an umbrella philosophy _N ≥ 1_	• Postpositivist in Phase 1 • Constructivist in Phase 2	• Constructivist in Phase 1 • Postpositivist in Phase 2	• Worldview may reflect the primary approach (e.g., postpositivist or constructivist) or pragmatism if concurrent _N ≥ 1_ • Constructivist for the qualitative component and postpositivist for the quantitative component if sequential	• Transformative worldview as an umbrella philosophy	• Pragmatism if concurrent • Constructivist for the _N ≤ 1_ qualitative component and postpositivist for the quantitative component if sequential
Level of interaction	Independent	Interactive	Interactive	Interactive	Interactive	Interactive
Priority of the strands	Equal emphasis	Quantitative emphasis	Qualitative emphasis	Either quantitative or qualitative emphasis	Equal, quantitative, or qualitative emphasis	Equal emphasis
Timing of the strands	Concurrent	Sequential: quantitative first	Sequential: qualitative first	Either concurrent or sequential	Either concurrent or sequential	Multiphase combination
Primary point of interface for mixing	• Interpretation if independent • Analysis if interactive	• Data collection	• Data collection	• Design level	• Design level	• Design level

Prototypical Characteristics	Convergent Design	Explanatory Design	Exploratory Design	Embedded Design	Transformative Design	Multiphase Design
Primary mixing strategies	Merging the two strands: • After separate data analysis • With further analyses (e.g., comparisons or transformations) of separate results	Connecting the two strands: • From quantitative data analysis to qualitative data collection • Use quantitative results to make decisions about qualitative research questions, sampling, and data collection in Phase 2	Connecting the two strands: • From qualitative data analysis to quantitative data collection • Use qualitative results to make decisions about quantitative research questions, sampling, and data collection in Phase 2	Embedding one strand within a design based on the other type: • Before, during, or after major component • Use secondary results to enhance planning, understanding, or explaining of primary strand	Mixing within a theoretical framework: • Merging, connecting, or embedding the strands within a transformative theoretical lens	Mixing within a program-objective framework: • Connecting and possibly merging and/or embedding within a programmatic objective

(Continued)

Table 3.3 (Continued)

Prototypical Characteristics	Convergent Design	Explanatory Design	Exploratory Design	Embedded Design	Transformative Design	Multiphase Design
Common variants	• Parallel databases • Data transformation • Data validation	• Follow-up explanations • Participant selection	• Theory development • Instrument development	• Embedded experiment • Embedded correlational design • Mixed methods case study • Mixed methods narrative research • Mixed methods ethnography	• Feminist lens • Disability lens • Socioeconomic class lens	• Large-scale program development and evaluation projects • Multilevel statewide studies • Single mixed methods studies that combine both concurrent and sequential phases

The Convergent Parallel Design

The most well-known approach to mixing methods is the convergent design. Scholars began discussing this design as early as the 1970s (e.g., Jick, 1979), and it is probably the most common approach used across disciplines. The convergent design was initially conceptualized as a "triangulation" design where the two different methods were used to obtain triangulated results about a single topic, but it often becomes confused with the use of triangulation in qualitative research, and researchers often use this design for purposes other than to produce triangulated findings. Since the 1970s, this design has gone by many names, including simultaneous triangulation (Morse, 1991), parallel study (Tashakkori & Teddlie, 1998), convergence model (Creswell, 1999), and concurrent triangulation (Creswell, Plano Clark, et al., 2003). Regardless of the name, the convergent design occurs when the researcher collects and analyzes both quantitative and qualitative data during the same phase of the research process and then merges the two sets of results into an overall interpretation.

N = 1 initial approach

The purpose of the convergent design. The purpose of the convergent design is "to obtain different but complementary data on the same topic" (Morse, 1991, p. 122) to best understand the research problem. The intent in using this design is to bring together the differing strengths and nonoverlapping weaknesses of quantitative methods (large sample size, trends, generalization) with those of qualitative methods (small sample, details, in depth) (Patton, 1990). This design is used when the researcher wants to triangulate the methods by directly comparing and contrasting quantitative statistical results with qualitative findings for corroboration and validation purposes. Other purposes for this design include illustrating quantitative results with qualitative findings, synthesizing complementary quantitative and qualitative results to develop a more complete understanding of a phenomenon, and comparing multiple levels within a system.

When to choose the convergent design. In addition to matching the design to the study's purpose, the following considerations also suggest when to use the convergent design:

- The researcher has limited time for collecting data and must collect both types of data in one visit to the field.
- The researcher feels that there is equal value for collecting and analyzing both quantitative and qualitative data to understand the problem.
- The researcher has skills in both quantitative and qualitative methods of research.
- The researcher can manage extensive data collection and analysis activities. In view of this, this design is best suited for team research or for the sole researcher who can collect limited quantitative and qualitative data.

Philosophical assumptions behind the convergent design. Since the convergent design involves collecting, analyzing, and merging quantitative and qualitative data and results at one time, it can raise issues regarding the philosophical assumptions behind the research. Instead of trying to "mix" different paradigms, we recommend that researchers who use this design work from a paradigm such as pragmatism to provide an "umbrella" paradigm to the research study. The assumptions of pragmatism (as discussed earlier in Chapter 2) are well suited for guiding the work of merging the two approaches into a larger understanding.

The convergent design procedures. The procedures for implementing a convergent design are outlined in the procedural flowchart in Figure 3.3. As indicated in the figure, there are four major steps in the convergent design. First, the researcher collects both quantitative data and qualitative data about the topic of interest. These two types of data collection are concurrent but separate—that is, one does not depend on the results of the other. They also typically have equal importance for addressing the study's research questions. Second, the researcher analyzes the two data sets separately and independently from each other using typical quantitative and qualitative analytic procedures. Once the two sets of initial results are in hand, the researcher reaches the point of interface and works to merge the results of the two data sets in the third step. This merging step may include directly comparing the separate results or transforming results to facilitate relating the two data types during additional analysis. In the final step, the researcher interprets to what extent and in what ways the two sets of results converge, diverge from each other, relate to each other, and/or combine to create a better understanding in response to the study's overall purpose.

Strengths of the convergent design. This design has a number of strengths and advantages:

- The design makes intuitive sense. Researchers new to mixed methods often choose this design. It was the design first discussed in the literature (Jick, 1979), and it has become a popular approach for thinking about mixed methods research.
- It is an efficient design, in which both types of data are collected during one phase of the research at roughly the same time.
- Each type of data can be collected and analyzed separately and independently, using the techniques traditionally associated with each data type. This lends itself to team research, in which the team can include individuals with both quantitative and qualitative expertise.

| Figure 3.3 | Flowchart of the Basic Procedures in Implementing a Convergent Design |

STEP 1

Design the Quantitative Strand:
- State quantitative research questions and determine the quantitative approach.

Collect the Quantitative Data:
- Obtain permissions.
- Identify the quantitative sample.
- Collect closed-ended data with instruments.

and

Design the Qualitative Strand:
- State qualitative research questions and determine the qualitative approach.

Collect the Qualitative Data:
- Obtain permissions.
- Identify the qualitative sample.
- Collect open-ended data with protocols.

STEP 2

Analyze the Quantitative Data:
- Analyze the quantitative data using descriptive statistics, inferential statistics, and effect sizes.

and

Analyze the Qualitative Data:
- Analyze the qualitative data using procedures of theme development and those specific to the qualitative approach.

STEP 3

Use Strategies to Merge the Two Sets of Results:
- Identify content areas represented in both data sets and compare, contrast, and/or synthesize the results in a discussion or table.
- Identify differences within one set of results based on dimensions within the other set and examine the differences within a display organized by the dimensions.
- Develop procedures to transform one type of result into the other type of data (e.g., turn themes into counts). Conduct further analyses to relate the transformed data to the other data (e.g., conduct statistical analyses that include the thematic counts).

STEP 4

Interpret the Merged Results:
- Summarize and interpret the separate results
- Discuss to what extent and in what ways results from the two types of data converge, diverge, relate to each other, and/or produce a more complete understanding.

Challenges in using the convergent design. Although this design is the most popular mixed methods design, it is also probably the most challenging of the major types of designs. Here are some of the challenges facing researchers using the convergent design as well as options for addressing them:

- Much effort and expertise is required, particularly because of the concurrent data collection and the fact that equal weight is usually given to each data type. This can be addressed by forming a research team that includes members who have quantitative and qualitative expertise, by including researchers who have quantitative and qualitative expertise on graduate committees, or by training single researchers in both quantitative and qualitative research. Considerations for team research were discussed in Chapter 1.
- Researchers need to consider the consequences of having different samples and different sample sizes when merging the two data sets. Different sample sizes may arise because the quantitative and qualitative data are usually collected for different purposes (generalization vs. in-depth description, respectively). Effective strategies, such as collecting large qualitative samples or using unequal sample sizes, are discussed in Chapter 6.
- It can be challenging to merge two sets of very different data and their results in a meaningful way. Researchers need to design their studies so that the quantitative and qualitative data address the same concepts. This strategy facilitates merging the data sets. In addition, Chapter 7 provides techniques for designing a discussion, building comparison displays, and using data transformation.
- Researchers may face the question of what to do if the quantitative and qualitative results do not agree. Contradictions may provide new insights into the topic, but these differences can be difficult to resolve and may require the collection of additional data. The question then develops as to what type of additional data to collect or to reanalyze: quantitative data, qualitative data, or both? Chapter 7 discusses the collection of additional data or the reexamination of existing data to address this challenge.

Convergent design variants. Design variants convey the variation found in researchers' use of the major designs. There are three common variants of the convergent design found in the literature:

- The **parallel-databases variant** is the common approach where two parallel strands are conducted independently and are only brought together during the interpretation. The researcher uses the two types of data to examine facets of a phenomenon, and the two sets of independent results are then synthesized or compared during the discussion. For example, Feldon and Kafai (2008) gathered qualitative ethnographic interviews along with quantitative survey responses and computer server logs and discussed how the two sets of results developed a more complete picture of youth's activities within online virtual communities.

- The **data-transformation variant** occurs when researchers implement the convergent design using an unequal priority, placing greater emphasis on the quantitative strand, and use a merging process of data transformation. That is, after the initial analysis of the two data sets, the researcher uses procedures to quantify the qualitative findings (e.g., creating a new variable based on qualitative themes). The transformation allows the results from the qualitative data set to be combined with the quantitative data and results through direct comparison, interrelation, and further analyses. The study of parental values by Pagano, Hirsch, Deutsch, and McAdams (2002) is an example of using this approach. They derived qualitative themes from the qualitative interview data and then scored the themes dichotomously as "present" or "not present" for each participant. These quantified scores were then analyzed with the quantitative data, using correlations and logistical regression to identify relationships between categories, as well as gender and racial differences.

- The **data-validation variant** is used when the researcher includes both open- and closed-ended questions on a questionnaire and the results from the open-ended questions are used to confirm or validate the results from the closed-ended questions. Because the qualitative items are an add-on to a quantitative instrument, the items generally do not result in a complete context-based qualitative data set. However, they provide the researcher with emergent themes and interesting quotes that can be used to validate and embellish the quantitative survey findings. For example, Webb, Sweet, and Pretty (2002) included qualitative questions with their quantitative survey measures in their study of the emotional and psychological impact of mass casualty incidents on forensic odontologists. They used the qualitative data to validate the quantitative results from the survey items.

The Explanatory Sequential Design

Most writings about mixed methods designs have emphasized sequential approaches, using design names such as sequential model (Tashakkori & Teddlie, 1998), sequential triangulation (Morse, 1991), and iteration design (Greene, 2007). Although these names apply to any sequential two-phase approach, we introduced specific names to distinguish whether the sequence begins quantitatively or qualitatively (Creswell, Plano Clark, et al., 2003). The explanatory design is a mixed methods design in which the researcher begins by conducting a quantitative phase and follows up on specific results with a second phase (refer back to Figure 3.1b). The second, qualitative phase is implemented for the purposes of explaining the initial results in more depth, and it is due to this focus on explaining results that is reflected in the design name. This design has also been called a qualitative follow-up approach (Morgan, 1998).

The purpose of the explanatory design. The overall purpose of this design is to use a qualitative strand to explain initial quantitative results (Creswell, Plano Clark, et al., 2003). For example, the explanatory design is well suited when the researcher needs qualitative data to explain quantitative significant (or nonsignificant) results, positive-performing exemplars, outlier results, or surprising results (Bradley et al., 2009; Morse, 1991). This design can also be used when the researcher wants to form groups based on quantitative results and follow up with the groups through subsequent qualitative research or to use quantitative results about participant characteristics to guide purposeful sampling for a qualitative phase (Creswell, Plano Clark, et al., 2003; Morgan, 1998; Tashakkori & Teddlie, 1998).

When to choose the explanatory design. This design is most useful when the researcher wants to assess trends and relationships with quantitative data but also be able to explain the mechanism or reasons behind the resultant trends. Other important considerations include

- The researcher and the research problem are more quantitatively oriented.
- The researcher knows the important variables and has access to quantitative instruments for measuring the constructs of primary interest.
- The researcher has the ability to return to participants for a second round of qualitative data collection.
- The researcher has the time to conduct the research in two phases.
- The researcher has limited resources and needs a design where only one type of data is being collected and analyzed at a time.
- The researcher develops new questions based on quantitative results, and they cannot be answered with quantitative data.

Philosophical assumptions behind the explanatory design. Since this study begins quantitatively, the research problem and purpose often call for a greater importance to be placed on the quantitative aspects. Although this may encourage researchers to use a postpositivist orientation to the study, we encourage researchers to consider using different assumptions within each phase—that is, since the study begins quantitatively, the researcher typically begins from the perspectives of postpositivism to develop instruments, measure variables, and assess statistical results. When the researcher moves to the qualitative phase that values multiple perspectives and in-depth description, there is a shift to using the assumptions of constructivism. The overall philosophical assumptions in this design change and shift from postpositivist to constructivist as researchers use multiple philosophical positions.

The explanatory design procedures. The explanatory design is probably the most straightforward of the mixed methods designs. Figure 3.4 provides an overview of the procedural steps used to implement a typical two-phase explanatory design. During the first step, the researcher designs and implements a quantitative strand that includes collecting and analyzing quantitative data. In the second step, the researcher connects to a second phase—the point of interface for mixing—by identifying specific quantitative results that call for additional explanation and using these results to guide the development of the qualitative strand. Specifically, the researcher develops or refines the qualitative research questions, purposeful sampling procedures, and data collection protocols so they follow from the quantitative results. As such, the qualitative phase depends on the quantitative results. In the third step, the researcher implements the qualitative phase by collecting and analyzing qualitative data. Finally, the researcher interprets to what extent and in what ways the qualitative results explain and add insight into the quantitative results and what overall is learned in response to the study's purpose.

Strengths of the explanatory design. The many advantages of the explanatory design make it the most straightforward of the mixed methods designs. These advantages include the following:

- This design appeals to quantitative researchers, because it often begins with a strong quantitative orientation.
- Its two-phase structure makes it straightforward to implement, because the researcher conducts the two methods in separate phases and collects only one type of data at a time. This means that single researchers can conduct this design; a research team is not required to carry out the design.
- The final report can be written with a quantitative section followed by a qualitative section, making it straightforward to write and providing a clear delineation for readers.
- This design lends itself to emergent approaches where the second phase can be designed based on what is learned from the initial quantitative phase.

Figure 3.4 Flowchart of the Basic Procedures in Implementing an Explanatory Design

STEP 1

Design and Implement the Quantitative Strand:
- State quantitative research questions and determine the quantitative approach.
- Obtain permissions.
- Identify the quantitative sample.
- Collect closed-ended data with instruments.
- Analyze the quantitative data using descriptive statistics, inferential statistics, and effect sizes to answer the quantitative research questions and facilitate the selection of participants for the second phase.

STEP 2

Use Strategies to Follow From the Quantitative Results:
- Determine which results will be explained, such as
  ○ significant results,
  ○ nonsignificant results,
  ○ outliers, or
  ○ group differences.
- Use these quantitative results to
  ○ refine the qualitative and mixed methods questions,
  ○ determine which participants will be selected for the qualitative sample, and
  ○ design qualitative data collection protocols.

STEP 3

Design and Implement the Qualitative Strand:
- State qualitative research questions that follow from the quantitative results and determine the qualitative approach.
- Obtain permissions.
- Purposefully select a qualitative sample that can help explain the quantitative results.
- Collect open-ended data with protocols informed by the quantitative results.
- Analyze the qualitative data using procedures of theme development and those specific to the qualitative approach to answer the qualitative and mixed methods research questions.

STEP 4

Interpret the Connected Results:
- Summarize and interpret the quantitative results.
- Summarize and interpret the qualitative results.
- Discuss to what extent and in what ways the qualitative results help to explain the quantitative results.

Challenges in using the explanatory design. Although the explanatory design is straightforward, researchers choosing this approach still need to anticipate challenges specific to this design. The explanatory design faces the following challenges:

- This design requires a lengthy amount of time for implementing the two phases. Researchers should recognize that the qualitative phase takes more time to implement than the quantitative phase. Although the qualitative phase can be limited to a few participants, adequate time must still be budgeted for the qualitative phase.
- It can be difficult to secure institutional review board (IRB) approval for this design, because the researcher cannot specify how participants will be selected for the second phase until the initial findings are obtained. Approaches to addressing this issue by tentatively framing the qualitative phase for the IRB and informing participants of the possibility that they will be contacted again are discussed in Chapter 6.
- The researcher must decide which quantitative results need to be further explained. Although this cannot be determined precisely until after the quantitative phase is complete, options such as selecting significant results and strong predictors can be considered as the study is being planned, as discussed in Chapters 6 and 7.
- The researcher must decide who to sample in the second phase and what criteria to use for participant selection. Chapter 6 explores approaches to using individuals from the same sample to provide the best explanations and criteria options, including the use of demographic characteristics, groups used in comparisons during the quantitative phase, and individuals who vary on select predictors.

Explanatory design variants. There are two variants of the explanatory design:

- The prototypical **follow-up explanations variant** is the most common approach for using the explanatory design. The researcher places the priority on the initial, quantitative phase and uses the subsequent qualitative phase to help explain the quantitative results. For example, Igo, Riccomini, Bruning, and Pope (2006) started by quantitatively studying the effect of different modes of note taking on test performance for middle school students with learning disabilities. Based on the quantitative results, the researchers conducted a qualitative phase that included gathering interviews and documents from the students to understand their note taking attitudes and behaviors to help explain the quantitative results.

- Although less common, the **participant-selection variant** arises when the researcher places priority on the second, qualitative phase instead of the initial quantitative phase. This variant has also been called a quantitative preliminary design (Morgan, 1998). This variant is used when the researcher is focused on qualitatively examining a phenomenon but needs initial quantitative results to identify and purposefully select the best participants. For example, May and Etkina (2002) collected quantitative data to identify physics students with consistently high and low conceptual learning gains. They then completed an in-depth qualitative comparison study of these two groups of students' perceptions of learning.

The Exploratory Sequential Design

As was depicted in Figure 3.1c, the exploratory design is also a two-phase sequential design that can be recognized because the researcher starts by qualitatively exploring a topic before building to a second, quantitative phase. This emphasis on exploration is reflected in the design name. In many applications of this iterative design, the researcher develops an instrument as an intermediate step between the phases that builds on the qualitative results and is used in the subsequent quantitative data collection. For that reason, this design has been referred to as the instrument development design (Creswell, Fetters, & Ivankova, 2004) and the quantitative follow-up design (Morgan, 1998).

The purpose of the exploratory design. The primary purpose of the exploratory design is to generalize qualitative findings based on a few individuals from the first phase to a larger sample gathered during the second phase. As with the explanatory design, the intent of the two-phase exploratory design is that the results of the first, qualitative method can help develop or inform the second, quantitative method (Greene et al., 1989). This design is based on the premise that an exploration is needed for one of several reasons: (1) measures or instruments are not available, (2) the variables are unknown, or (3) there is no guiding framework or theory. Because this design begins qualitatively, it is best suited for exploring a phenomenon (Creswell, Plano Clark, et al., 2003). This design is particularly useful when the researcher needs to develop and test an instrument because one is not available (Creswell, 1999; Creswell et al., 2004) or to identify important variables to study quantitatively when the variables are unknown. It is also appropriate when the researcher wants to generalize qualitative results to different groups (Morse, 1991), to test aspects of an emergent theory or classification

(Morgan, 1998), or to explore a phenomenon in depth and measure the prevalence of its dimensions.

When to choose the exploratory design. The exploratory design is most useful when the researcher wants to generalize, assess, or test qualitative exploratory results to see if they can be generalized to a sample and a population. In addition, the following considerations are relevant:

- The researcher and the research problem are more qualitatively oriented.
- The researcher does not know what constructs are important to study, and relevant quantitative instruments are not available.
- The researcher has the time to conduct the research in two phases.
- The researcher has limited resources and needs a design where only one type of data is being collected and analyzed at a time.
- The researcher identifies new emergent research questions based on qualitative results that cannot be answered with qualitative data.

Philosophical assumptions behind the exploratory design. Since the exploratory design begins qualitatively, the research problem and purpose often call for the qualitative strand to have greater priority within the design. Therefore, researchers generally work from constructivist principles during the first phase of the study to value multiple perspectives and deeper understanding. When the researcher moves to the quantitative phase, the underlying assumptions may shift to those of postpositivism to guide the need for identifying and measuring variables and statistical trends. Thus, multiple worldviews are used in this design, and the worldviews shift from one phase to the other phase.

The exploratory design procedures. The four major steps of the exploratory design are summarized in Figure 3.5. As this figure shows, this design starts with the collection and analysis of qualitative data to explore a phenomenon. In the next step, which represents the point of interface in mixing, researchers using this design build on the results of the qualitative phase by developing an instrument, identifying variables, or stating propositions for testing based on an emergent theory or framework. These developments connect the initial qualitative phase to the subsequent quantitative strand of the study. In the third step, the researcher implements the quantitative strand of the study to examine the salient variables using the developed instrument with a new sample of participants. Finally, the researcher interprets in what ways and to what extent the quantitative results generalize or expand on the initial qualitative findings.

Figure 3.5 Flowchart of the Basic Procedures in Implementing an Exploratory Design

STEP 1

Design and Implement the Qualitative Strand:
- State qualitative research questions and determine the qualitative approach.
- Obtain permissions.
- Identify the qualitative sample.
- Collect open-ended data with protocols.
- Analyze the qualitative data using procedures of theme development and those specific to the qualitative approach to answer the qualitative research questions and identify the information needed to inform the second phase.

STEP 2

Use Strategies to Build on the Qualitative Results:
- Refine quantitative research questions or hypotheses and the mixed methods question.
- Determine how participants will be selected for the quantitative sample.
- Design and pilot test a quantitative data collection instrument based on the qualitative results.

STEP 3

Design and Implement the Quantitative Strand:
- State quantitative research questions or hypotheses that build on the qualitative results, and determine the quantitative approach.
- Obtain permissions.
- Select a quantitative sample that will generalize or test the qualitative results.
- Collect closed-ended data with the instrument designed from quantitative results.
- Analyze the quantitative data using descriptive statistics, inferential statistics, and effect sizes to answer the quantitative and mixed methods research questions.

STEP 4

Interpret the Connected Results:
- Summarize and interpret the qualitative results.
- Summarize and interpret the quantitative results.
- Discuss to what extent and in what ways the quantitative results generalize or test the qualitative results.

Strengths of the exploratory design. Due to its two-phase structure and the fact that only one type of data is collected at a time, the exploratory design shares several of the same advantages as the explanatory design. Its specific advantages are as follows:

- Separate phases make the exploratory design straightforward to describe, implement, and report.
- Although this design typically emphasizes the qualitative aspect, the inclusion of a quantitative component can make the qualitative approach more acceptable to quantitative-biased audiences.
- This design is useful when the need for a second, quantitative phase emerges based on what is learned from the initial qualitative phase.
- The researcher can produce a new instrument as one of the potential products of the research process.

Challenges in using the exploratory design. There are a number of challenges associated with using the exploratory design:

- The two-phase approach requires considerable time to implement, potentially including time to develop a new instrument. Researchers need to recognize this factor and build time into their study's plan.
- It is difficult to specify the procedures of the quantitative phase when applying for initial IRB approval for the study. Providing some tentative direction in a project plan or planning to submit two separate applications for the IRB will be discussed further in Chapter 6.
- Researchers should consider using a small purposeful sample in the first phase and a large sample of different participants in the second phase to avoid questions of bias in the quantitative strand (see the discussion of sampling in Chapter 6).
- If an instrument is developed between phases, the researcher needs to decide which data to use from the qualitative phase to build the quantitative instrument and how to use these data to generate quantitative measures. In Chapter 6, we will discuss procedures for using qualitative themes, codes, and quotes to generate aspects of quantitative instruments.
- Procedures should be undertaken to ensure that the scores developed on the instrument are valid and reliable. In Chapter 6, we will review rigorous steps of instrument and scale development for this process.

Exploratory design variants. As with the explanatory design, the two main variants of the exploratory design are differentiated by the relative priority of the two strands:

- In the **theory-development variant**, the researcher places the priority on the initial qualitative phase with the ensuing quantitative phase playing a secondary role to expand on the initial results. The qualitative strand is conducted to develop an emergent theory or a taxonomy or classification system, and the researcher examines the prevalence of the findings and/or tests the theory with a larger sample (Morgan, 1998; Morse, 1991). This model is used when the researcher formulates quantitative research questions or hypotheses based on qualitative findings and proceeds to conduct a quantitative phase to answer the questions. For example, Goldenberg, Gallimore, and Reese (2005) described how they identified new variables and hypotheses about predictors of family literacy practices based on their qualitative case study. Then they conducted a quantitative path analysis study to test these qualitatively identified variables and relationships.

- Researchers using the exploratory design, however, often place the emphasis on the second, quantitative phase. In the **instrument-development variant**, the initial qualitative phase plays a secondary role, often for the purpose of gathering information to build a quantitative instrument that is needed for the prioritized quantitative phase. Using this model, Mak and Marshall (2004) initially qualitatively explored young adults' perceptions about the significance of the self to others in romantic relationships (i.e., how they perceive that they matter to someone else). Based on their qualitative results, they developed the Mattering to Romantic Others Questionnaire and administered it as part of the second, quantitative phase to test hypotheses based on the theoretical model of the formation and maintenance of perceived mattering.

The Embedded Design

The embedded design is a mixed methods approach where the researcher combines the collection and analysis of both quantitative and qualitative data within a traditional quantitative research design or qualitative research design (refer back to Figure 3.1d) (Caracelli & Greene, 1997; Greene, 2007). The collection and analysis of the second data set may occur before, during, and/or after the implementation of the data collection and analysis procedures traditionally associated with the larger design. In some embedded designs, one data set provides a supportive, secondary role in the study. For example, researchers embed a qualitative strand within quantitative experiments to support aspects of the experimental design (Creswell, Fetters, Plano Clark, & Morales, 2009). In other cases, the quantitative and qualitative approaches are combined and embedded within a traditional design or procedure. For example, in an embedded mixed methods case study, the researcher collects and analyzes both quantitative and qualitative data to examine a case. The researcher could also embed quantitative and qualitative approaches within a procedure such as social network analysis.

The purpose of the embedded design. The premises of this design are that a single data set is not sufficient, that different questions need to be answered, and that each type of question requires different types of data. In the case of the embedded experimental mixed methods design, researchers use it when they need to include qualitative data to answer a secondary research question within the predominantly quantitative study. In the experimental example, the investigator embeds qualitative data for several reasons, such as to improve recruitment procedures (e.g., Donovan et al., 2002), examine the process of an intervention (e.g., Victor, Ross, & Axford, 2004), or to explain reactions to participation in an experiment (e.g., Evans & Hardy, 2002a, 2002b). Notice that the purposes for including the qualitative data are tied to but different from the primary purpose of the experiment to assess whether a treatment has a significant effect. This distinguishes the embedded design from a convergent design where the researcher is using both methods to address a single overarching question.

When to choose the embedded design. The embedded design is appropriate when the researcher has different questions that require different types of data in order to enhance the application of a quantitative or qualitative design to address the primary purpose of the study. The following are additional considerations:

- The researcher has the expertise necessary to implement the planned quantitative or qualitative design in a rigorous way.
- The researcher is comfortable having the study be driven by either a quantitative or a qualitative primary orientation.
- The researcher has little prior experience with the supplemental method.
- The researcher does not have adequate resources to place equal priority on both types of data.

- The researcher identifies emergent issues related to the implementation of the primary quantitative or qualitative design, and insight into these issues can be obtained with a secondary data set.

Philosophical assumptions behind the embedded design. The embedded design is used to enhance the application of a traditional quantitative or qualitative design. The assumptions of this design are therefore established by the primary approach, and the other data set is subservient within that methodology. For example, if the primary design is experimental, correlational, longitudinal, or focused on instrument validation, then the researcher will most likely be working from postpositivist assumptions as the overarching paradigm. Likewise, if the primary design is phenomenological, grounded theory, ethnography, case study, or narrative, then the researcher will most likely be working from a constructivist paradigm. In either case, the supplemental method is used in service to the guiding approach.

The embedded design procedures. A good way to think about the procedures for the embedded design is to focus on the timing of the collection and analysis of the supplemental data relative to the primary strand of the study and the reasons for adding in the supplemental data. Sandelowski (1996) first introduced the notion of the supplemental strand occurring before, during, or after (or some combination) the primary strand, and we find this to be a useful framework for thinking about the embedded design no matter which approach is placed in the primary role. The researcher makes this procedural decision (before, during, after, or some combination) based on the purpose of the supplemental data within the larger design (Creswell et al., 2009). Therefore, embedded designs can use either a one-phase or a two-phase approach for the embedded strand, and the procedures reflect the issues relevant to the sequential or concurrent nature of the implementation.

Because the most common type of embedded design found in the literature occurs when researchers embed qualitative data within an experimental design, Figure 3.6 provides a general overview of the procedures for implementing qualitative data before, during, and/or after the intervention in an experiment. The general steps include (1) designing the overall experiment and deciding the reason why qualitative data need to be included, (2) collecting and analyzing qualitative data to enhance the experimental design, (3) collecting and analyzing quantitative outcome data for the experimental groups, and (4) interpreting how the qualitative results enhanced the experimental procedures and/or understanding of the experimental outcomes.

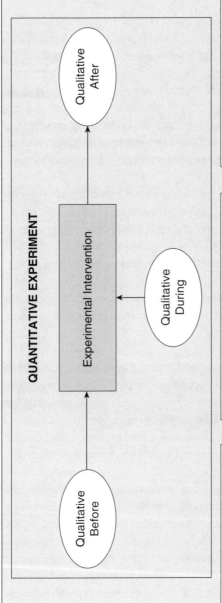

QUANTITATIVE EXPERIMENT

Qualitative Before → Experimental Intervention → Qualitative After

Qualitative During → Experimental Intervention

Implement the Qualitative Strand Before the Experiment:
- Decide the reason for the qualitative strand.
- State qualitative research questions, and determine the qualitative approach.
- Obtain permissions.
- Identify the qualitative sample.
- Collect open-ended data.
- Analyze the qualitative data using procedures of theme development and those specific to the qualitative approach.

Use the Qualitative Strand to Plan the Experiment, Such as:
- Refine recruitment procedures.
- Develop outcome measure.
- Develop intervention.

Implement the Qualitative Strand During the Experiment:
- Decide the reason for the qualitative strand.
- State qualitative research questions, and determine the qualitative approach.
- Obtain permissions.
- Identify the qualitative sample.
- Collect open-ended data.
- Analyze the qualitative data using procedures of theme development and those specific to the qualitative approach.

Use the Qualitative Strand to Understand the Experiment, Such as:
- Describe participants' experiences with the intervention.
- Describe the process.
- Describe treatment fidelity.

Implement the Qualitative Strand After the Experiment:
- Decide the reason for the qualitative strand.
- State qualitative research questions, and determine the qualitative approach.
- Obtain permissions.
- Identify the qualitative sample.
- Collect open-ended data.
- Analyze the qualitative data using procedures of theme development and those specific to the qualitative approach.

Use the Qualitative Strand to Explain the Experiment, Such as:
- Describe why outcomes occurred.
- Describe how participants respond to the results.
- Describe what long-term effects are experienced.

Strengths of the embedded design. There are several advantages specific to the embedded design:

- This design can be used when the researcher does not have sufficient time or resources to commit to extensive quantitative and qualitative data collection because one data type is given less priority than the other.
- By the addition of supplemental data, the researcher is able to improve the larger design.
- Because the different methods are addressing different questions, this design fits a team approach well, where members on the team can focus their work on one of the questions based on their interests and expertise.
- The focus on different questions means that the two types of results can be published separately.
- This design may be appealing to funding agencies that are less familiar with mixed methods research because the primary focus of the approach is on a traditional quantitative or qualitative design.

Challenges in using the embedded design. There are many challenges associated with the embedded design. The following are challenges and suggested strategies for dealing with them:

- The researcher needs to have expertise in the quantitative or qualitative design used in addition to expertise in mixed methods research.
- The researcher must specify the purpose of collecting qualitative (or quantitative) data as part of a larger quantitative (or qualitative) study. Researchers can state these as the primary and secondary purposes for the study. See Chapter 5 for examples for writing these primary and secondary purpose statements.
- The researcher must decide at what point in the experimental study to collect the qualitative data in relation to the intervention (i.e., before, during, after, or some combination). This decision should be made based on the intent for including the qualitative data (e.g., to shape the intervention, to explain the process of participants during treatment, or to follow up on results of the experimental trial). Chapter 6 provides more detail about these options at different phases of a project.
- It can be difficult to integrate the results when the two methods are used to answer different research questions. However, unlike the convergent design, the intent of the embedded design is not to merge two

different data sets collected to answer the same question. Researchers using an embedded design can keep the two sets of results separate in their reports or even report them in separate papers (see Chapter 8 for further discussion about these writing strategies).

- For during-intervention experimental approaches, the qualitative data collection may introduce potential treatment bias that affects the outcomes of the experiment. Suggestions for addressing this potential bias through collecting unobtrusive data are discussed in Chapter 6.

Embedded design variants. Conceptually, there are two variants of the embedded design based on whether one method is embedded as a supplement to a larger design or both methods are embedded in combination within a larger design or procedure. Many variations also exist within these two larger categories:

- The prototypical variant of the embedded design occurs when the researcher embeds a supplemental data set within a larger design to address different questions. The most common example is the **embedded-experiment variant**, which occurs when the researcher embeds qualitative data within an experimental trial. Other similar variants include the **embedded-correlational variant** (Harrison, 2005) and the **embedded instrument development and validation variant** (Plano Clark & Galt, 2009). For example, Hilton, Budgen, Molzahn, and Attridge (2001) gathered qualitative information (e.g., participant comments, open-ended responses, and observational field notes) as they pilot tested their instrument to provide additional evidence that the instrument measured meaningful client outcomes at a nursing center.
- Recently, scholars have also discussed hybrid designs where researchers embed both quantitative and qualitative data within traditional designs or procedures. These approaches result in variants, such as **mixed methods case studies** (Luck, Jackson, & Usher, 2006) and **mixed methods narrative research** (Elliot, 2005). In these examples, the case, or the narrative, becomes a placeholder for collecting both quantitative and qualitative data (Creswell & Tashakkori, 2007). Another example would be a **mixed methods ethnography** in which the researcher discusses the collection of both forms of data within an ethnographic design (Morse & Niehaus, 2009). The embedding of both forms of data may also take place within larger procedures, such as Neighborhood History Calendars, Life History Calendars, or geographic information systems (GIS) as discussed by

social demographers (Axinn & Pearce, 2006). For example, Skinner, Matthews, and Burton (2005) joined quantitative spatial data with qualitative ethnographic information within a GIS procedure to map the experiences of families meeting the needs of their children with disabilities.

The Transformative Design

A design that goes beyond the basic four mixed methods designs occurs when researchers conduct mixed methods research using a theoretical-based framework, such as a transformative worldview. A transformative-based theoretical framework is a framework for advancing the needs of underrepresented or marginalized populations. As discussed in Chapter 2, it involves the researcher taking a position, being sensitive to the needs of the population being studied, and recommending specific changes as a result of the research to improve social justice for the population under study. Some scholars discount ideological perspectives as a criterion for classifying mixed methods designs, arguing that they relate more to the content purpose of the study than the methods decisions of the study (e.g., Teddlie & Tashakkori, 2009). Others, however, have included transformative designs among the major mixed methods designs (Creswell, Plano Clark, et al., 2003; Greene, 2007; Greene & Caracelli, 1997; Mertens, 2003). Mertens (2003, 2009) specifically discussed ways in which a transformative perspective influences every stage of the research and design process. We do find researchers planning and naming their designs in ways that reflect the importance that they place on the use of a transformative perspective. As mentioned in Chapter 2, a number of mixed methods studies have been published that use a transformative lens drawn from a feminist theory, a racial or ethnic theory, a sexual orientation theory, or a disability theory (Mertens, 2009). For example, Lehan-Mackin (2007) classified her two-phase proposed study of unintended pregnancies in college-aged women as an "equivalent, sequential, transformative, mixed-methods study" (Abstract, para. 1). She planned her procedures so that implications for social contexts and policies that promote health disparities would result.

The purpose of the transformative design. The purpose of this design is to conduct research that is change oriented and seeks to advance social justice causes by identifying power imbalances and empowering individuals and/or communities—that is, the purpose for mixing methods in the transformative design is for value-based and ideological reasons more than for reasons related to methods and procedures (Greene, 2007). The purpose is to use the methods that are best suited for advancing the transformative goals (e.g., challenging the status quo and developing solutions) of the study.

When to choose the transformative design. This design should be used when the researcher determines that mixed methods is needed to address a transformative aim. The following are other considerations:

- The researcher seeks to address issues of social justice and call for change.
- The researcher sees the needs of underrepresented or marginalized populations.
- The researcher has a good working knowledge of theoretical frameworks used to study underrepresented or marginalized populations.
- The researcher can conduct the study without further marginalizing the population under study.

Philosophical assumptions behind the transformative design. The transformative paradigm provides the overarching assumptions behind the conduct of the transformative design (Mertens, 2003, 2007). This worldview, as discussed in Chapter 2 as the advocacy and participatory worldview, provides an umbrella paradigm to the project and includes political action, empowerment, collaborative, and change-oriented research perspectives.

The transformative design procedures. Depending on the specific contexts of an individual transformative study, the researcher may end up using procedures that are consistent with any of the four basic mixed methods designs already discussed. The difference is that the transformative paradigm and theoretical lens in use by the researcher has a "pervasive influence throughout the research process" (Mertens, 2003, p. 142). Mertens described ways in which this perspective influences five steps of the research process, including (1) defining the problem and searching the literature; (2) identifying the research design; (3) identifying data sources and selecting participants; (4) identifying or constructing data collection instruments and methods; and (5) analyzing, interpreting, reporting, and using results. In addition, Plano Clark and Wang (2010) identified several procedures for conducting mixed methods research in a multiculturally competent way by examining researchers' practices in 11 published studies. As suggested by these authors, Figure 3.7 summarizes some of the key considerations that transformative researchers need to consider as they design their mixed methods procedures. More details will be provided in Chapters 6 and 7 about the data collection and analysis procedures within a transformative design.

Figure 3.7   Flowchart of the Basic Considerations for Designing a Transformative Design

Defining the Problem and Searching the Literature:
- Deliberately search the literature for concerns of diverse groups and issues of discrimination and oppression.
- Allow the definition of the problem to arise from the community of concern.
- Build trust with community members.
- Resist deficit-based theoretical frameworks.
- Ask balanced—positive and negative—research questions.
- Develop questions that lead to transformative answers, such as questions focused on authority and relations of power in institutions and communities.

Identifying the Research Design:
- Use mixed methodologies to capture the complexity of the problem and respond to different stakeholder needs.
- Ensure that your research design respects ethical considerations of participants.
- Do not deny treatment to any groups if incorporating experimental procedures.

Identifying Data Sources and Selecting Participants:
- Focus on participants of groups associated with discrimination and oppression.
- Avoid stereotypical labels for participants.
- Recognize the diversity within the target population.
- Use sampling strategies that improve the inclusiveness of the sample to increase the probability that traditionally marginalized groups are adequately and accurately represented.

Identifying or Constructing Data Collection Instruments and Methods:
- Consider how the data collection process and outcomes will benefit the community being studied.
- Use methods to ensure that the research findings will be credible to that community.
- Design data collection to permit effective communication with community members.
- Use collection methods that are sensitive to the community's cultural contexts.
- Design the data collection to open up avenues for participation in the social change process.

Analyzing, Interpreting, Reporting, and Using Results:
- Be open to the results raising new hypotheses.
- Analyze subgroups (i.e., multilevel analyses) to examine the differential impact on diverse groups.
- Frame the results to help understand and elucidate power relationships.
- Report the results in ways to facilitate social change and action.

SOURCE: Adapted from D. M. Mertens (2003) and J. W. Creswell (2009c, pp. 67–68). Adapted with permission of SAGE Publications, Inc.

Strengths of the transformative design. Researchers may implement procedures consistent with any of the four basic mixed methods designs within their transformative designs. As such, the transformative design shares the same strengths previously discussed with these designs. In addition, the transformative design has the following advantages:

- The researcher positions the study within a transformative framework and an advocacy or emancipatory worldview.
- The research helps to empower individuals and bring about change and action.
- Participants often play an active, participatory role in the research.
- The researcher is able to use a collection of methods that produces results that are both useful to community members and viewed as credible to stakeholders and policy makers.

Challenges in using the transformative design. As with the strengths, the transformative design shares procedural challenges associated with the corresponding basic mixed methods designs. In addition, the transformative design has these further challenges:

- There is still little guidance in the literature to assist researchers with implementing mixed methods in a transformative way. One way to proceed is to review published mixed methods studies that employ a transformative lens (see Sweetman, Badiee, & Creswell, 2010).
- The researcher may need to justify the use of the transformative approach. This can be done by explicitly discussing the philosophical and theoretical foundations as part of the study proposal and report, as discussed in Chapter 2.
- The researcher must develop trust with participants and be able to conduct the research in a culturally sensitive way.

Transformative design variants. The variants of the transformative design are best described by the diverse theoretical frameworks used rather than by different methods decisions. For example, Sweetman, Badiee, and Creswell (2010) identified several transformative mixed methods studies in the literature and classified the variants by the theoretical lens in use. These studies used different theoretical lenses, including a feminist lens (e.g., Cartwright, Schow, & Herrera, 2006), a disability lens (e.g., Boland, Daly, & Staines, 2008), and a socioeconomic class lens (Newman & Wyly, 2006). Therefore, three variants of the transformative design are (1) the **feminist lens transformative variant**, in which the researcher frames the study using a feminist theoretical lens; (2) the **disability lens transformative variant**, in which the researcher frames the study using a disability theoretical lens; and (3) the socioeconomic class lens transformative variant, in which the researcher frames the study using a socioeconomic class theoretical lens.

The Multiphase Design

The multiphase design is an example of a mixed methods design that goes beyond the basic designs (convergent, explanatory, exploratory, and embedded). Multiphase designs occur when an individual researcher or team of investigators examines a problem or topic through an iteration of connected quantitative and qualitative studies that are sequentially aligned, with each new approach building on what was learned previously to address a central program objective. Early writings in the area referred to the sandwich design, which occurs when the researcher alternates the quantitative and qualitative methods across three phases (e.g., qualitative then quantitative then qualitative) (Sandelowski, 2003). Today, multiphase designs combine sequential and concurrent aspects and are most common in large funded studies that have numerous questions being investigated to advance one programmatic objective. Two primary examples of this design would be a multi-project funded mixed methods project involving numerous investigators and researchers for U.S. federal funding (e.g., a National Institutes of Health [NIH] or National Science Foundation [NSF] project) or a statewide evaluation study involving multiple levels of data collection and analysis as well as multiple studies.

The purpose of the multiphase design. The purpose of this design is to address a set of incremental research questions that all advance one programmatic research objective. It provides an overarching methodological framework to a multiyear project that calls for multiple phases to develop an overall program of research, or evaluation. For example, in the context of program evaluation, these multiple phases may be tied to phases for needs assessment, program development, and program evaluation testing.

When to choose the multiphase design. In addition to matching the design to the series of research questions, a multiphase design should be selected for the following considerations:

- The researcher cannot fulfill the long-term program objective of the study with a single mixed methods study.
- The researcher has experience in large-scale research (e.g., an evaluation background, a background in complex health science projects).
- The researcher has sufficient resources and funding to implement the study over multiple years.

- The researcher is part of a team that includes practitioners in addition to individuals with research expertise in both qualitative and quantitative research.
- The researcher is conducting a mixed methods project that is emerging, and new questions arise during different stages of the research project.

Philosophical assumptions behind the multiphase design. The philosophical assumptions that provide the foundation for a multiphase design will vary dependent on the specifics of the design. As a general framework, we suggest that the researcher use pragmatism as an umbrella foundation if strands are implemented concurrently and use constructivism for the qualitative component and postpositivism for the quantitative component if the strands are sequential. Since teams often implement this approach, it is common for different subgroups within the teams to be working from different assumptions and focusing on different aspects of the overall design. In addition to the importance of philosophical assumptions, multiphase designs also benefit from a strong theoretical perspective that provides a guiding framework for thinking about the substantive aspects of the study across the multiple phases.

The multiphase design procedures. The general procedures indicative of a multiphase design are depicted in Figure 3.8. As the figure illustrates, the multiphase design allows for each individual study to address a specific set of research questions that evolve to address a larger program objective. These procedures within a given study phase, or sequence of studies, often mirror the procedures for implementing one or more of the basic mixed methods designs. In addition, researchers utilizing a multiphase design also have to carefully state the research questions for each phase, which both contribute to the overall program of inquiry and build upon what has been learned in previous phases, and design procedures that build on the earlier findings and results.

Strengths of the multiphase design. This design has a number of strengths:

- The multiphase design incorporates the flexibility needed to utilize the mixed methods design elements required to address a set of interconnected research questions.
- Researchers can publish the results from individual studies while at the same time still contributing to the overall evaluation or research program.
- The design fits the typical program evaluation and development approach well.
- The researcher can use this design to provide an overall framework for conducting multiple iterative studies over multiple years.

Figure 3.8 Flowchart of the Basic Procedures in Implementing a Multiphase Design

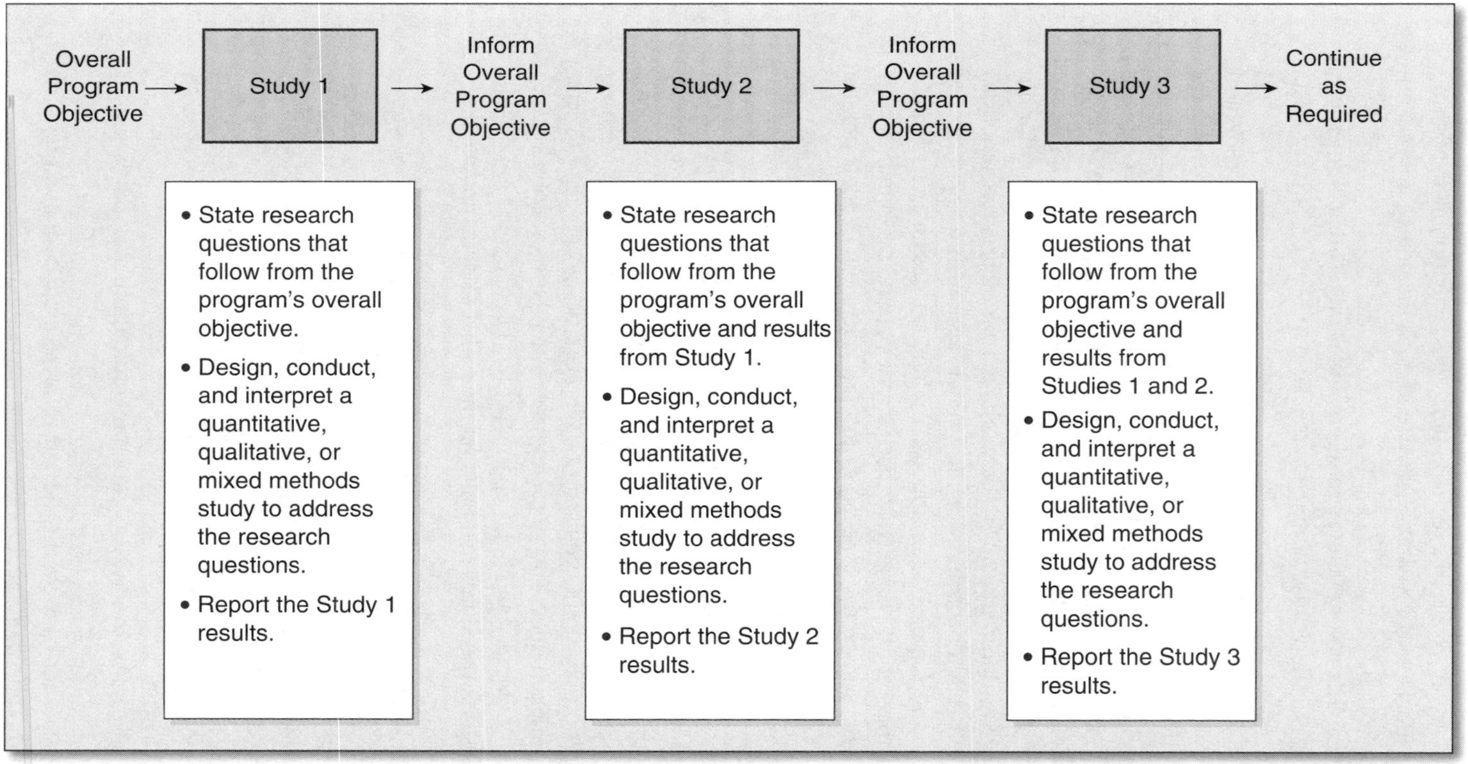

Overall
Program ⟶ Study 1
Objective

Inform
Overall
Program ⟶ Study 2
Objective

Inform
Overall
Program ⟶ Study 3
Objective

Continue
as
Required

Study 1

- State research questions that follow from the program's overall objective.
- Design, conduct, and interpret a quantitative, qualitative, or mixed methods study to address the research questions.
- Report the Study 1 results.

Study 2

- State research questions that follow from the program's overall objective and results from Study 1.
- Design, conduct, and interpret a quantitative, qualitative, or mixed methods study to address the research questions.
- Report the Study 2 results.

Study 3

- State research questions that follow from the program's overall objective and results from Studies 1 and 2.
- Design, conduct, and interpret a quantitative, qualitative, or mixed methods study to address the research questions.
- Report the Study 3 results.

SOURCE: Figure based on Creswell and Plano Clark (2007) and Morse and Niehaus (2009)

Challenges in using the multiphase design. While the multifaceted nature and flexibility of the multiphase design are its main strengths, they also represent the primary challenges:

- The researcher must anticipate the challenges generally associated with individual concurrent and sequential approaches within individual or subsequent phases.
- The researcher needs sufficient resources, time, and effort to successfully implement several phases over multiple years.
- The researcher needs to effectively collaborate with a team of researchers over the scope of the project, while also accommodating the potential addition and loss of team members.
- The researcher needs to consider how to meaningfully connect the individual studies in addition to mixing quantitative and qualitative strands within phases.
- Due to the practical focus of many multiphase designs for program development, the investigator needs to consider how to translate research findings into practice through developing materials and programs.
- The researcher may need to submit new or modified protocols to the IRB for each phase of the project.

Multiphase design variants. We are only beginning to think about how to classify variants of the multiphase designs. Examples can be difficult to identify, because they are frequently published as different projects across different journals. Considering the examples we do have from the literature, we suggest the following variants:

- **Large-scale program development and evaluation projects** may be the most common use of multiphase designs. These projects are often federally funded research programs in areas such as education and health services research where investigators conduct projects that require exploration, program development, program testing, and feasibility studies.
- **Multilevel statewide studies** utilize different methods and phases to examine different levels within a system, such as at the local, state, and national levels. For example, Teddlie and Yu (2007) discussed how multilevel projects focused on educational issues need to study five different levels: school systems, school districts, schools, teachers and classrooms, and students, with each level requiring different methods.
- A final variant includes **single mixed methods studies that combine both concurrent and sequential phases.** For example, Fetters, Yoshioka, Greenberg, Gorenflo, and Yeo (2007) reported their use of a combined design to study the practice of seeking consent for epidural anesthesia in advance of childbirth for Japanese women.

These researchers used a sequential approach to identify and explain the women's perspectives with a survey followed by interviews. They combined this sequential approach with a concurrent study of health professionals' perspectives by collecting quantitative and qualitative data in an e-mail survey.

● A MODEL FOR DESCRIBING A DESIGN IN A WRITTEN REPORT

Because many researchers and reviewers are currently unfamiliar with the different types of mixed methods designs, it is important to include a paragraph that introduces the design when writing about a study in proposals or research reports. This overview paragraph generally is placed at the beginning of the methods discussion and should address four topics. First, identify the type of mixed methods design. Next, give the defining characteristics of this design, including its level of interaction, timing, priority, and mixing decisions. Third, state the overall purpose or rationale for using this design for the study. Finally, include references to the mixed methods literature on this design. An example of an overview paragraph is included in Figure 3.9, along with comments that will assist in identifying these features within the paragraph.

Figure 3.9   A Sample Paragraph for Writing a Mixed Methods Design Into a Report

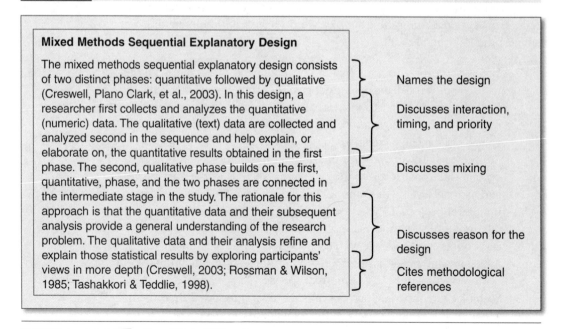

Mixed Methods Sequential Explanatory Design

The mixed methods sequential explanatory design consists of two distinct phases: quantitative followed by qualitative (Creswell, Plano Clark, et al., 2003). In this design, a researcher first collects and analyzes the quantitative (numeric) data. The qualitative (text) data are collected and analyzed second in the sequence and help explain, or elaborate on, the quantitative results obtained in the first phase. The second, qualitative phase builds on the first, quantitative, phase, and the two phases are connected in the intermediate stage in the study. The rationale for this approach is that the quantitative data and their subsequent analysis provide a general understanding of the research problem. The qualitative data and their analysis refine and explain those statistical results by exploring participants' views in more depth (Creswell, 2003; Rossman & Wilson, 1985; Tashakkori & Teddlie, 1998).

- Names the design
- Discusses interaction, timing, and priority
- Discusses mixing
- Discusses reason for the design
- Cites methodological references

SOURCE: Ivankova, Creswell, and Stick, 2006, p. 5.

SUMMARY

Like quantitative and qualitative research approaches, mixed methods research encompasses several different designs. The designs provide sound frameworks for collecting, analyzing, mixing, interpreting, and reporting quantitative and qualitative data to best address specific types of research purposes. There are four principles that researchers should consider as they design their mixed methods studies. First, mixed methods designs can be fixed from the start and/or emerge as the study is underway. Second, researchers should consider their approach to research design and weigh the use of a typology-based or dynamic approach. Third, researchers must match the design to their research problem and questions. Finally, researchers should articulate at least one reason why they are mixing methods.

Researchers designing a mixed methods study make four key decisions in choosing a mixed methods design: whether the strands will remain independent or be interactive; whether the two strands will have equal or unequal priority for addressing the study's purpose; whether the strands will be implemented concurrently, sequentially, or across multiple phases; and how the strands are to be mixed. Mixing involves making decisions as to the stage in the research in which mixing occurs and the specific strategies used in mixing (i.e., merging, connecting, embedding, or using a framework). These decisions, along with the underlying logic that is best suited to the research problem and practical considerations, are the foundation researchers should use in selecting a mixed methods design for their study.

Researchers can choose among six major mixed methods designs: convergent, explanatory, exploratory, embedded, transformative, or multiphase. These designs are suited for different purposes and often find their basis within different philosophical assumptions. They each include a specific set of procedures that provide the underlying logic of the approach. Researchers should carefully consider the challenges associated with their design choice and plan strategies for addressing these challenges. Within the different designs, we also find that there are common variants in addition to the design decisions that are most common within studies published in the recent literature.

ACTIVITIES

1. Reflect on the four principles of mixed methods design (using a design that is fixed and/or emergent, using a mixed methods design approach, matching the design to the problem, and stating the reason for mixing methods) in regards to a study you are planning. Briefly describe how these principles will be applied in your study.

2. Identify a substantive topic of interest to you. Describe how this topic could be studied using each of the major designs discussed in this chapter.

3. Which of the major design types will you use in your study? Write a one-paragraph overview that identifies this design; defines its level of interaction, priority, timing, and mixing; and conveys your reason for choosing it for your study.

4. What challenges are associated with your design choice? Write a paragraph that discusses the challenges that you anticipate occurring with your design and how you might address them.

ADDITIONAL RESOURCES TO EXAMINE

For additional discussions on the major types of mixed methods designs, consult the following resources:

Creswell, J. W., Plano Clark, V. L., Gutmann, M., & Hanson, W. (2003). Advanced mixed methods research designs. In A. Tashakkori & C. Teddlie (Eds.), *Handbook of mixed methods in social & behavioral research* (pp. 209–240). Thousand Oaks, CA: Sage.

Greene, J. C. (2007). *Mixed methods in social inquiry*. San Francisco: Jossey-Bass.

Mertens, D. M. (2003). Mixed methods and the politics of human research: The transformative-emancipatory perspective. In A. Tashakkori & C. Teddlie (Eds.), *Handbook of mixed methods in social & behavioral research* (pp. 135–164). Thousand Oaks, CA: Sage.

Morse, J. M., & Niehaus, L. (2009). *Mixed methods design: Principles and procedures*. Walnut Creek, CA: Left Coast Press.

Teddlie, C., & Tashakkori, A. (2009). *Foundations of mixed methods research*. Thousand Oaks, CA: Sage.

Look to these resources for further discussions of the interaction, timing, priority, and mixing decisions:

Bazeley, P. (2009). Integrating data analyses in mixed methods research [Editorial]. *Journal of Mixed Methods Research, 3*(3), 203–207.

Caracelli, V. J., & Greene, J. C. (1993). Data analysis strategies for mixed-method evaluation designs. *Educational Evaluation and Policy Analysis, 15*(2), 195–207.

Greene, J. C., Caracelli, V. J., & Graham, W. F. (1989). Toward a conceptual framework for mixed-method evaluation designs. *Educational Evaluation and Policy Analysis, 11*(3), 255–274.

See the following discussions for alternative approaches for mixed methods design:

Hall, B., & Howard, K. (2008). A synergistic approach: Conducting mixed methods research with typological and systemic design considerations. *Journal of Mixed Methods Research, 2*(3), 248–269.

Maxwell, J. A., & Loomis, D. M. (2003). Mixed methods design: An alternative approach. In A. Tashakkori & C. Teddlie (Eds.), *Handbook of mixed methods in social & behavioral research* (pp. 241–271). Thousand Oaks, CA: Sage.

CHAPTER 4

EXAMPLES OF MIXED METHODS DESIGNS

A fter considering the choice of a mixed methods design to address a research problem, a good next step is for researchers to examine examples of published studies in which the mixed methods design was applied in practice. We introduced search terms useful for locating mixed methods studies in Chapter 1. Now we need to consider how to read studies that represent the different mixed methods designs. Reading and under-standing mixed methods studies is facilitated by identifying the key design features used in studies, describing general designs using a notation system, and drawing diagrams that convey the detailed procedures of mixed meth-ods studies. Using these tools, we present six complete published studies and examine them for their mixed methods features as examples of the dif-ferent major mixed methods designs.

This chapter will address

- lessons to be learned from studying examples of published mixed methods studies;
- two tools—a notation system and diagrams—that can facilitate an understanding of published mixed methods studies;
- the design features helpful in reviewing a mixed methods study;
- six examples of mixed methods studies that each illustrate a different type of design; and
- the similarities and differences among the six examples.

● LEARNING FROM EXAMPLES OF MIXED METHODS RESEARCH

From our experience of reading and reviewing many hundreds of mixed methods studies, we have found great value in examining the practice of other researchers as they implement and report the mixed methods designs they used in their research studies. It is also helpful for researchers planning to use mixed methods to locate mixed methods studies published within their discipline in order to identify the language and designs that are common within their disciplinary context. By identifying studies that make use of a selected mixed methods design, researchers can cite these studies as examples of the design in the methods section of their proposals and reports. In addition, researchers who examine examples of mixed methods designs learn about different procedures used when conducting mixed methods research. They also are better able to anticipate challenges that can occur with a specific design. Published studies also provide models for how to write up and report the results of a specific mixed methods design (a topic we will discuss further in Chapter 8). To address this step, we have included an example of each of the major mixed methods designs in this book. First, however, we consider two tools that facilitate the design, communication, and review of mixed methods studies.

● USING TOOLS TO DESCRIBE MIXED METHODS DESIGNS

Two tools are helpful in reviewing published mixed methods studies: a notation system and diagrams for describing the procedures, methods, and products of mixed methods studies. The notation system and diagrams have a substantial history of use in the mixed methods literature. They are useful tools for designing and communicating the complexity inherent in mixed methods designs. Because of their extensive use in the literature and value for conveying mixed methods approaches, researchers need to be familiar with interpreting the information conveyed by these tools and to be comfortable using them to describe their own studies.

A Notation System

To facilitate the discussion of mixed methods design features, a notation system, first used by Morse (1991), has been expanded and appears extensively throughout the mixed methods literature. The common notations used from this system are summarized in Table 4.1. Morse's initial notation system used

Table 4.1 Summary of Notations Used to Describe Mixed Methods Designs

Notation	Example Application	What the Notation Indicates	Key Citations
Shorthand: Quan, Qual	Quan strand	Quantitative methods	Morse (1991, 2003)
Uppercase letters: QUAN, QUAL	QUAL priority	The qualitative methods are prioritized in the design.	Morse (1991, 2003)
Lowercase letters: quan, qual	qual supplement	The qualitative methods have a lesser priority in the design.	Morse (1991, 2003)
Plus sign: +	QUAN + QUAL	The QUAN and QUAL methods occur concurrently.	Morse (1991, 2003)
Arrow: →	QUAN → qual	The methods occur in a sequence of QUAN followed by qual.	Morse (1991, 2003)
Parentheses: ( )	QUAN(qual)	A method is embedded within a larger design or procedure or mixed within a theoretical or program-objective framework.	Plano Clark (2005)
Double arrows: →←	QUAL →← QUAN	The methods are implemented in a recursive process (QUAL → QUAN → QUAL → QUAN → etc.).	Nastasi et al. (2007)
Brackets: [ ]	QUAL → QUAN → [QUAN + qual]	Mixed methods [QUAN + qual] is used within a single study or project within a series of studies.	Morse & Niehaus (2009)
Equal sign: =	QUAN → qual = explain results	The purpose for mixing methods	Morse & Niehaus (2009)

"quan" to indicate the quantitative methods of a study and "qual" to indicate the qualitative methods. This shorthand aims to convey an equal status of the two methods (i.e., both abbreviations have the same number of letters and same format). The relative priority of the two methods within a particular study is indicated through the use of uppercase and lowercase letters—that is, prioritized methods are indicated with uppercase letters (i.e., QUAN and/or QUAL) and secondary methods with lowercase letters (i.e., quan and/or qual). In addition, the notation uses a plus (+) to indicate methods that occur at the same time and an arrow (→) to indicate methods that occur in a sequence. As shown in Table 4.1, several authors have expanded the notations to go beyond these basic elements. Plano Clark (2005) added the use of parentheses to indicate methods that are embedded within a larger framework. Nastasi et al. (2007) added double arrows (→←) to indicate methods that were implemented in a recursive fashion. More recently, Morse and Niehaus (2009) suggested the use of brackets ([ ]) to distinguish mixed methods projects in a series of studies and an equal sign (=) to indicate the purpose for combining the methods.

This shorthand notation can be very helpful for describing the overall design of a study. Consider the following examples of using this notation system for the four basic mixed methods:

- QUAN + QUAL = converge results: This notation indicates a convergent design in which the researcher implemented the quantitative and qualitative strands at the same time, both strands had equal emphasis, and the results of the separate strands were converged.
- QUAN → qual = explain results: This notation indicates an explanatory design in which the researcher implemented the two strands in a sequence, the quantitative methods occurred first and had a greater emphasis in addressing the study's purpose, and the qualitative methods followed to help explain the quantitative results.
- QUAL → quan = generalize findings: This notation indicates an exploratory design in which the researcher implemented the two strands in a sequence, the qualitative methods occurred first and had a greater emphasis in addressing the study's purpose, and the quantitative methods followed to assess the extent to which the initial qualitative findings generalize to a population.
- QUAN (+ qual) = enhance experiment: This notation indicates an embedded design in which the researcher implemented a secondary qualitative strand within a larger quantitative experiment, the qualitative methods occurred during the conduct of the experiment, and the qualitative strand enhanced the conduct and understanding of the experiment.

Procedural Diagrams

Building from this notation system, procedural diagrams have been used to convey the complexity of mixed methods designs. Such diagrams were introduced by Steckler, McLeroy, Goodman, Bird, and McCormick (1992) and have been adopted by many authors (e.g., Morse & Neihaus, 2009; Tashakkori & Teddlie, 2003b). These diagrams use geometric shapes (boxes and ovals) to illustrate the steps in the research process (i.e., data collection, data analysis, interpretation) and arrows made with solid lines (→) to show the progression through these steps. They incorporate details about specific procedures and products (e.g., specific product reports that might go to a funding agency) that go beyond the level of information conveyed by the mixed methods notation system. Ivankova, Creswell, and Stick (2006) studied the use of procedural diagrams and suggested 10 guidelines for drawing diagrams for mixed methods designs so that they can be easily and conveniently constructed. These guidelines are listed in Figure 4.1 and are applied in the development of diagrams that appear throughout the remainder of this chapter.

| Figure 4.1 | Ten Guidelines for Drawing Procedural Diagrams for Mixed Methods Studies |

1. Give a title to the diagram.

2. Choose either a horizontal or a vertical layout for the diagram.

3. Draw boxes for the quantitative and qualitative stages of data collection, data analysis, and interpretation of the study results.

4. Use uppercase or lowercase letters to designate the relative priority of the quantitative and qualitative data collection and analysis.

5. Use single-headed arrows to show the flow of procedures in the design.

6. Specify procedures for each stage of quantitative and qualitative data collection and analysis.

7. Specify expected products or outcomes of each procedure in quantitative and qualitative data collection and analysis.

8. Use concise language for describing the procedures and products.

9. Make your diagram simple.

10. Limit your diagram to a single page.

SOURCE: Adapted from Ivankova et al. (2006, p. 15) with permission of SAGE Publications, Inc.

◗ EXAMINING THE DESIGN FEATURES OF MIXED METHODS STUDIES

In Chapter 1, we defined mixed methods research as collecting and analyzing both quantitative and qualitative data, mixing the data, and using a design to frame the procedures. We now add further details to these steps and present a checklist in Figure 4.2 that aids the process of reviewing mixed methods studies by identifying the features of the mixed methods design used. Note that although a few items speak to the substantive content of the study, our attention is focused on the methods' decisions that occurred during the conduct of the study. Specifically, we recommend using the following steps to examine a study's mixed methods design:

- Assess the study's content topic. The content topic is the general issue being studied. It is usually named within the study's title and identified within the abstract.
- Note the philosophical and theoretical foundations. If explicitly addressed, the philosophical and theoretical foundations for a study are often discussed in the background or literature review section of an article. These foundations provide the larger perspectives that the author is using to guide the study.
- Identify the study's content purpose by locating the purpose statement. The purpose statement is the passage in which the author states the specific intent of the study. It is generally found within the introduction section of the article and often at the very end of this section. It usually includes a phrase such as "the purpose of this study is" or "the primary aim of this study was."
- Identify the samples used for the quantitative and qualitative strands. Sampling procedures and the size of the two samples are stated in the methods section of an article. Information about the quantitative and qualitative samples may be discussed together or in separate paragraphs.
- Identify the data collection procedures for the quantitative and qualitative strands. Data collection is described in the methods section of an article, and the quantitative and qualitative data collection procedures are often discussed in separate paragraphs.
- Identify the data analysis procedures for the quantitative and qualitative strands. Data analysis procedures are also discussed in the methods section of an article and, like data collection, are often discussed separately for each data type. In some studies, the data analysis techniques may have to be inferred from the results.

- Assess the author's reason for using mixed methods research. The reason for using a mixed methods approach may be found in one of several places. It may be discussed in close proximity to the study's purpose statement in the introduction or as part of the description of the methods. Some authors may highlight the reason in the final section as part of the discussion of the study's findings.

- Determine the relative priority of the quantitative and qualitative strands. There are two possibilities for the relative priority of the strands for addressing the study's purpose: equal or unequal (where either the quantitative or qualitative strand has greater emphasis). Many authors will explicitly discuss the study's priority in the introduction or methods section or indicate it in a diagram. If not explicitly stated, then judgments about a study's priority might be made based on the study's overarching framework or philosophical foundations, the extensiveness of the databases and sophistication of the analytic procedures, and the language used in the study's title and purpose.

- Determine the timing of the quantitative and qualitative strands. There are three possibilities for the timing of the two strands: They are implemented concurrently in one phase, sequentially in two phases, or combined in multiple phases or projects. The timing of the methods will be described in the methods section of a study and in a procedural diagram, if provided.

- Determine the point of interface between the quantitative and qualitative strands. There are four possible points of interface between the study's strands: at the point of interpretation, at the point of data analysis, at the point of data collection, and at the level of the design. The point of interface often has to be inferred by how the authors report their results and describe their methods. Authors typically convey the point of interface in the results section. In addition, the point of interface may be discussed in other sections of a study, such as in the purpose statement, in the methods section, in a diagram, or in the final conclusion of the study. Judgments about the point(s) of interface are made based on when the two strands directly interact with each other.

- Determine how the quantitative and qualitative strands were mixed. In all mixed methods studies, the author should make interpretations in the discussion section about what was learned from combining the two strands. In addition, there are overall procedures for mixing the quantitative and qualitative strands of an interactive mixed methods study: merging the results of the two data sets, connecting from the results of one type of data to the collection of the other, and embedding the two types of data within a larger design, theoretical framework, or program objective framework. Ideally, the author will discuss in the methods section how the strands were mixed, but in many studies, this process must be inferred by how the quantitative and qualitative results relate to each other as found in the results and interpretation sections.

- Identify the overall mixed methods design by using the mixed methods notation system. Examine how the different methods were implemented within the study, considering the priority, timing, point(s) of interface, and mixing of the two methods. Use the notation system to describe the overall mixed methods approach and to name the corresponding design using the classifications introduced in Chapter 3.
- Draw a one-page diagram of the flow of activities that occurred in the study. Consider the main activities of data collection, data analysis, mixing, and interpretation of results for both the quantitative and qualitative strands. Sketch out how these activities occurred in the study. Refer to Figure 4.1 for guidelines about drawing this diagram.

Figure 4.2   Checklist for Reviewing the Features of a Mixed Methods Study

_____   Assess the study's content topic.

_____   Note the philosophical and theoretical foundations.

_____   Identify the study's content purpose.

_____   Determine whether the author conducted a quantitative strand that included selecting a sample, collecting quantitative data, and analyzing the quantitative data.

_____   Determine whether the author conducted a qualitative strand that included selecting a sample, collecting qualitative data, and analyzing the qualitative data.

_____   Assess the reasons for collecting both quantitative and qualitative data.

_____   Determine the relative priority of the quantitative and qualitative strands for addressing the study's purpose. Use (1) equal or (2) unequal.

_____   Determine the timing of the quantitative and qualitative strands. Use (1) concurrent, (2) sequential, or (3) multiphase combination.

_____   Determine the point of interface between the quantitative and qualitative strands. Use (1) interpretation, (2) data analysis, (3) data collection, or (4) design level.

_____   Determine how the author mixed the two strands. Use (1) merged, (2) connected, (3) embedded, (4) within a theoretical framework, or (5) within a program-objective framework.

_____   Identify the overall mixed methods design.

_____   Provide the notation of the design using the mixed methods notation system.

_____   Draw a diagram of the flow of activities that occurred during the study.

SIX EXAMPLES OF MIXED METHODS DESIGNS ●

To facilitate our discussion of mixed methods, we have included six complete studies in this book (see Appendixes A, B, C, D, E, and F). These studies represent examples of mixed methods research from the health, social, education, and evaluation sciences. In addition, each study reports the application of a different mixed methods design.

The six articles included in the appendixes are

- Wittink, M. N., Barg, F. K., & Gallo, J. J. (2006). Unwritten rules of talking to doctors about depression: Integrating qualitative and quantitative methods. *Annals of Family Medicine, 4*(4), 302–309. (See Appendix A.)
- Ivankova, N. V., & Stick, S. L. (2007). Students' persistence in a Distributed Doctoral Program in Educational Leadership in Higher Education: A mixed methods study. *Research in Higher Education, 48*(1), 93–135. (See Appendix B.)
- Myers, K. K., & Oetzel, J. G. (2003). Exploring the dimensions of organizational assimilation: Creating and validating a measure. *Communication Quarterly, 51*(4), 438–457. (See Appendix C.)
- Brady, B., & O'Regan, C. (2009). Meeting the challenge of doing an RCT evaluation of youth mentoring in Ireland: A journey in mixed methods. *Journal of Mixed Methods Research, 3*(3), 265–280. (See Appendix D.)
- Hodgkin, S. (2008). Telling it all: A story of women's social capital using a mixed methods approach. *Journal of Mixed Methods Research, 2*(3), 296–316. (See Appendix E.)
- Nastasi, B. K., Hitchcock, J., Sarkar, S., Burkholder, G., Varjas, K., & Jayasena, A. (2007). Mixed methods in intervention research: Theory to adaptation. *Journal of Mixed Methods Research, 1*(2), 164–182. (See Appendix F.)

At this time, read these different articles and examine their application of the different mixed methods features using the checklist provided in Figure 4.2. After reading each of the six articles and identifying the features, read the commentary provided in the following sections. This commentary analyzes and reviews the important mixed methods features reported in each of the sample studies. In addition to this commentary, we have included diagrams of the procedures reported in each of the articles using figures provided by the authors themselves when available or figures that we created based on the procedures described within the articles.

Study A: An Example of the Convergent Parallel Design (Wittink, Barg, & Gallo, 2006)

The convergent design involves collecting and analyzing two independent strands of qualitative and quantitative data in a single phase; merging the results of the two strands; and then looking for convergence, divergence, contradictions, or relationships between the two databases. The Wittink et al. (2006) study illustrates the major features of this design.

Wittink et al. (2006) were interested in the contexts surrounding the determination of patients' depression status by primary care physicians with a focus on the patients' views of the interactions they have with their physicians. The purpose of their study was to develop a better understanding of the occurrence of concordance and discordance between patients' and physicians' assessments of a patient's depression status for older adult patients.

To address their study's purpose, the researchers selected a sample made up of all participants in a larger research study (the Spectrum Study) who self-identified as depressed ($N = 48$). The databases assembled for this study then included quantitative and qualitative data collected for each of these 48 individuals. In terms of the quantitative data, the researchers gathered three measures of participants' depression status: a physician's rating, a patient's self-rating, and the participant's score on a standardized measure of depressive symptoms (known as the CES-D). The researchers also gathered several other measures from each participant, including demographic characteristics and assessments of anxiety, hopelessness, health status, and cognitive functioning. When analyzing the quantitative data, the researchers identified whether the patient and physician ratings were concordant (agreed with each other) or discordant (disagreed with each other) for each participant and then calculated descriptive statistics and group comparisons to see whether significant differences existed for the concordant and discordant groups in terms of the other variables of interest.

The researchers also included qualitative semistructured interviews about patients' perceptions of their encounters with their physicians. The interviews were transcribed, and the research team analyzed the texts using constant comparative strategies for theme development. This analysis was independent from the quantitative analysis, as the researchers purposefully did not have access to the quantitative information as they completed the qualitative analysis. Four major themes emerged to describe patients' interactions with their physicians: (1) my doctor just picked it up, (2) I'm a good patient, (3) they just check out your heart and things, and (4) they'll just send you to a psychiatrist. These themes provided a typology for classifying participants based on how they discussed the interactions.

Wittink et al. (2006) described that they needed both types of data in order to develop a more complete understanding. When explaining their mixed methods approach, they wrote, "This design allowed us to link the themes regarding how patients talk to their physicians with personal characteristics and standard measures of distress" (p. 303). Therefore, in order to relate these two different types of information, they selected and analyzed their quantitative and qualitative data sets concurrently and separately from each other. Both types of data appeared equally important for addressing the study's purpose. After the initial separate analyses, they merged the two sets of results in an interactive way so that the point of interface occurred during the analysis and the interpretation. They further analyzed the data to develop a matrix (see Table A.3 in Appendix A) that brought together the qualitative findings (four groups derived from the qualitative themes) with the quantitative results (concordance of depression ratings and other important variables). The information contained within the cells of the table show the descriptive statistics of the variables for each of the qualitatively derived groups for purposes of comparison among the different qualitative perspectives. The researchers concluded with a brief discussion of how the comparisons across the two data sets provided a better understanding of the study's topic.

This study is an example of a convergent mixed methods design. The notation of the study's design can be written as QUAN + QUAL = complete understanding. Although the authors did not provide a diagram of their procedures, we developed one, and it is presented in Figure 4.3. The quantitative data collection and analysis appear on the left side of the figure, and the qualitative data collection and analysis appear on the right side. As shown in this diagram, the quantitative and qualitative strands were implemented during the same phase of the research process and appeared to have an equal emphasis within the study. These two data types and their results were then merged with a comparison matrix and into one overall interpretation, as depicted in the two ovals, which indicate these points of interface between the strands.

Figure 4.3 Diagram for a Study That Used the Convergent Design

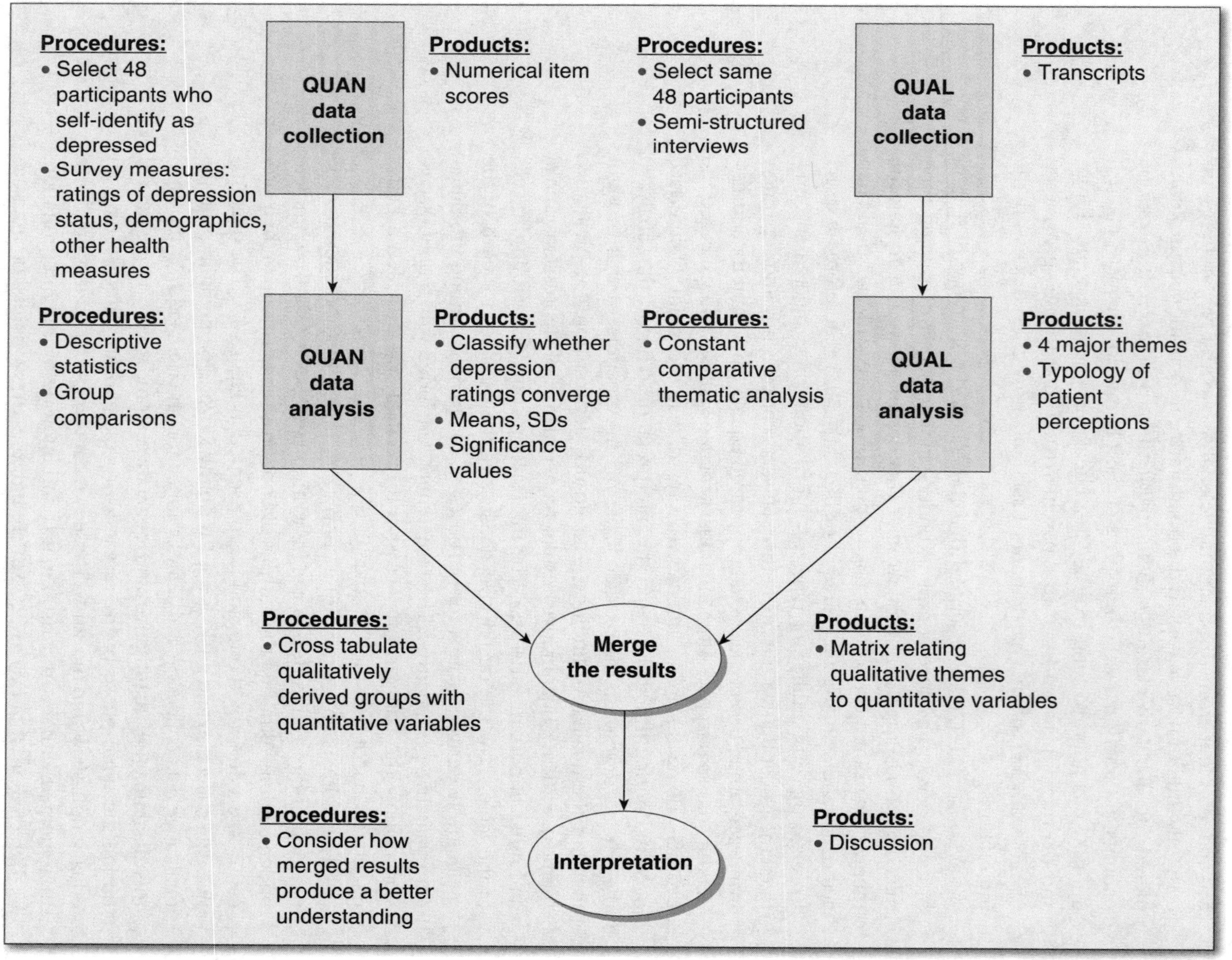

Procedures:
- Select 48 participants who self-identify as depressed
- Survey measures: ratings of depression status, demographics, other health measures

Products:
- Numerical item scores

Procedures:
- Select same 48 participants
- Semi-structured interviews

Products:
- Transcripts

QUAN data collection

QUAL data collection

Procedures:
- Descriptive statistics
- Group comparisons

Products:
- Classify whether depression ratings converge
- Means, SDs
- Significance values

Procedures:
- Constant comparative thematic analysis

Products:
- 4 major themes
- Typology of patient perceptions

QUAN data analysis

QUAL data analysis

Procedures:
- Cross tabulate qualitatively derived groups with quantitative variables

Products:
- Matrix relating qualitative themes to quantitative variables

Merge the results

Procedures:
- Consider how merged results produce a better understanding

Products:
- Discussion

Interpretation

NOTE: Diagram based on Wittink et al. (2006). *SD* = indicates standard deviation.

Study B: An Example of the Explanatory Sequential Design (Ivankova & Stick, 2007)

The explanatory design is implemented in two distinct phases. The first phase involves collecting and analyzing quantitative data. Based on a need to further understand the quantitative results, the researcher implements a second, qualitative phase that is designed to help explain the initial quantitative results. The study by Ivankova and Stick (2007) illustrates the major features of the explanatory design.

Ivankova and Stick (2007) studied the issue of students' persistence within the discipline of higher education. Building on three major theories about student persistence, they chose to study the persistence of doctoral students in one distributed doctoral program in educational leadership. Specifically, their purpose was to identify factors that contribute to students' persistence in the program and to explore participant views about these factors.

The researchers implemented their study in two phases, starting with a quantitative strand. First, they approached all students who had been or were currently enrolled in the program, and 207 agreed to participate in the study. Using a cross-sectional survey design, the researchers developed and administered an online survey to the participants that measured nine predictor variables suggested by theories of student persistence. The responding students represented four groups related to persistence in the program: (1) beginning, (2) matriculated, (3) graduated, and (4) withdrawn or inactive. The analysis of the quantitative data resulted in descriptions of the demographic characteristics of the four groups and identified five variables that significantly discriminated the four different groups defined by their level of persistence.

The researchers conducted a second, qualitative phase after completing the quantitative phase. Using the quantitative results, they identified individuals within the sample that had scores that were typical of the average scores for each group. They purposefully selected four "typical" individuals (one per group) and conducted an in-depth case study of each person's experiences in and perceptions of the program. The primary form of data collection was one-on-one interviews using a protocol that was developed to explore the factors found to be significant in the quantitative phase. Other forms of qualitative data gathered included electronic interviews, written responses, and documents. The analysis first examined the data for description and themes within each case, and this was followed by a cross-case analysis to identify important themes about persistence across the four cases.

Ivankova and Stick (2007) noted that one method alone is not sufficient to capture the trends and details of complex situations such as student persistence in this program. They went on to describe the purpose for their mixing in the following statement: "Thus, the quantitative data and results provided a general picture of the research problem, while the qualitative data and its analysis refined and explained those statistical results by exploring the participants' views regarding their persistence in more depth" (p. 97).

The researchers needed to first identify the general picture and statistically significant results before they knew what quantitative results needed to be further explored with a qualitative strand. As such, the study used sequential timing with the quantitative methods being implemented in the first phase and the qualitative methods following in a second phase. The authors noted that the qualitative phase was prioritized because "it focused on in-depth explanations of the results obtained in the first, quantitative, phase, and involved extensive data collection from multiple sources and two-level case analysis" (p. 97). The primary point of interface occurred at the point of qualitative data collection during the second phase. The authors connected the phases by using the results of the quantitative phase to inform the sampling plan and interview protocol used in the qualitative phase. They also connected the results during the interpretation by discussing a major quantitative result and then how a follow-up qualitative result helped to explain the statistical result in more depth.

Based on the implemented design features, the notation for the study can be written as quan → QUAL = explain significant factors. Since the study was conducted in two phases with the second, qualitative phase dependent on the results of the initial quantitative phase, this study is an example of the explanatory mixed methods design. Its two-phase timing and points of mixing are highlighted in the diagram developed by the authors and reproduced in Figure 4.4. The data collection and analysis procedures of the initial quantitative phase are described in the first two rectangular boxes. The connections to the qualitative phase through case selection and interview protocol development are shown in the oval (the first point of interface). Then, the procedures in the second, qualitative phase are described in the next two rectangular boxes. The diagram concludes with another oval indicating the second point of interface and how the authors interpreted the overall mixed methods results.

Figure 4.4 Diagram for a Study That Used the Explanatory Design

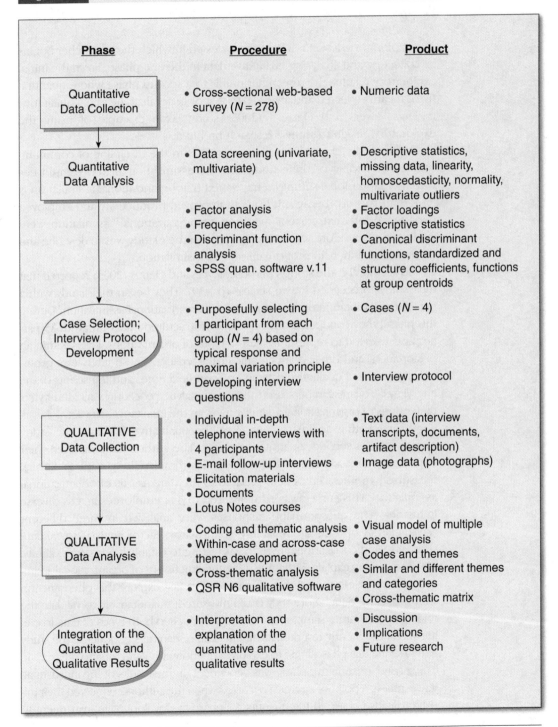

SOURCE: Reprinted from Ivankova and Stick (2007, p. 98) with permission of Springer Science+Business Media, Inc.

Study C: An Example of the Exploratory
Sequential Design (Myers & Oetzel, 2003)

The exploratory design is a two-phase design in which the researcher begins by collecting and analyzing qualitative data in the first phase. From the initial exploratory results, the researcher builds to a second phase where quantitative data are collected and analyzed to test or generalize the initial qualitative findings. Myers and Oetzel's (2003) study is an example of using the exploratory design to study a research problem.

Myers and Oetzel (2003) are researchers in the discipline of communications. The topic of their study was the assimilation of new employees within organizational settings. They stated that organizational assimilation is important to study because it leads to improved productivity and employee persistence but that current measures of organizational assimilation were inadequate. Therefore, the overall purpose of their study was to describe and measure the dimensions of organizational assimilation.

To accomplish the study's purpose, Myers and Oetzel (2003) reported that their study "proceeded in two stages" (p. 439). They began their study with a qualitative exploration of the dimensions of organizational assimilation. During this phase, they conducted one-on-one semistructured interviews with 13 participants selected to represent different types of organizations, levels within an organization, and other demographic characteristics. These interviews generated two types of qualitative data: interviewer field notes and transcripts of the interviews. The researchers used thematic analytic procedures to identify six dimensions of organizational assimilation from the qualitative data set.

After creating an instrument from the qualitative findings, the study moved into its second, quantitative phase. The authors administered their Organizational Assimilation Index (OAI) instrument along with additional measures hypothesized as being related to the dimensions of organizational assimilation. This survey was administered to 342 employees across diverse industries. The questionnaire responses were analyzed in three different ways: analysis of the scale reliability, confirmatory factor analysis to validate subscales, and correlational hypothesis testing to establish construct validity.

The authors explained that the different dimensions of organizational assimilation were unknown and that they needed to first explore this phenomenon with qualitative data before they could measure it quantitatively to validate the findings with a larger sample. Therefore, they needed both types of data to create and subsequently test an instrument. The researchers conducted the study in two sequential phases: first to explore a phenomenon and then to measure it. The second, quantitative phase was dependent on the results of the initial, qualitative phase. A point of interface occurred when the authors connected their initial, qualitative phase to the quantitative phase by developing an instrument to measure organizational assimilation. Building from their qualitative findings, the authors developed 61 survey items to represent the six dimensions of organizational assimilation. This instrument was then implemented in the second phase. In the final discussion, they noted specific qualitative findings and then discussed the extent to which the quantitative results validated the findings. Because of the authors' emphasis on developing and validating a quantitative instrument, this study seemed to emphasize the quantitative aspects, thus demonstrating the overall importance of the quantitative data in this study.

The notation for this study can be written as qual → QUAN = validate exploratory dimensions by designing and testing an instrument. The authors used two connected phases to implement this study's methods in an exploratory mixed methods design. As depicted in Figure 4.5, the design began with qualitative data collection and analysis to explore a phenomenon (the first two boxes of the diagram). From this initial phase, an instrument was developed at a point of interface (note the "develop an instrument" oval in Figure 4.5). The researchers used this instrument to collect quantitative data in a second phase (the next two boxes in the diagram) and concluded by interpreting what was learned across the two phases.

Study D: An Example of the Embedded Design (Brady & O'Regan, 2009)

The embedded design involves collecting and analyzing at least one type of data within a design framework generally associated with the other type of data, such as when a researcher chooses to embed a qualitative strand within a quantitative experiment. The purpose of the embedded data is to enhance the conduct or interpretation of the larger design. Brady and O'Regan's (2009) study illustrates the major features of this design.

In their 2009 article, Brady and O'Regan reported on the design of their mixed methods study to evaluate the Big Brothers Big Sisters (BBBS) mentoring program within the Ireland context. The purpose of this study was twofold: to assess the impact of the BBBS program for Irish youth and to examine the process and implementation of the program. The authors provided a rich discussion of how a pragmatic foundation and dialectical position allowed them to be open to adding a qualitative component to an overall experimental design. They also described how their guiding theoretical framework, Rhodes's Model of Mentoring, provided a means for thinking about the combination of these different aspects.

This study's methods were shaped by a quantitative, randomized controlled trial (RCT) for testing whether the BBBS program has a significant impact on youth outcomes. In order to satisfy the needs to study the program

Figure 4.5 Diagram for a Study That Used the Exploratory Design

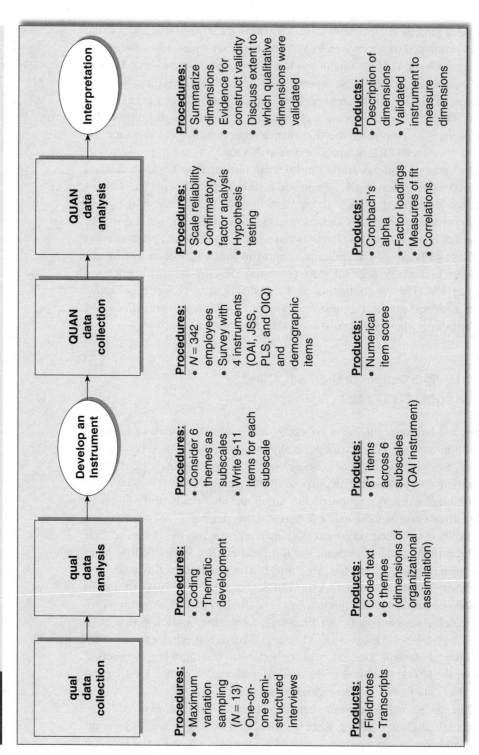

qual data collection

Procedures:
• Maximum variation sampling (N = 13)
• One-on-one semi-structured interviews

Products:
• Fieldnotes
• Transcripts

qual data analysis

Procedures:
• Coding
• Thematic development

Products:
• Coded text
• 6 themes (dimensions of organizational assimilation)

Develop an Instrument

Procedures:
• Consider 6 themes as subscales
• Write 9–11 items for each subscale

Products:
• 61 items across 6 subscales (OAI instrument)

QUAN data collection

Procedures:
• N = 342 employees
• Survey with 4 instruments (OAI, JSS, PLS, and OIQ) and demographic items

Products:
• Numerical item scores

QUAN data analysis

Procedures:
• Scale reliability
• Confirmatory factor analysis
• Hypothesis testing

Products:
• Cronbach's alpha
• Factor loadings
• Measures of fit
• Correlations

Interpretation

Procedures:
• Summarize dimensions
• Evidence for construct validity
• Discuss extent to which qualitative dimensions were validated

Products:
• Description of dimensions
• Validated instrument to measure dimensions

NOTE: Diagram based on Myers and Oetzel (2003). OAI indicates Organizational Assimilation Index. OIQ indicates Organizational Identification Questionnaire. JSS indicates Job Satisfaction Scale. PLS indicates Propensity to Leave Scale.

where it was well established and to not deny services to youth in need, the researchers had to compromise with a sample size of 164 participating youth randomly assigned to either the program or a treatment-as-usual control condition. The study also included the participating youth's parents, mentors, and teachers. The researchers gathered measures of outcomes predicted by the theoretical model at baseline and times 12 and 18 months later. The quantitative data analysis included regression and structural equation modeling (SEM) analyses based on the hypothesized relationships within the mentoring model.

The researchers also described the design of the qualitative component of their study, which was needed to make the quantitative approach acceptable for stakeholders and to address questions related to process, feasibility, and implementation. They gathered information on the experiences and perceptions of youth, mentors, parents, and staff. Specifically, they planned to purposefully select 12 mentoring pairs and interview the corresponding individuals at the time the mentoring relationship was formed and again 6 or more months later. In addition, the team collected observations, case file documents, and focus groups with program staff. The qualitative analysis focused on thematic development across the cases and perspectives.

Although the researchers started with a planned rigorous experimental study as their "primary" focus, they found they needed to address ethical, feasibility, and methodological issues associated with using this design that could not be handled with a purely quantitative design. The researchers therefore embedded a "secondary" qualitative strand within the quantitative experimental design to address the issues, such as stakeholder concerns with the research, questions about the process and implementation in addition to the outcomes of the program, and issues at both the individual and programmatic levels. Therefore, the primary point of interface occurred at the level of the design. The strands in this study can be considered to have an interactive relationship because the decisions about the qualitative strand depended on the quantitative experimental design within which they were implemented. Specifically, the qualitative methods played a supplemental role to examine issues of process and implementation of the experiment. The researchers also linked the two concurrent strands through their theoretical model, with neither strand depending on the results of the other strand. The two methods were utilized to address different research questions within the overarching experimental design. In addition, the research team combined the quantitative impact data and the qualitative case study data through the theoretical model to enhance understanding of how the intervention was experienced in the Irish context.

This study is an example of an embedded mixed methods design. The notation of the study's design can be written as QUAN (+ qual) = enhance experiment. The authors provided a detailed diagram of their procedures, which is reproduced in Figure 4.6. This diagram indicates the primary

Figure 4.6   Diagram for a Study That Used the Embedded Design

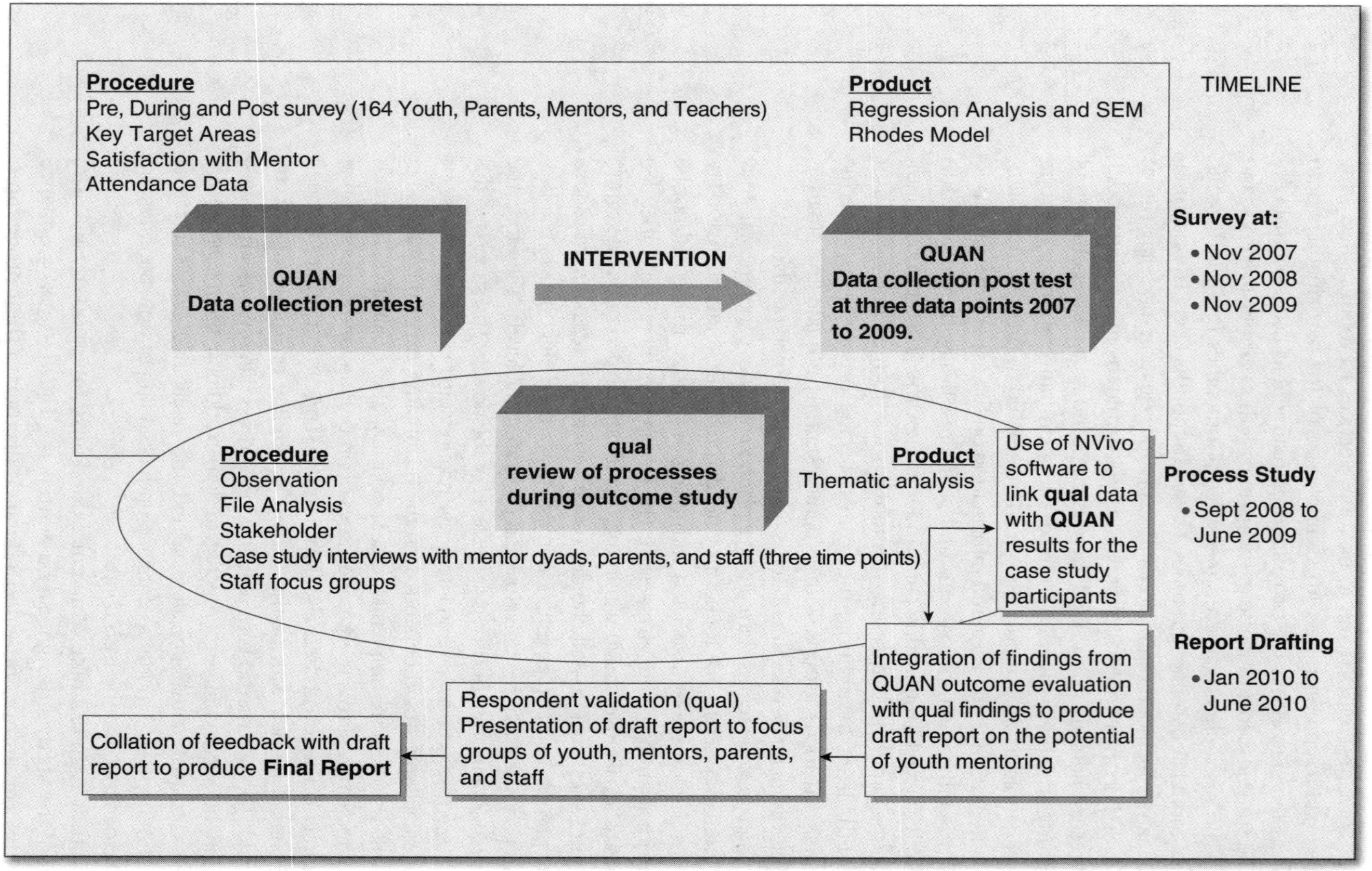

Procedure
Pre, During and Post survey (164 Youth, Parents, Mentors, and Teachers)
Key Target Areas
Satisfaction with Mentor
Attendance Data

QUAN
Data collection pretest

INTERVENTION

QUAN
Data collection post test at three data points 2007 to 2009.

Product
Regression Analysis and SEM Rhodes Model

Procedure
Observation
File Analysis
Stakeholder
Case study interviews with mentor dyads, parents, and staff (three time points)
Staff focus groups

qual
review of processes during outcome study

Product
Thematic analysis

Use of NVivo software to link **qual** data with **QUAN** results for the case study participants

Integration of findings from QUAN outcome evaluation with qual findings to produce draft report on the potential of youth mentoring

Respondent validation (qual)
Presentation of draft report to focus groups of youth, mentors, parents, and staff

Collation of feedback with draft report to produce **Final Report**

TIMELINE

Survey at:
• Nov 2007
• Nov 2008
• Nov 2009

Process Study
• Sept 2008 to June 2009

Report Drafting
• Jan 2010 to June 2010

SOURCE: Reprinted from Brady and O'Regan (2009, p. 277) with permission of Sage Publications, Inc.

quantitative strand in the large rectangular box at the top of the figure. This box indicates the major procedures for the RCT, including the intervention and the preintervention and postintervention measures. The secondary qualitative strand to examine the processes during the intervention is indicated in the large oval, which is shown concurrent with the intervention and experimental procedures. The large area of overlap between the rectangular box and oval conveys the interface at the design level. The diagram also indicates ways in which the researchers planned to link outcome results to the case study results for individuals and for the program and to report combined results in the final project report. Note that this diagram also includes a timeline for the different components on the right side.

Study E: An Example of the Transformative Design (Hodgkin, 2008)

The transformative design is used when the researcher frames a mixed methods study within a transformative theoretical perspective to help address injustices or bring about change for an underrepresented or marginalized group. The qualitative and quantitative strands of the study can proceed concurrently or sequentially or both. Hodgkin's (2008) article describes her application of the main features of a transformative design.

Hodgkin's 2008 article discussed how her study was located within the transformative research paradigm and specifically was framed from a feminist theoretical lens. She was interested in the topic of social capital and was focused on understanding women's social capital and challenging the lack of gender sensitivity. Stemming from her transformative and feminist perspectives, she described that the purpose of her research was to highlight gender inequality by identifying differences in the social capital profiles of men and women and explaining why these differences existed for women.

Hodgkin (2008) started with a quantitative strand using cross-sectional survey procedures to identify whether men and women have different social capital profiles. Using simple random sampling procedures, she gathered mailed survey responses from 1,431 individuals in a regional city of Australia, including 998 females. In designing the survey instrument, Hodgkin described specifically locating a measure of social capital that was shown to be sufficiently sensitive to gender issues by including scales related to social, community, and civic participation. She analyzed the quantitative data using multivariate analyses to compare men and women and found significant differences on three scales of participation, with women scoring higher in informal social participation, social participation in groups, and community group participation. Hodgkin concluded that the quantitative data provided evidence of gendered patterns of social, civic, and community participation.

Hodgkin (2008) next conducted a qualitative phase to explain why women had different social capital profiles from men. She was unable to select participants from their quantitative responses due to ethical considerations related to keeping the survey data confidential. Therefore, she used cluster random sampling to select a subsample of women who had completed the quantitative survey. She conducted two in-depth one-on-one interviews with each woman participant one week apart, asking each one to record written reflections in a diary for the week between the two interview sessions. These written reflections of women's activities were discussed as part of the second interview. Interviews related to women's participation were completed until saturation was reached with 12 participants. Hodgkin conducted narrative analysis of the participants' stories that emerged through the qualitative data. Three themes tied to motherhood emerged as explaining different reasons for participation: wanting to be a "good mother," wanting to avoid social isolation, and wanting to be a good citizen.

Hodgkin (2008) stated that there was a need to challenge the lack of gender sensitivity in the study of social capital and to use mixed methods to build a comprehensive picture and produce data types that would be deemed acceptable by those who needed to change. The relationship between the strands was interactive, because she first identified whether and in what ways social capital profiles differed between men and women. Then she sought to explain the identified differences. The study used sequential timing with the quantitative methods being implemented in the first phase and the qualitative methods following in a second phase. The author implied that both methods were equally important for developing an understanding that could promote change. She connected from the quantitative strand to the qualitative strand in two ways. She used a subsample from the first phase in the second phase and designed the qualitative data collection protocols to follow up on initial quantitative results, thereby connecting at the point of data collection. In conclusion, she interpreted how the quantitative results identified differences in involvement between the genders and how the qualitative findings explained why women became involved.

This study is an example of a transformative mixed methods design. Although the author used the procedures of the explanatory design, these procedures were implemented within a larger transformative theoretical framework that shaped the design decisions. Although the author did not provide a diagram of her procedures, we developed one as shown in Figure 4.7. The larger transformative framework is illustrated with the dashed line encircling the methods, which provides an overview of the transformative aims for the different components of the design and indicates the mixing within a

Figure 4.7 Diagram for a Study That Used the Transformative Design

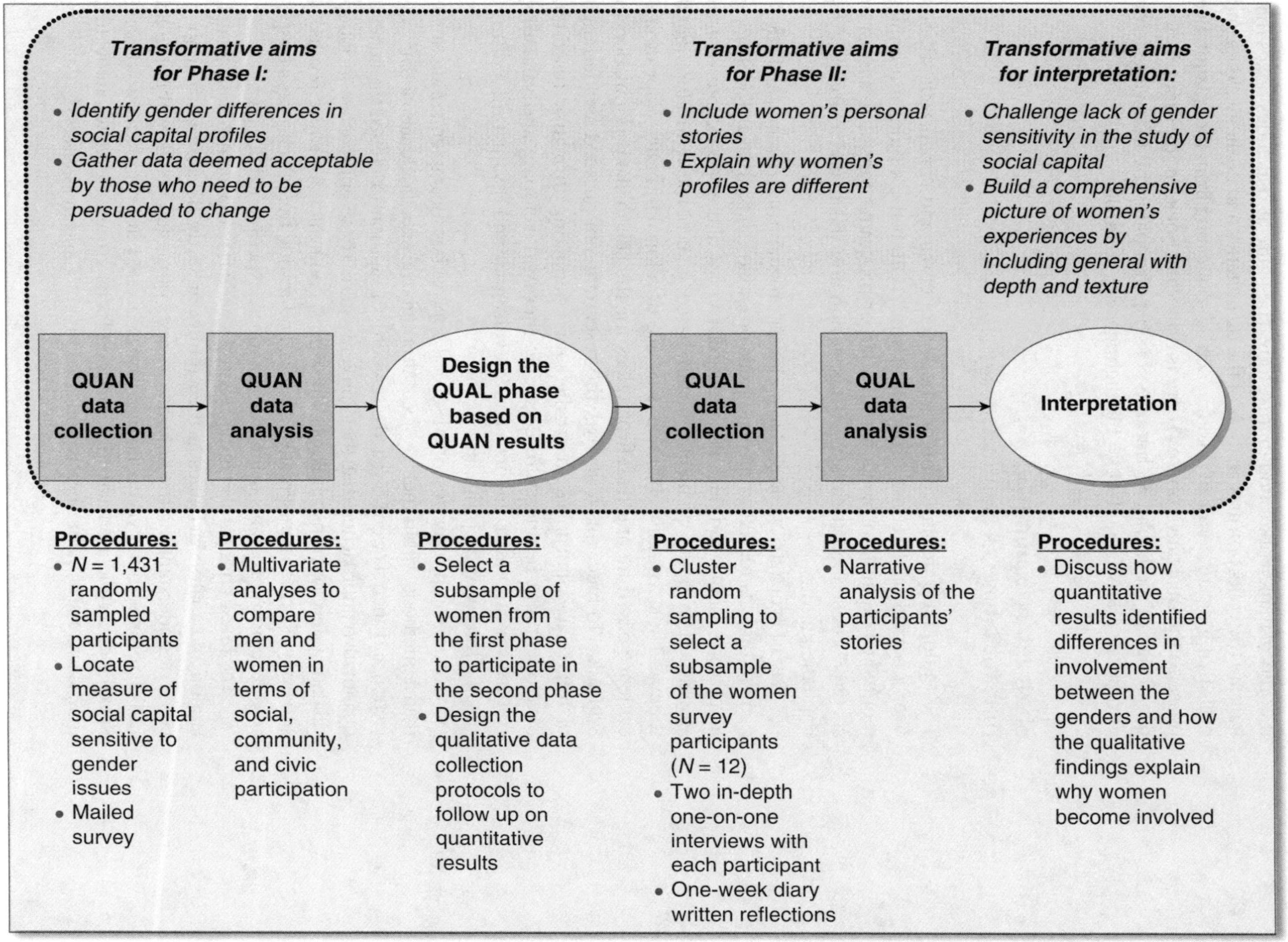

Transformative aims for Phase I:
- *Identify gender differences in social capital profiles*
- *Gather data deemed acceptable by those who need to be persuaded to change*

Transformative aims for Phase II:
- *Include women's personal stories*
- *Explain why women's profiles are different*

Transformative aims for interpretation:
- *Challenge lack of gender sensitivity in the study of social capital*
- *Build a comprehensive picture of women's experiences by including general with depth and texture*

QUAN data collection

Procedures:
- $N = 1,431$ randomly sampled participants
- Locate measure of social capital sensitive to gender issues
- Mailed survey

QUAN data analysis

Procedures:
- Multivariate analyses to compare men and women in terms of social, community, and civic participation

Design the QUAL phase based on QUAN results

Procedures:
- Select a subsample of women from the first phase to participate in the second phase
- Design the qualitative data collection protocols to follow up on quantitative results

QUAL data collection

Procedures:
- Cluster random sampling to select a subsample of the women survey participants ($N = 12$)
- Two in-depth one-on-one interviews with each participant
- One-week diary written reflections

QUAL data analysis

Procedures:
- Narrative analysis of the participants' stories

Interpretation

Procedures:
- Discuss how quantitative results identified differences in involvement between the genders and how the qualitative findings explain why women become involved

NOTE: Diagram based on Hodgkin (2008).

theoretical framework found in this study. The figure also includes boxes that indicate the collection and analysis of the quantitative and qualitative data and ovals that indicate points of mixing, such as connecting from the quantitative phase to the qualitative phase. There is no formal notation for a transformative design in the literature, but one possible example is the following: feminist theory (QUAN → QUAL) = highlight gender inequality.

Study F: An Example of the Multiphase Design (Nastasi et al., 2007)

The multiphase design combines both sequential and concurrent strands over a period of time in a program of study. Often the strands are implemented as multiple projects within a larger program of inquiry. Nastasi et al. (2007) described their use of this design and the major features of its implementation.

Nastasi et al. (2007) engaged in multiyear programmatic research and development projects related to mental health promotion in Sri Lanka. The guiding frameworks for this study included the Participatory Culture-Specific Intervention Model and a model of mental health based in ecological–developmental theory. Their overall objective was to develop culturally appropriate evidence-based mental health practices for the school-aged population in Sri Lanka. To meet this objective, the research team pursued a wide range of interrelated purposes that called for conducting formative research, developing and testing culture-specific theory, developing and validating culture-specific instruments, and developing and evaluating culture-specific intervention programs.

The research team described several approaches for implementing quantitative methods within their project. Although the specific details of data collection and analysis were detailed elsewhere, the authors discussed the general quantitative approaches that they implemented. These approaches included validating developed psychological measures, confirming formative results by surveying a larger representative sample, and testing the effectiveness of specific developed programs using true and quasi-experimental designs.

The research team also implemented a wide range of qualitative data collection and analysis activities in their multiyear study. Because of the importance of understanding the cultural contexts of mental health within the Sri Lanka setting, much of qualitative research used an ethnographic design. Specific data collection activities included focus group interviews, individual interviews, participant observations, documents, and field notes.

Nastasi et al. (2007) argued that their goal of developing culturally appropriate evidence-based mental health practices required a recursive and integrative combination of quantitative and qualitative methods, which is an example of mixing within a program objective framework. They needed qualitative methods to identify cultural contexts that helped guide program development and adaptation of program to new contexts and this required quantitative methods to test cultural models and program effectiveness. At times these methods were interactive, such as when quantitative methods were used to validate a psychological measure developed based on qualitative results. This dependent relationship was most strongly seen as the program moved from one sequential phase to the next. In addition, it was likely that there were times that the methods were independent when they were implemented concurrently, such as when the authors merged the two types of information to understand the acceptability, integrity, and effectiveness of intervention methods. The authors did not specifically discuss the priority of the two approaches. Although it was possible that an individual phase could have one method prioritized over the other, it was clear from looking over the full research process that the two methods played equally important roles in addressing the study's objective. The authors described many ways that they mixed the quantitative and qualitative strands throughout the project, such as designing a quantitative strand to test the effectiveness of a program adapted based on a qualitative strand (i.e., connecting) and combining both methods to examine acceptability of a program (i.e., merging).

This large-scale, multiyear evaluation project was an example of a multiphase mixed methods design. The study was implemented over multiple phases, and the quantitative and qualitative methods were conducted sequentially across phases and also concurrently within some phases. There is no simple notation for describing this study because of its iterative and recursive nature of implementing the methods, and in fact, the authors introduced a new notation of double arrows ($\rightarrow\leftarrow$) to convey the recursive nature of the process. A simpler start to describing this overall study might look something like this: QUAL $\rightarrow$ QUAN $\rightarrow$ [QUAN + QUAL] . . . = program development. Better yet, the authors' diagram of the process shown in Figure 4.8 conveys extensive detail of the multiple phases. This figure outlines the many phases involved in the program development process, where each circle represents the use of at least one qualitative and/or quantitative strand. In addition to this figure, the authors also provided a table detailing the concurrent and sequential interactions of data at different phases in the project.

Figure 4.8 Diagram for a Study That Used the Multiphase Design

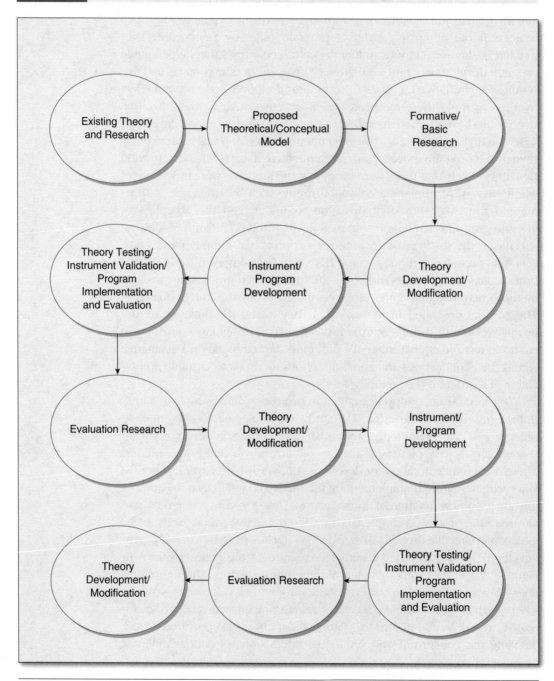

SOURCE: Reprinted from Nastasi et al. (2007, p. 166) with permission of SAGE Publications, Inc.

SIMILARITIES AND DIFFERENCES ● AMONG THE SAMPLE STUDIES

The main features of the six mixed methods studies discussed in this chapter are summarized in Table 4.2. The similarities and differences among the information within this table highlight many of the important features of mixed methods research and the different approaches for applying mixed methods research.

First, it is interesting to note that these six example studies represent diverse disciplines, examine different research topics, and incorporate different philosophical and theoretical perspectives. Their diversity is also reflected in the fact that they were drawn from different disciplines and conducted for different purposes. Wittink et al. (2006) examined how patients and physicians communicate about older adults' depression status, using both qualitative and quantitative information. Ivankova and Stick (2007) identified and explained predictors of student persistence in one distributed doctoral program. Myers and Oetzel (2003) explored and validated the dimensions of organizational assimilation. Brady and O'Regan (2009) wanted to evaluate the process and implementation of their mentoring program as part of their experiment to test its impact. Hodgkin (2008) wanted to challenge gender inequality in the study of social capital. Finally, Nastasi et al. (2007) worked to develop culturally appropriate mental health practices in Sri Lanka.

Each of these studies included samples of individuals for the quantitative and qualitative strands, although they used different strategies. For example, Wittink et al. (2006) used the same sample (same individuals and same sample size) for both strands. Hodgkin (2008) and Ivankova and Stick (2007) both selected a smaller subsample of individuals that participated in their quantitative phase to participate in the qualitative phase. Myers and Oetzel (2003) selected a small sample for their qualitative phase and then selected a large sample of different individuals for the quantitative phase.

Each of these studies collected at least one form of quantitative data and at least one form of qualitative data. Quantitative cross-sectional survey and experimental approaches were used across the six studies. The quantitative data were collected using various types of structured questionnaires or measurement instruments. The forms of qualitative data collected across these studies included one-on-one interviews, focus group interviews, observations, written responses, and researcher field notes.

Each study also included procedures for analyzing the quantitative and qualitative data. The quantitative procedures presented descriptive analyses, group comparisons, reliability checks, confirmatory factor analyses, correlational analyses, and multivariate analyses. Qualitative analytic procedures included developing description, thematic analyses, and narrative story development.

Table 4.2 A Comparison of the Example Mixed Methods Studies

	Wittink, Barg, and Gallo (2006)	Ivankova and Stick (2007)	Myers and Oetzel (2003)	Brady and O'Regan (2009)	Hodgkin (2008)	Nastasi et al. (2007)
Content area and field of study	• Depression status (mental health)	• Persistence of doctoral students (higher education studies)	• Organizational assimilation (organization studies)	• BBBS mentoring program (youth studies)	• Gender and social capital (sociology)	• Mental health promotion in Sri Lanka (evaluation)
Philosophical foundations	• Not explicitly discussed	• Not explicitly discussed	• Not explicitly discussed	• Pragmatic and dialectical	• Transformative paradigm	• Participatory Culture-Specific Intervention Model
Theoretical foundations (social science or advocacy)	• Not explicitly discussed	• Three major theories of students' persistence (social science)	• Theories of organizational assimilation stages (social science)	• Rhodes's Model of Mentoring (social science)	• Feminist theory (advocacy)	• Model of mental health based in ecological–developmental theory (social science)
Content purpose	• To understand concordance and discordance between physicians and patients about depression status	• To understand student persistence in one distributed doctoral program by identifying and exploring factors that predict student persistence	• To describe and measure the dimensions that describe new employee assimilation	• To evaluate the impact, as well as the process and implementation, of the BBBS program for youth in Ireland	• To highlight gender inequality by identifying differences in the social capital profiles of men and women and explaining why these differences exist for women	• To develop culturally appropriate evidence-based mental health practices in Sri Lanka

	Wittink, Barg, and Gallo (2006)	Ivankova and Stick (2007)	Myers and Oetzel (2003)	Brady and O'Regan (2009)	Hodgkin (2008)	Nastasi et al. (2007)
Quantitative strand:						
Sample	$N = 48$ individuals that self-identified as depressed in a larger study	$N = 207$ students across four matriculation status groups	$N = 342$ employees across industries	$N = 164$ participating youth and their parents, mentors, and teachers	$N = 1,431$ randomly sampled participants ($n = 403$ male; $n = 998$ female)	Samples selected as appropriate for each phase
Data collection	• Cross-sectional survey design • Measures of depression (physician rating, patient self-report, and standardized depression scale, CES-D) • Demographics • Other health measures (e.g., anxiety, health status, and cognitive functioning)	• Cross-sectional survey design • Online survey to assess predictor variables	• Questionnaire including multiple scales to measure six dimensions of OAI as well as OIQ, JSS, and PLS.	• RCT design • Gather pretest and posttest measures over 3 years • Measures assess satisfaction with mentor and attendance data	• Cross-sectional survey design • Locate measure of social capital sufficiently sensitive to gender issues • Mail survey that assesses social capital in terms of social, community, and civic participation	• Data collection as appropriate for each phase • Approaches include instrument validation techniques and experimental designs

(Continued)

Table 4.2 (Continued)

	Wittink, Barg, and Gallo (2006)	Ivankova and Stick (2007)	Myers and Oetzel (2003)	Brady and O'Regan (2009)	Hodgkin (2008)	Nastasi et al. (2007)
Data analysis	• Descriptive statistics • Group comparisons	• Descriptive statistics • Discriminant function analysis	• Scale reliability • Confirmatory factor analysis • Correlational tests	• Regression analysis • SEM analysis	• Multivariate analyses to compare men and women	• Data analysis as appropriate for each phase
Qualitative strand:						
Sample	• Same individuals (N = 48) that self-identified as depressed	• Purposefully selected four individuals from the quantitative sample who were typical of the four matriculation status groups	• Purposefully selected 13 individuals for maximum variation	• Select program stakeholders including youth, mentors, parents, and staff	• Cluster random sampling to select a subsample of the women survey participants (N = 12)	• Samples selected as appropriate for each phase
Data collection	• Semistructured interviews	• Multiple case study design • Telephone interviews, electronic interviews, open-ended questionnaire responses, and program-related documents	• One-on-one semistructured interviews • Researcher field notes	• Interviews focused around 12 mentoring pairs • Case file documents • Focus groups with program staff • Observations	• Two in-depth one-on-one interviews with each woman • Participants' written reflections in a 1-week diary	• Data collection as appropriate for each phase, such as focus group interviews, participant observation, documents, and field notes

	Wittink, Barg, and Gallo (2006)	Ivankova and Stick (2007)	Myers and Oetzel (2003)	Brady and O'Regan (2009)	Hodgkin (2008)	Nastasi et al. (2007)
Data analysis	• Thematic analysis	• Within-case descriptive and thematic analysis • Cross-case thematic analysis	• Thematic analysis	• Thematic analysis	• Narrative analysis of the participants' stories	• Data analysis as appropriate for each phase
Mixed methods features:						
Reason for mixing methods	• Need to relate quantitative measures of depression and characteristics with qualitative descriptions of patient experiences with physicians to develop a more complete picture	• Need to obtain general statistical picture of predictors of persistence and to explore participant views in-depth to explain the statistical results	• Need quantitative data to validate qualitative findings	• Need to address ethical, feasibility, and methodological issues associated with using a RCT to study the program impact	• Need to challenge the lack of gender sensitivity in the study of social capital using methods that will build a comprehensive picture using data that will be deemed acceptable by those who need to be persuaded to change	• Need qualitative methods to identify cultural contexts that help guide program development and adaptation of program to new contexts and need quantitative methods to test cultural models and program effectiveness
Priority of the strands	• Equal	• Qualitative priority	• Quantitative priority	• Quantitative priority	• Equal	• Equal

(Continued)

Table 4.2 (Continued)

	Wittink, Barg, and Gallo (2006)	Ivankova and Stick (2007)	Myers and Oetzel (2003)	Brady and O'Regan (2009)	Hodgkin (2008)	Nastasi et al. (2007)
Timing of the strands	• Concurrent	• Sequential: Quantitative followed by qualitative	• Sequential: Qualitative followed by quantitative	• Concurrent	• Sequential: Quantitative followed by qualitative	• Sequential and concurrent
Primary points of mixing (point of interface)	• Data analysis • Interpretation	• Data collection • Interpretation	• Data collection • Interpretation	• Design	• Design • Data collection • Interpretation	• Design • Interpretation
Mixing of the strands	• Merge: Developed a matrix that related qualitatively derived groups to quantitative scores • Interpretation: Discussed how comparisons across the two data sets	• Connect: Used quantitative results to select participants and develop interview protocol for the qualitative phase • Interpretation: Described specific quantitative	• Connect: Used qualitative findings to inform development of an instrument for the quantitative phase • Interpretation: Discussed the extent to which	• Embed: Qualitative strand is embedded within the quantitative experiment • Merge: Quantitative and qualitative data at the individual level	• Theoretical framework: Gathered both types of data within a feminist lens • Connect: Used a subsample in the second phase • Designed the qualitative data collection	• Program-objective framework: Promote mental health • Connect: Used quantitative strand to test the effectiveness of a program adapted based on a qualitative strand

	Wittink, Barg, and Gallo (2006)	Ivankova and Stick (2007)	Myers and Oetzel (2003)	Brady and O'Regan (2009)	Hodgkin (2008)	Nastasi et al. (2007)
	provide a better understanding	results and discussed how the qualitative findings help to explain the results	quantitative results validated qualitative findings	• Merge: Impact results and case study results in regards to the guiding theoretical model	protocols to follow up on quantitative results • Interpretation: Discussed quantitative differences in involvement between the genders and how the qualitative findings explain why women become involved	• Merge: Used both methods to examine acceptability of a program • Embed: Used process evaluation procedures within a summative evaluation

Mixed methods design:

	Wittink, Barg, and Gallo (2006)	Ivankova and Stick (2007)	Myers and Oetzel (2003)	Brady and O'Regan (2009)	Hodgkin (2008)	Nastasi et al. (2007)
Mixed methods design type	Convergent	Explanatory	Exploratory	Embedded	Transformative	Multiphase
Notation	QUAN + QUAL = complete understanding	quan → QUAL = explain significant factors	qual → QUAN = validate exploratory dimensions	QUAN (+ qual) = enhance experiment	Feminist theory (QUAN → QUAL) = highlight gender inequality	QUAL → QUAN → [QUAN + QUAL]... = program development

NOTE: BBBS indicates Big Brothers Big Sisters. CES-D indicates Center for Epidemiologic Studies Depression Scale. OAI indicates Organizational Assimilation Index. OII indicates Organizational Identification Questionnaire. JSS indicates Job Satisfaction Scale. PLS indicates Propensity to Leave Scale. RCT indicates randomized controlled trial. SEM indicates structural equation modeling.

The authors of each study offered their reasons for collecting both quantitative and qualitative forms of data. Wittink et al. (2006) needed to directly relate both types of data to best understand the problem. Ivankova and Stick (2007) needed to collect qualitative data to explain their initial quantitative results. Myers and Oetzel (2003) wanted to validate qualitative findings by developing an instrument based on an initial exploration of their topic before they tried to measure it. Brady and O'Regan (2009) needed qualitative data to address ethical, feasibility, and methodological issues as part of their experimental trial. Hodgkin (2008) needed a combination of methods to challenge gender inequality with data types deemed acceptable by those who need to be persuaded to change and that conveyed the big picture with the personal story. Nastasi et al. (2007) needed a combination of methods over multiple years to identify cultural contexts that help guide program development and adaptation of program to new contexts and to test cultural models and program effectiveness.

In line with the different purposes, the studies employed different priority and timing among the strands. The strands were equally important for addressing the overall purpose (e.g., Wittink et al., 2006) or had an unequal priority, either prioritizing the qualitative strand (e.g., Ivankova & Stick, 2007) or the quantitative strand (e.g., Brady & O'Regan, 2009). Likewise, the timing varied to include concurrent (i.e., Brady & O'Regan, 2009; Wittink et al., 2006), sequential (i.e., Hodgkin, 2008; Ivankova & Stick, 2007; Myers & Oetzel, 2003), or a combination of multiple phases (i.e., Nastasi et al., 2007).

These studies all mixed their quantitative and qualitative strands, but they did so at different points and in different ways. Wittink et al. (2006) merged the two data sets during data analysis by interrelating the two sets of findings and combining them in a table. Hodgkin (2008), Ivankova and Stick (2007), and Myers and Oetzel (2003) both mixed by connecting two sequential phases in which the second phase data collection built off the first phase results. In their respective studies, Hodgkin (2008) and Ivankova and Stick (2007) identified key results from their quantitative data and used them to direct their qualitative phase. Myers and Oetzel (2003) developed an instrument based on their qualitative findings, which they then used to gather quantitative data in their second phase. Brady and O'Regan (2009) mixed by embedding qualitative data within their experimental trial design to examine the process and implementation of the intervention from stakeholders' perspectives. Hodgkin (2008) also mixed by positioning her use of mixed methods within a feminist theoretical framework. Nastasi et al. (2007) mixed within a program objective framework. Among the multiple phases, they also mixed by merging, connecting, and embedding strands within their larger program of inquiry.

Each of these studies represents a different mixed methods design, and we can highlight the methodological differences across the designs by examining the different patterns that emerge from the stated notations and the diagrams pictured in Figures 4.3 through 4.8. Notice how the designs differ in their timing. Figures 4.3 and 4.6 depict both methods being implemented concurrently in one phase, whereas Figures 4.4, 4.5, and 4.7 depict the methods being implemented in a definite sequence. Figure 4.8 encompasses both sequential and concurrent aspects of the multiphase design. The designs also differ in terms of the relative priority given to the different forms of data, as shown by the QUAN and QUAL letters appearing in uppercase or lowercase letters. The diagrams also highlight other important features of the designs, such as the importance of a transformative perspective in Figure 4.7 and the recursive and iterative nature of a multiyear development project in Figure 4.8.

These six studies illustrate many important features of mixed methods research that we enumerated as the core characteristics of mixed methods research in Chapter 1. These studies illustrate the collection of two types of data; the analysis of both data sets; reasons for collecting the two types of data; the point of interface, relative emphasis, and timing between the two strands; and how the two types of data are mixed. It is important to pay attention to these features when examining published articles, and it can be very helpful to use the shorthand notation and to draw a diagram to present and organize the procedures when reading others' studies or designing one's own study.

SUMMARY

It is important to locate and review examples of mixed methods research from the literature to enhance one's knowledge about the application of different designs. Because of their inherent complexity and the numerous important methodological features, mixed methods designs can be communicated through a notation system and diagrams that are drawn using standard conventions and that identify the specific procedures and products for each stage of the research process. These diagrams also convey the timing and relative emphasis of the two methods. Reading published mixed methods studies require that key features be examined. These features include summarizing the content by assessing the topic, identifying relevant philosophical and theoretical perspectives, and locating the purpose statement. The features also include analyzing the use of the two methods, such as identifying the samples used for the quantitative and qualitative strands, the types of qualitative and quantitative data collection, and the types of analysis procedures. Identifying features that further highlight the use of mixed methods include noting the reason given for collecting both types of data and determining the points of interface, relative priority, and timing of the quantitative and qualitative strands as well as how the two strands are mixed. Ultimately, these features work together to indicate the overall mixed methods design used in a study. Researchers wanting to read and design mixed methods studies will benefit from having the skills to locate key features within study reports to identify models for mixing methods and the ability to draw diagrams that communicate a study's methods.

ACTIVITIES

1. Locate an example study in the literature that uses the same mixed methods design that you are planning. Make a list of different ways that you can learn from and use this study in your own work.

2. Carefully read the mixed methods study you located. Using the checklist in Figure 4.2, review this study by considering the items included in this list.

3. Using the mixed methods notation system described in Table 4.1, write a shorthand notation that identifies the overall design used in the study.

4. Using the rules for drawing procedural diagrams (see Figure 4.1) and the sample figures in this chapter, draw a diagram representing the mixed methods features of the article you located.

5. Using the mixed methods notation system, the rules for drawing diagrams (see Figure 4.1), and the sample figures in this chapter, draw a diagram representing the study you are working to design.

ADDITIONAL RESOURCES TO EXAMINE

For additional discussions on drawing procedural diagrams for mixed methods studies, see the following resources:

Ivankova, N. V., Creswell, J. W., & Stick, S. (2006). Using mixed methods sequential explanatory design: From theory to practice. *Field Methods, 18*(1), 3–20.

Morse, J. M., & Niehaus, L. (2009). *Mixed methods design: Principles and procedures.* Walnut Creek, CA: Left Coast Press.

Tashakkori, A., & Teddlie, C. (2003). The past and future of mixed methods research: From data triangulation to mixed model designs. In A. Tashakkori & C. Teddlie (Eds.), *Handbook of mixed methods in social and behavioral research* (pp. 671–701). Thousand Oaks, CA: Sage.

For examples of mixed methods studies using different designs, see the following published collections:

Plano Clark, V. L., & Creswell, J. W. (2008). *The mixed methods reader.* Thousand Oaks, CA: Sage.

Weisner, T. S. (Ed.). (2005). *Discovering successful pathways in children's development: Mixed methods in the study of childhood and family life.* Chicago: University of Chicago Press.

CHAPTER 5

INTRODUCING A MIXED METHODS STUDY

A fter we learn the characteristics of mixed methods research, assess the preliminary considerations, select a research design, and review studies, we may begin the more detailed process of designing and conducting a mixed methods study. This chapter discusses how the beginning of a mixed methods study might be shaped. It starts with designing a title for your mixed methods study. We realize that this may be an unusual place to begin. However, the title becomes a focusing device to help shape the study, and it can be stated in "draft" form and then revised as the project proceeds. Writing an introduction to the study follows this step. It includes discussing the research problem leading to a need for the study followed by a purpose statement and research questions. The title, as well as the introductory sections that follow, are scripted using mixed methods ideas, and the idea of a mixed methods research question will catch some off guard as this type of question is not traditionally included in research methods texts. It is, however, an important step in good mixed methods research as it ties together the overall purpose of the study and the methods to follow.

This chapter will address

- writing a mixed methods title that reflects a type of mixed methods design,
- developing an introductory section that highlights the research problem leading to the study,

- scripting a purpose statement that includes the elements of an appropriate mixed methods statement and that relates to a type of mixed methods design, and
- writing a mixed methods research question (as well as quantitative and qualitative research questions) that fits the type of design being used in the study.

● WRITING A MIXED METHODS TITLE

Many researchers do not pay much attention to titles or simply draft them late in a study when one is needed. In contrast, our approach is to emphasize the importance of titles. They serve as important placeholders in a research study and help to keep researchers focused on the primary aim of their study. We also see this preliminary title as a work in progress that can be shaped as the project proceeds.

Qualitative and Quantitative Titles

Before discussing a mixed methods title, we consider the elements of good titles in general and then the aspects that differentiate qualitative and quantitative titles. In general, titles need to convey basic information about a study so that other researchers can easily grasp the meaning of a study when it is referenced in the literature. Typically, titles are short, often containing 12 words or fewer. Good titles reflect four major components: the major subject area or topic being researched, the general research approach, the participants, and the site or place where the research will take place.

For **qualitative study titles**, researchers may state a question or use literary words, such as metaphors or analogies. Qualitative titles include several components: the central phenomenon (or concept) being examined, the participants, and the site at which the study will occur. In addition, a qualitative title might include the type of qualitative research being used, such as ethnography or grounded theory. Qualitative titles do not suggest a comparison of groups or a relationship among variables. Instead, they explore one idea (the central phenomenon) for an in-depth understanding. These sample titles illustrate these components:

- "Campus Response to a Student Gunman" (Asmussen & Creswell, 1995)
- "Waiting for a Liver Transplant" (Brown, Sorrell, McClaren, & Creswell, 2006)

- "How Rural Low-Income Families Have Fun: A Grounded Theory Study" (Churchill, Plano Clark, Prochaska-Cue, Creswell, & Ontai-Grzebik, 2007)

For quantitative study titles, in contrast, investigators compare groups or relate variables. In fact, the primary variables are evident in the title, as well as the participants and possibly the site for the research study. Words in a title, such as "a comparison of" or "the relationship between" or "prediction," signal quantitative studies. Sometimes researchers mention the theory being tested, the prediction being made in the study, or the foreshadowed results. As with qualitative titles, quantitative titles are short and concise. Three examples of quantitative titles are listed here:

- "Factors That Predict a Positive Working Alliance for Education Students in a Mentoring Project" (Harrison, 2005)
- "Affirmation of Personal Values Buffers Neuroendocrine and Psychological Stress Responses" (Creswell et al., 2005)
- "Academic Performance Gap Between Summer-Birthday and Fall-Birthday Children in Grades K–8" (Oshima & Domaleski, 2006)

Clearly, the titles for qualitative and quantitative studies reflect some basic differences between qualitative and quantitative research, such as the study of a single phenomenon versus multiple variables, the language of exploration versus explanation and relationships, and theory development as opposed to theory testing. Given these differences, how would one write a mixed methods title that combines elements of both qualitative and quantitative research?

Mixed Methods Titles

It is important to write a specifically worded title that conveys the point that mixed methods research was used. Mixed methods titles provide reviewers with an introduction to this form of research. They foreshadow the use of mixed methods and the type of mixed methods design that the researcher will use. They also give increased visibility to mixed methods as a distinct approach in the social and human sciences. Since many see mixed methods as a new approach to research, we can highlight its use by incorporating words that denote this form of inquiry in the title. Here are some basic components of a good mixed methods title:

- It is short and succinct.
- It mentions the major topic being addressed, the participants in the study, and the location or site of the project.

- It includes the words *mixed methods* to highlight the overall approach being used. This practice is becoming increasingly used, as shown in a few of the examples provided next.
- It is neutral in that it does not include terms associated with either quantitative or qualitative research. An exception to this is when there is a priority given to one approach. The best practice is to first write the title in a neutral form and then revise it later when the type of mixed methods design is firmly in place and the relative emphasis given to quantitative or qualitative is known.
- It contains words that suggest the specific type of mixed methods design used in the study. If the type of design is still emerging at the time of designing the study, the title can later be revised to reflect the type of design after decisions are made about the type.

In addition, for each major type of mixed methods design, additional considerations come into play. For a convergent design, we recommend writing a title that is neutral in its orientation toward either quantitative (i.e., an explanation) or qualitative (i.e., an exploration) forms of research. Because the basic feature of this design is to merge both quantitative and qualitative data, we do not want the title to lean in one direction or the other. The leaning comes through in the words used that denote either a qualitative or quantitative orientation. For example, examples of qualitative words might be "explore," "meaning," "discover," "generate," or "understanding." Quantitative words might be "predict," "relationship," "comparison," "correlates," and "factors." These words should be left out of the titles or, alternatively, both qualitative and quantitative words might be included. The following examples of titles for studies using the convergent design convey these perspectives.

- The Predictive Validity of an ESL Placement Test: A Mixed Methods Approach (Lee & Greene, 2007)

In this example, there is one topic being studied: the predictive validity of a placement test. No words were used that lean the title in either a quantitative or qualitative direction. In addition, the words *mixed methods* were included to designate it as a mixed methods study. In this next example of a convergent design, the authors neutralized the qualitative and quantitative words by inserting both. They also included the words *mixed methods*.

- Closed and Open-Ended Question Tools in a Telephone Survey About "The Good Teacher": An Example of a Mixed Methods Study (Arnon & Reichel, 2009)

- "In Their Own Words and by the Numbers: A Mixed-Methods Study of Latina Community College Presidents" (Muñoz, 2010)

In the first example, the reader was introduced to both the quantitative and qualitative orientation through words such as *closed* and *open-ended* in the title, and, in the second example, through the words *own words* and *numbers*. These would be other ways to write the title for a convergent design. Another approach would be to specify both quantitative and qualitative approaches in the title:

- "The Meanings of Self-Ratings of Health: A Qualitative and Quantitative Approach" (Idler, Hudson, & Leventhal, 1999)

In an explanatory design, with an emphasis on explaining the initial quantitative phase with qualitative data, the emphasis in the title is often placed on the first, quantitative phase. The following example illustrated this approach. It made explicit the quantitative followed by qualitative sequence of the study.

- "Multimethod Measurement of High-Risk Drinking Locations: Extending the Portal Survey Method With Follow-Up Telephone Interviews" (Kelley-Baker, Voas, Johnson, Furr-Holden, & Compton, 2007)

In an exploratory design, we see different models for how to design the title. One is to begin with qualitative words, because the study starts with a qualitative exploration. Another is to emphasize what the study leads up to, such as a quantitative survey comparing groups, as in the case of the instrument development type of design. An example of starting with a qualitative exploration using the word *perceptions* was illustrated in the first example to follow and building up to the development of an instrument that assessed similarities and differences among different clusters of participants shown in the second example.

- "Shoppers' Perceptions of Retail Developments: Suburban Shopping Centres and Night Markets in Singapore" (Ibrahim & Leng, 2003)
- "Similarities and Differences in Teachers' Practical Knowledge About Teaching Reading Comprehension" (Meijer, Verloop, & Beijaard, 2001)

In an embedded design, we also suggest that the words *mixed methods* be included in the title. The title should reflect the use of embedded data and possibly the reason for the use of the embedded data. In the two examples that follow, both of the studies were intervention trials with a qualitative component.

- "Improving Design and Conduct of Randomised Trials by Embedding Them in Qualitative Research: ProtecT (Prostate Testing for Cancer and Treatment) Study" (Donovan et al., 2002)
- "Reactions of Participants to the Results of a Randomized Controlled Trial: Exploratory Study" (Snowdon, Garcia, & Elbourne, 1998)

In a transformative design, we would expect to see the theoretical framework being advanced in the title as a major topic of interest and wording incorporated to suggest an injustice or a need of a specific group. In the first example, a theory in sociology is emphasized, and, in the second example, the injustice of "myths" in college student–athlete cultures is being studied:

- "Bourdieu's Reflexive Sociology as a Theoretical Basis for Mixed Methods Research: An Application to Complementary and Alternative Medicine" (Fries, 2009)
- "Understanding Community-Specific Rape Myths: Exploring Student Athlete Culture" (McMahon, 2007)

A title for a multiphase design needs to capture the spirit of the many phases of a project. The title could also emphasize that the program is going through an evaluation consisting of many phases. These illustrations suggest such an orientation:

- "A Participatory Program Evaluation of a Systems Change Program to Improve Access to Information Technology by People With Disabilities" (Mirza, Anandan, Madnick, & Hammel, 2006)
- "Research in Action: Using Positive Deviance to Improve Quality of Health Care" (Bradley et al., 2009)

● STATING THE RESEARCH PROBLEM IN THE INTRODUCTION

After the researcher writes the title, framing it within both mixed methods research and the type of design, the next section to be developed is the "statement of problem," which introduces a study. This is true whether the study is a proposal, a journal article, a manuscript for conference presentation, or a dissertation or thesis. The **statement of the problem** conveys a specific problem, or issue, that needs to be addressed and the reasons why the problem is important to study. We will first review the basic parts that go into a statement of the problem section and then discuss how elements of mixed methods research can be included in this statement.

Topics in a Statement of the Problem Section

The structure for beginning a research study and introducing the problem includes several components—the topic, the research problem, the literature, the deficiencies, and the audiences (see Creswell, 2009c).

- Introduce the topic. Begin with a paragraph that identifies the topic of the study in a way that will appeal to a wide readership. This paragraph might begin with statistics about the problem, call for more research about the topic, or provide a thought-provoking question.
- Identify the "problem." Discuss the problem, or issue, that leads to a need for the study. To write this section, consider beginning with the words *an issue faced by* or *a current problem is*. Further, consider drafting this problem from one or two standpoints. The first would look at the problem from the perspective of an issue that exists in the day-to-day working world or lives of individuals. Perhaps, for example, students are at risk today because of crime in the schools or senior citizens feel disempowered because of health issues. These are "real-life" problems, and they deserve to be studied. The second standpoint would consider a problem related to a need for research on a topic. This need may arise because of a gap in the existing body of knowledge or a need to extend the current research to a new population or to new variables. An ideal problem statement might include several statements that convey both a real-life problem in our society and a weakness or gap in the literature.
- Discuss the research that has addressed this problem. In this section, indicate the published literature on this problem. Think in terms of groups of studies rather than individual studies (that are discussed in a literature review section) to advance broad trends in the literature. How could the present literature be organized and summarized? Identify the major themes of each group of studies to give readers a general understanding of existing trends. In this review, draw on quantitative, qualitative, and mixed methods research studies.

- Indicate deficiencies in the literature and how your study will fill this gap and in what way. These gaps may be content areas not addressed or flaws in the research methods (e.g., all of the studies have been quantitative studies, so we do not hear the voices of participants through qualitative studies). If the section in which the problem is stated already addresses these gaps, there is no need to repeat information but focus on how the study will add to the literature and make an important contribution.
- Discuss what audiences will benefit from addressing this gap or deficiency. Several audiences might be specifically identified who will profit from your study—audiences such as researchers, policy makers, administrators, teachers, providers, and others. It is useful to cite several audiences that will profit from your study and enumerate the ways that they might each benefit.

End the introduction (statement of the problem) with the purpose statement and research questions or hypotheses. These topics will be addressed later in this chapter.

Integrate Mixed Methods Into the Statement of the Problem

How does mixed methods research fit into this introduction? Although the sections included in an introduction do not necessarily relate to the methods or design used in a study, it is useful to foreshadow the type of mixed methods design even in the opening passages of the introduction. One way to do this is to relate the type of mixed methods design to be used to the deficiencies passage in the introduction. Specifically, the choice of a mixed methods design is partly based on a need arising out of the literature. Examine Table 5.1. Here we identify examples of the needs in the literature that each mixed methods design might address. You could include the arguments for the mixed methods design with the other deficiencies in the literature that you mention in your introduction and effectively foreshadow the type of design that will be developed later in your study.

An example of how the deficiency addressed by a type of mixed methods design can be integrated into an introductory statement follows:

> In a study of leadership styles, the literature has discussed transformational leadership, trait-based leadership, and person–situation leadership. These studies have all been quantitative investigations that describe leadership behaviors but do not incorporate the voices of participants to describe the meaning behind the different types of leadership behaviors. One issue that arises, then, is that the quantitative results are inadequate to describe and explain the leaders' experiences. (This issue implies that a need exists for an explanatory design.)

Table 5.1	Deficiencies in the Literature Related to the Different Mixed Methods Designs

Type of Mixed Methods Design	A Need Exists in the Literature to...
Convergent design	develop a complete understanding by collecting both quantitative and qualitative data, because each provides a partial view.
Explanatory design	not only obtain quantitative results but to explain such results in more detail, especially in terms of detailed voices and participant perspectives because little is known about the mechanisms behind the trends.
Exploratory design	explore a topic because variables are unknown and to assess the extent that the detailed results from a few participants generalize to a population.
Embedded design	to examine outcomes through experimental methods and processes by obtaining detailed views from participants through qualitative data.
Transformative design	lift up the voices of participants and to develop a call for action using data sources that can challenge injustices and provide evidence that is acceptable to stakeholders.
Multiphase design	to meet an overall objective through projects that develop over time with many phases.

[handwritten margin notes: LMS; N=1]

DEVELOPING THE PURPOSE STATEMENT ●

A mixed methods purpose statement can also include language to suggest a mixed methods design. Before we turn to useful scripts for writing this statement, it might be helpful to review the key elements of both quantitative and qualitative purpose statements (see Creswell, 2009c).

Qualitative and Quantitative Purpose Statements

A **qualitative purpose statement** conveys the overall qualitative purpose of the study and includes a central phenomenon, the participants, the research site for the study, and the type of qualitative design in the study. It begins with words such as *the purpose of this study* or *the intent of this study*. The statement also contains words denoting the one concept being explored in a qualitative study. This concept is called the central phenomenon. The writer includes action verbs to indicate an exploration of this central phenomenon. Words such as *describe*, *understand*, *explore*, and *develop* convey this exploration and the emerging understanding of the central phenomenon that will develop during the study. Because a qualitative study conveys multiple perspectives of participants, the qualitative purpose statement should not contain leading or directional words that convey a stance, such as "positive," "useful," or "predicts." The qualitative inquirer takes a nondirectional stance. Also, some reference might be made to the type of qualitative design or methods used in the study, such as an ethnography, or case study, or a grounded theory study. Finally, the qualitative purpose statement can also contain information about the individuals or sites that will be involved in the project.

An example of a qualitative purpose statement follows that begins with the purpose, identifies the type of qualitative design, uses an action verb phrase, specifies the central phenomenon, and mentions the participants and the location for the study:

> The purpose of this ethnographic study is to explore the culture-sharing behaviors and language of the homeless in a soup kitchen in a large, Eastern city. (Evident in this qualitative purpose statement is the lack of directional words and words relating variables or comparing groups.)

In a **quantitative purpose statement**, the researcher conveys the overall quantitative purpose of the study and presents the variables in the study, the participants, and the site for the research. The use of directional language and variables are central features. Writers specify their independent and dependent variables and typically order them left to right from independent to dependent. They begin with phrases such as *the purpose of the study* or the *intent of the study* and may identify the theory being tested in the study. Phrases that connect the variables, such as *the relationship between* or *a comparison of,* reflect the relationship among the variables in the study. As with qualitative research, the quantitative purpose statement might include the type of methods that will be employed and refer to the participants and the site for the study. This example illustrates these elements in a good quantitative purpose statement:

> The purpose of this correlational study will be to test sex-role theory, which predicts that males will be more conditioned than females to aggressive roles in college.

Mixed Methods Purpose Statements

We have found it useful to provide specific scripts for writing mixed methods purpose statements, because the purpose statement is the most important statement in a research project. If this statement is not clear, then a reader has difficulty understanding the entire study. Clear purpose statements are important in all types of research, but the need for clarity is especially important in a mixed methods project in which many elements of qualitative and quantitative research need to come together. Two elements that go into a mixed methods purpose statement are the qualitative purpose statement and the quantitative purpose statement, which need to be made explicit. Because the mixed methods statement contains these elements, it is not always necessary in a mixed methods study to state three purpose statements—quantitative, qualitative, and mixed methods—but having a mixed methods statement is essential. This statement is typically placed at the end of an introduction in a journal article. Alternatively, in a proposal for funding, it is found in a study aim section at the beginning of the proposal, and it is often presented as a separate section in a dissertation or thesis project.

A mixed methods purpose statement conveys the overall purpose of the mixed methods study, and it includes the intent of the study, the type of mixed methods design, quantitative and qualitative purpose statements, and the reasons for collecting both quantitative and qualitative data. The specific elements are as follows:

- Include the overall intent (the content aim) of the project in the first sentence. Begin with words such as *this study addresses, the purpose of this study is, the study aim is,* or *the intent of this study is*.
- Identify the type of mixed methods design using the full name (e.g., explanatory sequential design), so that the reader is introduced to the specific type of methods that will be used. Provide a brief definition of the type of design.
- Incorporate specific quantitative and qualitative purpose statements that indicate the type of data to be collected as well as the participants and the site for the two strands of the study.
- State the reasons for collecting both forms of data that match the rationale for the type of design (see Chapter 3).

[handwritten margin note: Mixed methods purpose statement]

An example of a script that illustrates these points is included in Figure 5.1. This example presents a model script for a convergent design, and it includes the intent of the study, the type of design and a brief description of the design, a quantitative and qualitative purpose statement, and a rationale for collecting

Figure 5.1	An Example of a Purpose Statement Script for a Convergent Design

Intent — This mixed methods study will address _____ [overall content aim of the study].

Design type — A convergent parallel mixed methods design will be used, and it is a type of design in which qualitative and quantitative data are collected in parallel, analyzed separately, and then merged. In this study, _____ [quantitative data] will be used to test the theory of

Quantitative and qualitative data and purposes — _____ [the theory] that predicts that _____ [independent variables] will _____ [positively, negatively] influence the _____ [dependent variables] for _____ [participants] at _____ [the research site]. The qualitative data _____ [type of qualitative data, such as interviews] will explore _____ [the central phenomenon] for _____ [participants] at _____ [research site]. The reason for

Rationale — collecting both quantitative and qualitative data is to converge [or compare results, validate results, corroborate results] the two forms of data to bring greater insight into the problem than would be obtained by either type of data separately.

both forms of data using the specific design. To use this script, researchers fill in the blanks with information from their own study and keep the elements of the script in order. In this way, it provides a complete, detailed mixed methods purpose statement.

An example of this convergent design script is the statement we designed in collaboration with workshop participants at the Qualitative International Conference at Edmonton, Canada, in February 2005. Here is the script that we developed, with slight changes to fit our model:

The intent of this study is to learn about the food choices of First Nations women with Type 2 diabetes. The purpose of this convergent parallel mixed methods study will be to converge both quantitative (numeric) and qualitative (text or image) data. In this approach, survey data will be used to measure the relationship between the factors (e.g., family backgrounds) and food choices. At the same time in the study, food choices will be explored using interviews and participant observations with First

Nations women with Type 2 diabetes in northern Manitoba. The reason for collecting both quantitative and qualitative data is to compare the results from two different perspectives.

In an <u>explanatory design,</u> the reason for the exploratory follow-up is mentioned in the purpose statement between the initial quantitative phase and the second qualitative phase. The order of phases—from quantitative to qualitative—highlights the sequence of procedures used in this design. Also, the second, qualitative phase is tentatively stated, because the central phenomenon and perhaps the participants and site cannot be clearly specified until the first, quantitative phase of the study has been completed.

This study will address _____ [content aim of the study]. An explanatory sequential mixed methods design will be used, and it will involve collecting quantitative data first and then explaining the quantitative results with in-depth qualitative data. In the first, quantitative phase of the study, _____ [quantitative instrument] data will be collected from _____ [participants] at _____ [research site] to test _____ [name of theory] to assess whether _____ [independent variables] relate to _____ [dependent variables]. The second, qualitative phase will be conducted as a follow up to the quantitative results to help explain the quantitative results. In this exploratory follow-up, the tentative plan is to explore _____ [the central phenomenon] with _____ [participants] at _____ [research site].

explanatory design script

A student in one of our mixed methods classes provided an example of this purpose statement as a class project:

The intent of this study is to examine Latino adolescents' perspectives on family conflict. The purpose of this two-phase, explanatory mixed methods study will be to obtain statistical quantitative results from a sample and then follow up with a few individuals to probe or explain those results in more depth. In the first phase, quantitative hypotheses will address the relationship of acculturation and family conflict with Latino adolescents at their respective middle school and/or high school in Southern California. In the second phase, qualitative semi-structured interviews will be used in a multiple case study to explore aspects of family conflict with 4 individuals representing different combinations (from the quantitative results) at a Middle School and a High School. (Cerda, 2005)

In an exploratory design purpose statement, the reason for collecting quantitative follow-up data is placed after the description of the qualitative phase of the study, and it serves as a bridge between the first and second phase of a study. Also, the second-phase quantitative research questions and hypotheses cannot be specified until the qualitative phase is completed. If readers need these elements in the quantitative phase to be specified, they can be stated as "tentative" statements.

This study addresses _____ [content aim of the study]. The purpose of this exploratory sequential design will be to first qualitatively explore with a small sample and then to determine if the qualitative findings generalize to a large sample. The first phase of the study will be a qualitative exploration of _____ [the central phenomenon] in which _____ [types of data] will be collected from _____ [participants] at _____ [research site]. From this initial exploration, the qualitative findings will be used to develop measures that can be administered to a large sample. In the tentatively planned quantitative phase, _____ [instrument data] will be collected from _____ [participants] at _____ [research site].

An example of this purpose statement is drawn from another student paper in one of our mixed methods classes.

This study will address language brokering (children serving in the role of interpreters) among immigrant families. The purpose of this two-phase, exploratory mixed methods study will be to explore participant views with the intent of using this information to develop and test an instrument with a Latino sample from a Midwestern city. The first phase will be a qualitative exploration of what it means for Latino parents to have their son or daughter serve in the role of the language broker or interpreter/translator by collecting interview data from a sample of 20 Latino parents from a mentoring program at a Midwestern university. Because there are no existing instruments to assess language brokering, an instrument needs to be developed based on the qualitative views of participants. Statements and/or quotes from this qualitative data will then be developed into an instrument so that a series of hypotheses can be tested that relate to parents' views about language brokering for a group of 60 Latino parents whose children participate in an after school program for Latino students (elementary to high school) at the Hispanic Community Center at a Midwestern city. (Morales, 2005)

In an embedded design purpose statement, the basic components need to be in place: the intent of the study, a description of the design, a quantitative and qualitative purpose statement, and the reason for the design. In addition, several other elements related to the embedded aspect of this design need to be added: the nature of the major design, the types of data that will go into the major design, and how these types of data will be embedded.

This mixed methods study will address _____ [overall content aim of the study]. An embedded design will be used in which _____ [qualitative data, quantitative data] are embedded within a major design _____ [intervention trial, case study, or other design]. The quantitative data will be used to test the theory that predicts that _____ [independent variable] will influence _____ [positively, negatively] the _____ [dependent variable] for _____ [participants] at _____ [research site]. The _____ [type of qualitative, quantitative data] will be embedded in this larger design _____ [intervention trial, case study] _____ [before, during, or after] for the purpose of _____ [rationale for the embedded data set]. The qualitative data will explore _____ [the central phenomenon] for _____ [participants] at _____ [site].

Embedded design script

An example of the use of this script is found in the purpose statement designed in a mixed methods workshop:

The primary intent of this investigation will be to test a case management intervention enhanced by automated pharmacy and clinical information to improve blood pressure control in Veterans [Affairs] hospitals. The objectives will be to improve blood pressure control among patients with hypertension through more appropriate use of medication, and to augment case management through the use of electronic pharmacy and clinical data for more effective treatment of uncontrolled hypertension. The research design of the study will be an embedded mixed methods intervention design, and it will involve collecting qualitative data before and during the intervention phases of the study. In the initial qualitative phase of the study, the investigators will collect qualitative data to explore potential barriers to the intervention before the intervention begins. Then during the trial, qualitative data will be collected to understand the patient experiences with the intervention. At the baseline, at multiple points during the trial, and at the conclusion, quantitative data will be collected on several survey and patient clinical data outcomes. (Creswell, 2005)

In a transformative design, the basic features of a purpose statement need to be included: the intent of the study, mention of the transformative design, a quantitative and qualitative purpose statement, and the reason for the design. Further, it is important to specify the transformative lens being used, why it is being used, and the elements it brings to the mixed methods study. The mixed methods procedures in this design could be either concurrent or sequential data collection.

This mixed methods study will address _____ [overall content aim of the study]. A transformative design will be used in which _____ [type of theoretical lens] will provide an overarching framework for the study. This lens is being used for the following reason _____ [state reason], and it has the following elements _____ [aspects of the lens]. The study will include both quantitative and qualitative data gathered _____ [concurrently or sequentially]. The quantitative data will be used to test the theory that predicts that _____ [independent variable] will influence _____ [positively, negatively] the _____ [dependent variable] for _____ [participants] at _____ [research site]. The qualitative data will explore _____ [the central phenomenon] for _____ [participants] at _____ [site].

transformative design script

The following example from a published journal article illustrates a good transformative purpose statement:

The study uses a sequential transformative mixed methods research design to explain how political advertising fails to engage college students. Qualitative focus groups examined how college students interpret the value of political advertising to them, and a quantitative manifest content analysis concerning ad framing of more than 100 ads from the 2004 presidential race revealed why focus group participants felt so alienated by political advertising. (Parmelee, Perkins, & Sayre, 2007, p. 183)

In a multiphase design, the purpose statement needs to advance the idea that there are multiple phases (or multiple projects) in the program of inquiry, that they will unfold over time, and that they will involve both concurrent and sequential components (or one or the other). It also needs to include the concurrent and sequential components in the order in which they will be undertaken in the study, as well as the basic elements of intent, type of design, types of data, and the reason for the design.

This mixed methods study will address _____ [the intent or program objective of the study]. In this multiphase design, there will be several phases (or projects) conducted over time. These phases are _____ [identify the phases]. The types of data collected in each phase will be_____ [mention the qualitative and quantitative data phases] and the _____ [type of data] will be collected _____[concurrently/sequentially] in different phases of the study. The reason for using a multiphase design is _____ [rationale for the design].

multiphase design script

The following example from a funding proposal illustrates a multiphase design purpose statement:

The purpose of this 5-year, mixed-methods international study is to explore enacted stigma behaviors among indigenous, Asian-ancestry, and European-ancestry adolescents in school environments in Canada, New Zealand, and the U.S., to develop cross-cultural measures of enacted stigma for adolescent health surveys, and to examine the association of types of stigma and HIV risk behaviors among adolescents. The specific aims are: I. To compare the prevalence of HIV risk behaviors associated with sexual orientation and other stigmatized identities among youth in existing large-scale school-based surveys, and to identify both the existing indirect measures of stigma that are risk factors plus the protective factors significantly associated with the HIV risk behaviors. II. To identify the prevalence of HIV risk behaviors and associated risk and protective factors among indigenous adolescents—American Indian (U.S.), First Nations (Canada), Maori (New Zealand)—as well as youth of Asian ancestry in each country, and to compare the patterns among adolescents of similar ethnic backgrounds in the 3 countries. III. To explore among adolescent and adult key informants the ways stigma is understood, assigned, and enforced in the school environment, and to compare the patterns within the three countries. This exploration will be focused primarily on stigma based on sexual orientation status, but other types of stigmatized identities will be examined to understand the similarities and differences of how stigma is enacted, and the potential utility of generic stigma measures. IV. Within each country, to elicit explanatory models from adolescents and youth workers on the survey findings of HIV risk behaviors and stigma, and to tap suggested strategies for reducing stigma and addressing sexual risk behaviors in culturally appropriate ways among GLBQ youth. V. Incorporating the findings of aims I-IV, to develop, pilot, and psychometrically evaluate universal and country-specific culturally competent items and scales, for population-based adolescent health surveys, that measure perceived and enacted stigma in school, to allow cross-cultural comparisons of the effects of stigma among adolescents. (Saewyc, 2003)

WRITING RESEARCH QUESTIONS AND HYPOTHESES

Research questions and hypotheses narrow the purpose statement into specific questions and predictions that will be examined in the study. In a mixed methods study, qualitative, quantitative, and mixed methods questions are presented. First, we will review the basic components of qualitative and quantitative questions.

Qualitative Questions and Quantitative Questions and Hypotheses

Qualitative research questions focus and narrow the qualitative purpose statement and are stated as questions, not hypotheses. These questions typically include a central question and several subquestions. The subquestions take the topic of the central question and ask questions related to a small number of aspects of the central question. Thus, subquestions usually involve no more than five to seven questions.

The central question and subquestions are concise, open-ended questions that begin with words such as *what* or *how* to suggest an exploration of the central phenomenon. Although the beginning word *why* can be found in published studies, this word implies a quantitative orientation of cause and effect, an explanation of why something occurred. Such an explanation is contrary to the nature of qualitative research, which looks for an in-depth understanding of a central phenomenon, not for explanations. As with the qualitative purpose statement, the qualitative research questions focus on a single concept or phenomenon. There may be no need to include information about the participants and the research site for the study, because that is already included in the qualitative purpose statement. Here is an example of a qualitative central question and subquestions from an article about a campus response to a gunman incident:

- What happened? (central question)
- Who was involved in response to the incident? (subquestion)
- What themes of response emerged during the 8-month period that followed the incident? (subquestion)

- What theoretical constructs helped us understand the campus response, and what constructs were unique to this case? (subquestion) (Asmussen & Creswell, 1995, p. 576)

Quantitative research questions and hypotheses narrow the purpose statement through research questions (that relate variables) or through hypotheses (that make predictions about the results of relating variables). Hypotheses are typically chosen when the literature or past research provides some indication about the predicted relationship among the variables (e.g., men will display more aggression than women when considered in terms of sex role stereotypes). If predictions are made, then the researcher has the additional consideration of whether to write the prediction as a null hypothesis ("there is no significant difference") or as a directional hypothesis ("men will display more aggression than women"). Directional hypotheses seem more popular today, and they are more definitive about the anticipated results than a null hypothesis.

Whether the researcher writes hypotheses or research questions (typically, there will not be both in the same quantitative study), the investigator narrows the purpose statement so that it indicates specific variables to test. These variables are then related to each other or compared for one or more groups. The most rigorous hypotheses and questions follow from a theory in which other researchers have tested the relationships among variables. Here are examples of quantitative research hypotheses and a question:

- There is no significant difference between the effects of verbal instructions, rewards, and no reinforcement on learning spelling among fourth grade children. (a null hypothesis)
- Fourth grade children perform better on spelling tests when they receive verbal instructions than when they receive rewards or no reinforcement. (directional hypothesis)
- What is the relationship between instructional approach and spelling achievement for fourth grade students? (research question)

Mixed Methods Research Questions

How would mixed methods questions differ from qualitative and quantitative research questions? Readers may not have an immediate answer to this question, because the use of mixed methods questions has been little discussed in the mixed methods literature (exceptions include Onwuegbuzie & Leech, 2006; Plano Clark & Badiee, in press; and Tashakkori & Creswell, 2007a).

Mixed methods research questions are questions in a mixed methods study that address the mixing or integration of the quantitative and qualitative data. They are necessary in a mixed methods study, because both quantitative and qualitative data collection are central to this form of inquiry and they raise distinct questions in addition to the qualitative or quantitative questions. As research questions, mixed methods questions need to be answered (just as quantitative hypotheses or qualitative research questions need to be answered), and, in a results and discussion section, the mixed methods researcher needs to provide answers to the questions. As a new type of question, mixed methods question(s) often remain implicit in articles and proposals. Our recommendation, however, is that this question be made explicit and clearly stated.

Plano Clark and Badiee (in press) have provided some guidance as to how researchers might state questions in a mixed methods study. They address three dimensions: (1) when in the process of conducting a mixed methods study, the research questions are generated, (2) how multiple questions in a mixed methods study might be linked or kept separate, and (3) the specific rhetorical style of writing the questions.

First, in terms of when the research questions are generated in a mixed methods study, they feel that the questions might be predetermined and based on the literature, practice, personal tendencies or field, or disciplinary considerations. This approach may be used in a convergent design when the data collection is set in advance. It is also a recommended procedure for graduate students in designing a mixed method study that have committee members (and institutional review board [IRB] committees) requiring the specific statements of questions before the study begins. However, the questions might also be emergent and occur during the design, data collection, data analysis, or interpretation of the study. The emerging approach is consistent with traditional qualitative approaches, and this form of question may occur in sequential and multiphase designs. Way, Stauber, Nakkula, and London (1994) described the questions that emerged from their unexpected quantitative result that students from two schools had different patterns of substance used as predictors for depression. Christ (2007) also illustrated how new questions emerged within an exploratory, longitudinal mixed methods study. Taking advantage of unforeseen circumstances of a budget reduction at one of his study sites, he added new questions to a third phase of his study.

Research questions in mixed methods can be linked conceptually or framed so that they are independent of each other (Plano Clark & Badiee, in press). For example, they can be stated independently of each other in which the researcher writes two or more research questions in which one question does not depend on the results of the other question or dependently, in which one question depends on the other. The independent type of questioning often occurs in a concurrent design in which two separate and distinct strands of data (qualitative and quantitative) are collected. Multiphase, transformative, and embedded designs with convergent approaches also would fit this model. For example, Brady and O'Regan's (2009) study provided a good example of independent questions when they asked, "What is the impact of the BBBS [Big Brothers Big Sisters] program on participating youth? How is the program experienced by stakeholders?" (p. 273). The first question was addressed through surveys of youth that related to the impact of the mentoring program while the second question was answered through interviews with stakeholders. The dependent type of questioning often occurs in sequential types of designs, such as the explanatory design, the exploratory design, or sequential procedures in the embedded, transformative, and multiphase designs. Biddix (2009) provided a useful example of the dependent type of questions when two questions were asked: "(1) What career paths for women lead to the community college SSAO (Senior Student Affairs Officer)? (2) What influences path decisions to change jobs or institutions?" (p. 3). In the first phase of the study, SSAO resumes were the primary source of data while the second phase consisted of interviews with SSAOs.

The style of writing research questions into a mixed methods study might assume several forms (Plano Clark & Badiee, in press). The researcher could provide an overarching mixed methods question that does not indicate a specific quantitative or qualitative approach. For example, Igo, Kiewra, and Bruning (2008) asked this question: "How do different copy-and-paste note-taking interventions affect college students' learning of Web-based text ideas?" (p. 150). In this example, the word *how* calls attention to the qualitative component of the study and the words *affect* and *interventions* relate to the quantitative component.

The researcher could pose a hybrid or double-barreled question with two specific parts and use the quantitative approach to address one part and the qualitative approach to address the other part. For example, in a federally funded project, Kruger (2006) posed a doubled-barreled purpose statement that might have been phrased as a mixed methods hybrid question: "The purpose of the R21 mixed-methods exploratory study is to develop and test a family-nurse care coordination intervention for families" (Abstract, para.1). In this statement, the word *develop* was more open-ended and thus more implicitly qualitative whereas the word *test* demonstrated a quantitative approach.

The researcher could pose separate quantitative and qualitative questions for the quantitative and qualitative strands of the study. For example, Webster (2009) had two quantitative and two qualitative questions, and his approach can be illustrated by two questions here:

Is there a statistically significant difference in nursing student empathy, as measured by the Interpersonal Reactivity Index (IRI), after a psychiatric nursing clinical experience? (a quantitative question)

What are student perceptions of working with mentally ill clients during a psychiatric nursing clinical experience? (a qualitative question) (pp. 6–7)

The researcher could pose a question about the integration of the databases in their mixed methods study. We call this a "mixed methods research question," and its form relates specifically to the type of mixed methods design. We recommend this approach to writing a mixed methods question into a study.

This last point needs further elaboration, because it builds directly on our discussion about types of research designs. Mixed methods questions that relate to the integration or mixing of the databases can be written in several ways: with a method-focus, a content-focus, or some combination of content and method. A **method-focused mixed methods research question** is a research question about mixing the quantitative and qualitative data in a mixed methods study in which the researcher writes to focus on the methods of the mixed methods design, for example,

- To what extent do the qualitative results confirm the quantitative results?

On the other hand, a **content-focused mixed methods research question** is a research question about mixing the quantitative and qualitative data in a mixed methods study in which the researcher makes explicit the content of the study and implies the research methods. For example,

- How do the perspectives of adolescent boys support the results that their self-esteem changes during the middle school years?

A final example is a **combination mixed methods question**, which is a research question about mixing the quantitative and qualitative data in a mixed methods study in which the researcher makes explicit both the methods and the content of the study. In this model, we see that the content of the study is included as well as the methods of the design. For example,

- What results emerge from comparing the exploratory qualitative data about boys' self-esteem with outcome quantitative instrument data measured on a self-esteem instrument?

Of these three models for writing a mixed methods research question, we would recommend the combination model, because it is most complete. However, we would not rule out the method- or the content-focused models given the inclinations of certain researchers or reviewers to emphasize more methods or more content in their studies. Also, writing the method question helps to think about how the methods will be combined or linked in a mixed methods study.

These three types of mixed methods research questions—method, content, or some combination—can be related now to the types of research designs we have discussed. Table 5.2 provides examples of the three types of mixed methods questions for each type of design. Certainly, variations could be presented in all of these examples, and we have chosen the content area of self-esteem for boys in middle schools as a common topic of all of these hypothetical questions so that comparisons among them might be easily made. The convergent design mixed methods question needs to convey that the two databases are being merged, while the explanatory design question addresses the use of the qualitative data to help explain the quantitative results. The exploratory design question illustrates how the initial qualitative findings will be generalized to a larger sample through the quantitative data collection and analysis. The embedded design question indicates how the embedded data can help provide a supportive role for the major form of data. The transformative design examples show that the mixed methods question can be written from an explanatory, exploratory, or a convergent design model, but they need to include some of the language intended by this design to address inequities, to bring about transformation, or to change injustices in our society. The multiphase design mixed methods research questions combine both sequential and concurrent types of designs, and we have chosen to label the different studies in the multiphase design with the labels of Study 1 and Study 2. We have also chosen to illustrate the follow-up study as a qualitative study, but the various studies in Study 1, Study 2, and other studies in the overall project or program of inquiry might be either qualitative, quantitative, or mixed in orientation.

Table 5.2 Type of Design and Examples of Method-Focused, Content-Focused, and Combination Mixed Methods Research Questions

Type of Design	Method-Focused Mixed Methods Questions	Content-Focused Mixed Methods Questions	Combination Mixed Methods Questions (Methods and Content)
Convergent design	To what extent do the quantitative and qualitative results converge?	To what extent do self-esteem ratings agree with the views of self-esteem by middle school boys?	To what extent do the quantitative results on self-esteem agree with the focus group data on self-esteem for middle school boys?
Explanatory design	In what ways do the qualitative data help to explain the quantitative results?	In what ways do the views of middle school boys about their self-esteem explain what they reported about their self-esteem?	In what ways do the interview data reporting the views of middle school boys about their self-esteem help to explain the quantitative results about self-esteem reported on the surveys?
Exploratory design	In what ways do the quantitative results generalize the qualitative findings?	Are the views of middle school boys about their self-esteem generalizable to many middle school boys?	Are the themes about self-esteem from middle school boys generalizable to a sample of a population of middle school boys?
Embedded design	How do the qualitative findings provide an enhanced understanding of the quantitative results?	How do the views of middle school boys help to develop a treatment program or explain the outcomes of a program designated to improve self-esteem?	How do the interview data with middle school boys help to design a treatment program and to explain the results of the intervention trial aimed at testing a program for self-esteem improvement?

Type of Design	Method-Focused Mixed Methods Questions	Content-Focused Mixed Methods Questions	Combination Mixed Methods Questions (Methods and Content)
Transformative design	How do the qualitative findings provide an enhanced understanding of the quantitative results in order to explore inequalities?	How do the views of middle school boys help to develop a treatment program or explain the outcomes of a program designated to improve self-esteem in order to explore how after-school programs marginalize middle school boys?	How do the interview data with middle school boys help to design a treatment program and to explain the results of the intervention trial aimed at testing a program for self-esteem improvement in order to explore how after-school programs marginalize middle school boys?
Multiphase design	Include combinations of the previous questions at different phases in the project so that an overall research goal is addressed.	Include combinations of the previous questions in different phases of the project so that an overall research goal is addressed.	Include combinations of the previous questions in different phases of the project so that an overall research goal is addressed.

Finally, when designing mixed methods questions, we offer several overall recommendations (Plano Clark & Badiee, in press):

1. When writing mixed research questions, select the format (questions, aims, and/or hypotheses) that matches the norms of your audience. If there is a choice of format, use the question format to highlight their importance within the conduct of mixed methods research.

2. Use consistent terms to refer to variables/phenomena examined across multiple questions.

3. Use a combination of question types to (a) convey the larger question guiding the study, (b) state the specific subquestions associated with quantitative and qualitative methods, and (c) include a mixed methods research question that directs and foreshadows how and why the strands will be mixed.

4. Relate the question style and content to the specific mixed methods design being used. For example, dependent questions should be associated with either a sequential design or sequential procedures (such as use in the data transformation variant of the convergent design).

5. If the questions are independent, list them in their order of importance. If the questions are dependent, list them in order of what has to be answered first.

6. Determine whether the study is best addressed with predetermined and/or emergent questions. Even if starting with predetermined questions, be open to the possibility of emergent questions. When questions emerge, explicitly discuss the process by which they emerged and the considerations that led to posing new questions.

SUMMARY

A mixed methods study begins with a mixed methods title. In the study's introduction section, the research problem is highlighted, the problem is narrowed into a purpose statement, and the purpose statement is further refined into research questions or hypotheses. With each component of this introduction, the researcher foreshadows a mixed methods approach and a type of mixed methods design so that the study is rigorous, interconnected, and evaluated as a mixed methods project.

The title to a mixed methods study should contain the words *mixed methods* to signal the type of approach that will be used. The title also needs to be framed as a neutral or nondirectional title if the study gives equal priority to both quantitative and qualitative data or it can lean in the direction of either quantitative or qualitative if the priority of the study is weighted in one direction or the other. The introduction to a study can also foreshadow mixed methods research. In the model provided in this chapter, in which the researcher begins with a topic, the problem, the literature, the deficiencies, and the audience, the reason or reasons for conducting mixed methods research can be inserted into the deficiencies section as a shortcoming in the existing literature. The mixed methods purpose statement needs to be crafted to highlight the type of mixed methods design, the forms of data to be collected, and the basic reason(s) for gathering both forms of data. Scripts have been provided in this chapter to help design purpose statements that relate to the designs in Chapter 3. Finally, the research questions or hypotheses narrow the purpose statement. We provide examples of qualitative and quantitative research questions and add specifically worded mixed methods questions. In a mixed methods study, the researcher can advance quantitative questions or hypotheses, qualitative questions, and a mixed methods question. The mixed methods question is important to include in the introduction, because it highlights the mixing of data and promotes the view of mixed methods as an integral part of the research, not as an add-on. Several options are available for writing research questions into a mixed methods study, and we recommend including a mixed methods question framed from a method orientation, a content orientation, or a combination of these two.

ACTIVITIES

1. Look at the titles of published mixed methods studies, and evaluate them in terms of (a) the inclusion of terms that refer to mixed methods research (e.g., quantitative and qualitative, integrated, mixed methods) and (b) whether the wording in the title accurately reflects the type of design.

2. Do the introductions presented in mixed methods studies published in the journal literature reflect the reason(s) for using mixed methods research? Take one or two mixed methods studies and look closely at their introductions. Label the parts: (a) the topic, (b) the research problem, (c) the literature, (d) the deficiencies in the literature, and (e) the audience. Also label the section (possibly the deficiencies) in which the authors suggest a need for a mixed methods study.

3. Write a good mixed methods purpose statement. First, decide on the type of design best suited for your study (see Chapter 3). Then, using the script provided in this chapter, fill in the blanks. Did the script work for you? For others reviewing your study?

4. Write a mixed methods research question. Again, for the type of design best suited for your study, examine Table 5.2 and select the mixed methods question that needs to be written. Consider a methods-focused, a content-focused, or a combination mixed methods question. Adapt the wording to fit your particular study.

ADDITIONAL RESOURCES TO EXAMINE

For the elements that go into introductions, writing purpose statements, and posing research questions, see the following resources:

Creswell, J. W. (2009c). *Research design: Qualitative, quantitative, and mixed methods approaches*. (3rd ed.) Thousand Oaks, CA: Sage.

For the importance of creative, tentative titles that are continually revised as the research proceeds, see the following resource:

Glesne, C., & Peshkin, A. (1992). *Becoming qualitative researchers: An introduction*. White Plains, NY: Longman.

For a good overview of the importance of writing purpose statements, see the following resource:

Locke, L. F., Spirduso, W. W., & Silverman, S. J. (2000). *Proposals that work: A guide for planning dissertations and grant proposals* (4th ed.). Thousand Oaks, CA: Sage.

For additional resources on developing and writing mixed methods research questions, see the following resources:

Onwuegbuzie, A. J., & Leech, N. L. (2006). Linking research questions to mixed methods data analysis procedures. *The Qualitative Report, 11*(3), 474–498. Retrieved from http://www.nova.edu/ssss/QR/QR11–3/onwuegbuzie.pdf

Plano Clark, V. L., & Badiee, M. (in press). Research questions in mixed methods research. In A. Tashakkori & C. Teddlie (Eds.), *SAGE handbook of mixed methods in social & behavioral research* (2nd ed.). Thousand Oaks, CA: Sage.

Tashakkori, A., & Creswell, J. W. (2007). Exploring the nature of research questions in mixed methods research [Editorial]. *Journal of Mixed Methods Research, 1*(3), 207–211.

CHAPTER 6

COLLECTING DATA IN MIXED METHODS RESEARCH

T he basic idea of collecting data in any research study is to gather information to address the questions being asked in the study. In mixed methods research, the data collection procedure consists of several key components: sampling, gaining permissions, collecting data, recording the data, and administering the data collection. Data collection is more than simply collecting data; it involves several interconnected steps. Moreover, in mixed methods research, the data collection needs to proceed along two strands: qualitative and quantitative. Each strand needs to be fully executed with persuasive and rigorous approaches. Finally, there are certain decisions that need to be made when collecting mixed methods data within each of the specific research designs addressed in this book. This chapter, then, first turns to the more general procedures of data collection found in both qualitative and quantitative research and then considers how data collection might proceed within each of the six mixed methods research designs.

This chapter will address

- the procedures for quantitative and qualitative data collection in a research study and
- the specific decisions that arise in data collection for each of the six types of mixed methods designs.

● PROCEDURES IN COLLECTING QUALITATIVE AND QUANTITATIVE DATA

As mentioned in our definition of mixed methods in Chapter 1, complete qualitative and quantitative research methods need to be part of a mixed methods study, which includes the process of collecting data. As stated in their review of mixed methods procedures to study the Tamang Family Research Project in Nepal, Axinn and Pearce (2006) mentioned: "Thus, an integration of ethnographic and survey techniques must not be an excuse for doing less than a complete job with each of the components" (p. 73). In designing a mixed methods study, we recommend that the researcher advance a qualitative strand that includes "persuasive" qualitative data collection procedures and a quantitative strand that incorporates "rigorous" quantitative procedures. We use different terms—*persuasive* and *rigorous*—for the thoroughness of this aspect of research in order to respect the distinct terms that qualitative and quantitative researchers often use. What would these procedures entail? Table 6.1 provides the elements of these data collection procedures organized into the key components found in data collection: using sampling procedures, obtaining permissions, collecting information, recording the data, and administering the procedures.

The discussion that follows outlines the major steps in each of these data collection procedures. It is not meant to replace the more detailed information available in many research methods texts, such as those recommended as additional reading at the end of this chapter. Also, as mentioned earlier, the skills of qualitative and quantitative data collection are needed to conduct mixed methods research, and so they are important to review at this time. In this discussion, we highlight specific components that need to be addressed in order to have a complete mixed methods study. They have been gleaned from reviewing many mixed methods studies and writing about the detailed procedures in both quantitative and qualitative research (e.g., Creswell, 2008b; Plano Clark & Creswell, 2010).

Using Sampling Procedures

To address a research question or hypothesis, the researcher engages in a sampling procedure that involves determining the location or site for the research, the participants who will provide data in the study and how they will be sampled, the number of participants needed to answer the research questions, and the recruitment procedures for participants. These steps in sampling apply both to qualitative and quantitative research, although there are fundamental differences in how they are addressed—especially in terms of the sampling approach and the sample size.

Table 6.1	Recommended Qualitative and Quantitative Data Collection Procedures for Designing Mixed Methods Studies

Persuasive Qualitative Data Collection Procedures	Procedures in Data Collection	Rigorous Quantitative Data Collection
• Identify the site(s) to be studied. • Identify the participants for the study. • Note the sample size. • Identify the purposeful sampling strategy to enroll participants and why it was chosen (inclusion criteria). • Discuss recruitment strategies for participants.	Using sampling procedures	• Identify the site(s) to be studied. • Identify the participants for the study. • Note the sample size, the way it was determined, and how it provides sufficient power. • Identify the probabilistic or nonprobabalistic sampling strategy. • Discuss recruitment strategies for participants.
• Discuss permissions needed to study the sites and participants. • Obtain institutional review board approvals.	Obtaining permissions	• Discuss permissions needed to study the sites and participants. • Obtain institutional review board approvals.
• Discuss the types of data to be collected (open-ended interviews, open-ended observations, documents, audiovisual materials). • Indicate the extent of data collection. • State the interview questions to be asked.	Collecting information	• Discuss the types of data to be collected (instruments, observations, quantifiable records). • Discuss reported scores for validity and reliability for instruments used.
• Mention what protocols will be used (interview protocols, observational protocols). • Identify recording methods (e.g., audio recordings, field notes).	Recording the data	• State what instruments or checklists will be used and provide examples.
• Identify anticipated data collection issues (e.g., ethical, logistical).	Administering the procedures	• State how procedures will be standardized. • Identify anticipated ethical issues.

In qualitative research, the inquirer purposefully selects individuals and sites that can provide the necessary information. **Purposeful sampling** in qualitative research means that researchers intentionally select (or recruit) participants who have experienced the central phenomenon or the key concept being explored in the study. A number of purposeful sampling strategies are available, each with a different purpose (see Creswell, 2008b). One of the more common strategies is maximal variation sampling, in which diverse individuals are chosen who are expected to hold different perspectives on the central phenomenon. The criteria for maximizing differences depends on the study, but it might be race, gender, level of schooling, or any number of factors that would differentiate participants. The central idea is that if participants are purposefully chosen to be different in the first place, then their views will reflect this difference and provide a good qualitative study in which the intent is to provide a complex picture of the phenomenon. Another approach is to use extreme case sampling of individuals who provide unusual, troublesome, or enlightened cases. In contrast, a researcher might use homogeneous sampling of individuals who have membership in a subgroup with distinctive characteristics. As a study develops, the researcher may sample individuals who can shed light on the phenomenon being studied using sampling strategies that emerge during initial data collection.

In terms of the number of participants, rather than select a large number of people or sites, the qualitative researcher identifies and recruits a small number that will provide in-depth information about the central phenomenon or concept being explored in the study. The qualitative idea is not to generalize from the sample (as in quantitative research) but to develop an in-depth understanding of a few people—the larger the number of people, the less detail that typically can emerge from any one individual. Many qualitative researchers do not like to constrain research by giving definitive sizes of samples, but the numbers may range from 1 or 2 people, as in a narrative study, to 20 or 30 in a grounded theory project (see Creswell, 2007). Typically, when cases are studied, a small number is used, such as 4 to 10. The sample size relates to the question and the type of qualitative approach used, such as narrative, phenomenology, grounded theory, ethnography, or case study research (Creswell, 2007).

On the other hand, the intent of **probabilistic sampling** in quantitative research is to select a large number of individuals who are representative of the population or who represent a segment of the population. Ideally, individuals are randomly chosen from the population so that each person in the population has a known chance of being selected. Probabilistic sampling involves randomly choosing individuals based on a systematic procedure, such as the use of a random numbers table. Nonprobabilistic sampling involves selecting individuals who are available and can be studied. For example, a researcher may need to select all students in one classroom because they are available, recognizing that the sample is not representative of the population of all students in classrooms or even of the students in classrooms in the one school being studied. In addition, the investigator may want certain characteristics represented in the sample that may be out of proportion in the larger population. For example, more females than males may be in the population, and a random sampling procedure would, logically, oversample females. In this situation, the researcher first stratifies the population (e.g., females and males) and then randomly samples within each stratum. In this way, an equal number of participants on the stratification characteristic can be represented in the final sample chosen for data collection.

The sample size needed for a rigorous quantitative study is typically quite large. The sample needs to be large enough to meet the requirements of statistical tests. The sample needs to be a good estimate for the parameters of the population (reducing sampling error and providing adequate power). To determine the adequate sample size, we recommend that researchers use sample size formulas available in research methods textbooks, such as power analysis formulas for experiments (e.g., Lipsey, 1990) or sampling error formulas for surveys (e.g., Fowler, 2008).

Gaining Permissions

Researchers require permission to collect data from individuals and sites. This permission often needs to be sought from multiple individuals and levels in organizations, such as from individuals who are in charge of sites, from people providing the data (and their representatives, such as parents), and from campus-based institutional review boards (IRBs).

Obtaining access to people and sites requires obtaining permissions from individuals in charge of sites. Sometimes this involves individuals at different levels, such as the hospital administrator, the medical director, and the staff participating in the study. These levels of permissions are required regardless of whether the study is qualitative or quantitative. However, because qualitative data collection involves spending time at sites and the sites may be places not typically visited by the public (e.g., soup kitchens for the homeless), researchers need to find a gatekeeper, an individual in the organization supportive of the proposed research who will, essentially, "open up" the organization. Qualitative research is well known for the collaborative stance of its researchers, who seek to involve participants in many aspects of research. The opening up of an organization may also be necessary for quantitative studies in hard-to-visit organizations, such as the FBI or other governmental agencies.

To conduct research sponsored by a university or college, researchers must seek and obtain approvals from campus-based IRBs. These boards have been established to protect the rights of individuals participating in research studies and to assess the risk and potential harm of the research to these individuals. Researchers need to obtain the permission of the appropriate board and guarantee that the rights of participants will be protected. Failure to do so can have negative consequences for the university or college, such as withdrawal of federal funds. Typically, obtaining permission from an IRB involves filing an application, presenting information about the level of risk and harm, and guaranteeing that rights will be protected. The researcher guarantees protection of rights by stating them in writing and having the participants (or a responsible adult, if the participant is a minor) sign a form (i.e., an informed consent form) before they provide data. Researchers may not present or publish their findings if permissions were not obtained before the start of the data collection.

In qualitative research, procedures for how the inquirer will collect data and protect the gathered information need to be stated in detail, because the research often involves asking personal questions and collecting data in places where individuals live or work. The information collected from observing families at home, for example, may place individuals at particular risk. When behaviors are videotaped, participants are at risk of having unwanted behaviors disclosed. In quantitative research, individuals need to also provide the researcher with permission to complete instruments or have their behavior observed and checked off. Often this research does not take place in the individuals' homes or workplaces, and it is less obtrusive and less likely to put individuals at risk of harm. If the research involves manipulating the conditions experienced by participants, such as in an experiment, then the details of the treatment procedures and potential risks and benefits need to be carefully considered and described.

Collecting Information

There are many types of qualitative and quantitative data that can be collected in a mixed methods study. Researchers need to examine and weigh each option so they can determine what sources of data will best answer the research questions or hypotheses. Some forms of data cannot be easily categorized into qualitative or quantitative data, such as patient records in which both text in the form of providers' notes and numeric data in the form of results from screening tests reside side by side. The basic distinction we make between qualitative and quantitative data is that qualitative data consist of information obtained on **open-ended questions** in which the researcher does not use predetermined categories or scales to collect the data. Indeed, the participants provide information based on questions that do not restrict the participants' options for responding. In contrast, quantitative data are collected on **closed-ended questions** based on predetermined response scales, or categories. A quantitative questionnaire, for example, illustrates how a researcher identifies questions and asks participants to rate their answers to the questions on a scale.

In qualitative research, the types of data researchers can collect are much more extensive than the types in quantitative research. Some forms of qualitative data may be decided before a study begins and others will emerge. Qualitative types of data might be broadly organized into text data (i.e., words) or images (i.e., types of pictures). These two broad forms can, in turn, be categorized in terms of types of information that researchers typically collect: open-ended interviews (e.g., one-on-one interviews, phone interviews, e-mail interviews, focus groups), open-ended observations, documents (private and public), and audiovisual materials (e.g., videotapes, photographs, sounds). Options for sources of qualitative data continue to expand, and more recent forms include text messages, blogs and wikis, e-mails, and various forms of eliciting information (such as through interviews) using artifacts, pictures, and videotapes. Because qualitative data collection is so labor intensive, qualitative researchers often make a point to mention the extensive nature of their data collection (e.g., 3,000 pages of transcripts from the interviews, multiple observations of a setting over a 6-month period). In addition, qualitative researchers convey the primary interview questions asked—if interviews were collected during a study—to illustrate what information was obtained from participants. This is similar to the practice of quantitative researchers who include complete reproductions of their instruments in appendixes when reporting their results in journal articles.

In quantitative research, the forms of data have been reasonably stable over the years. Investigators collect quantitative data using instruments that measure individual performance (e.g., aptitude tests) or individual attitudes (e.g., attitudes toward self-esteem scales). They also gather structured interview and observational data in which the response categories are determined before the data collection, and the scores are recorded on scales in a closed-ended fashion. They collect factual information in the form of numbers from census data, attendance reports, and progress summaries. Other newer forms of quantitative data include biomedical tests (such as tracking eye movements or brain responses), geographical information systems (GIS) spatial data, and computer-based tracking data (such as from server logs). Again, as with the forms of qualitative data, mixed methods researchers need to assess which quantitative data types will best address their research questions or hypotheses.

Recording the Data

The approach we take to data collection involves systematically gathering information and recording it in such a way that it can be preserved and analyzed by a single researcher or a team of researchers. For qualitative data collection, forms for recording the information need to be developed. If interview data are collected, then an **interview protocol** is needed that includes questions to be asked during an interview and space for recording information gathered during the interview. This protocol also provides space to record essential data about the time, day, and place of the interview. In many cases, the researcher audiotapes the qualitative interviews and later transcribes the interviews, and the protocol becomes a backup system for recording information. Having an interview protocol helps keep the researcher organized, and it provides a record of information in the event that the recording devices do not work. An **observational protocol** also provides a useful way of organizing an observation. On this form, the researcher records a description of events and processes observed, as well as reflective notes about emerging codes, themes, and concerns that arise during the observation. Recording forms can also be developed for reviewing documents and for recording image data, such as photographs.

In quantitative research, the investigator selects an instrument to use, modifies an existing instrument, or develops an original instrument. If an existing instrument is selected, researchers need to identify one for which there is evidence that past use resulted in scores showing high validity and reliability. Alternatively, for structured observations, the researcher will use a proven checklist to record information. For documents with numeric data, the researcher often develops a form for recording information that summarizes the data. Data collected through computer-based methods need to be carefully recorded and organized within secure electronic files.

Administering the Procedures

Administering the procedures of data collection involves the specific actions taken by the researcher for gathering the data. In qualitative research, much discussion in the literature is directed toward reviewing and anticipating the types of issues likely to arise "in the field" that will yield less-than-adequate data. Issues, such as the time to recruit participants, the researcher's role in observing, the adequate performance of recording equipment, the time to locate documents, and the details of the proper placement of videotaping equipment, illustrate the types of concerns that need to be addressed. Also, the researcher needs to enter sites in a way that is respectful and does not disrupt the flow of activities. Ethical issues, such as providing reciprocity to participants for their willingness to provide data, handling sensitive information, and disclosing the purposes of the research, apply to both qualitative and quantitative research.

Administering the data collection in quantitative research involves attending to these ethical issues. In addition, the procedures of quantitative data collection need to be administered with as little variation as possible so that bias is not introduced into the process. Standardized procedures should exist for collecting data on instruments, on checklists, and from public documents. If more than one investigator is involved in data collection, training should be provided so that the procedure is administered in a standard way each time.

DATA COLLECTION IN MIXED METHODS ●

It is essential to know the general procedures of collecting data in qualitative and quantitative research, because mixed methods builds on these procedures. Before turning to specific mixed methods designs and their data collection procedures, we present several general guidelines for collecting both forms of data in mixed methods research:

- The purpose of data collection in a mixed methods study is to develop answers to the research questions (Teddlie & Yu, 2007). Mixed methods researchers cannot lose sight of this objective and should continually ask themselves whether their data will provide answers to the questions.
- Mixed methods research involves collecting both quantitative and qualitative data. Because multiple sources of data are collected, the mixed methods researcher needs to be familiar with the array of qualitative and quantitative data collection procedures. We encourage mixed methods procedures that involve creative qualitative data collection (e.g., use of photos to elicit information) and the careful selection of quantitative instruments that do not extend beyond those needed to answer the research questions.
- In sampling, it is possible to have a combined form of random (quantitative) and purposeful (qualitative) sampling. For example, Teddlie and Yu (2007) discussed a stratified purposive sampling procedure in which the researcher first stratifies the potential participants based on certain dimensions using procedures consistent with probabilistic sampling and then purposefully selects a small number of cases from each stratum.
- Not much has been written specifically about mixed methods data collection procedures except for the discussion of sampling strategies by Teddlie and Yu (2007), who admitted that there is no widely accepted typology of mixed methods sampling strategies.

- We stress the importance of detailing the data collection procedures in the methods section of a mixed methods study report. This allows readers and reviewers to understand the procedures used and make judgments about their quality. In addition, detailed reports of the procedures help others learn about mixed methods research and understand the often complex interweaving of qualitative and quantitative data collection efforts.
- The different types of mixed methods designs raise specific types of decisions and issues for the data collection procedures. Primarily, these decisions relate to sampling and sampling strategies, the types of questions asked during data collection, permission questions associated with securing approvals from IRBs, and acknowledging and respecting the participants. An overview of the types of decisions and our recommendations related to each of the six designs highlighted in this book is shown in Table 6.2.

It is to these issues that we now turn our attention by discussing each of the six designs highlighted in this book.

Convergent Design

In the convergent design, the data collection involves collecting both quantitative and qualitative data concurrently, analyzing the information separately, and then merging the two databases. Ideally this design prioritizes the two types of information equally, but researchers also use variants where there is a quantitative or qualitative priority for addressing the study's purpose. Within this overall process, researchers must make decisions related to sampling and data collection forms. Important **data collection decisions for the convergent design** include who will be selected for the two samples, the size of the two samples, the design of the data collection questions, and the format and order of the different forms of data collection.

Decide whether the two samples will include different or the same individuals. There are two options for selecting individuals to participate in the quantitative and qualitative strands of a convergent study: the samples can include different individuals or the same individuals. Different individuals may be used when the researcher is trying to synthesize information on a topic from different levels of participants. For example, Schillaci et al. (2004) included quantitative survey data collected from a random sample of households and qualitative ethnographic data collected from individuals (e.g., physicians, nurses, clerical staff, and patients) at

Table 6.2	Types of Mixed Methods Designs, Decisions, and Recommendations for Data Collection	
Type of Mixed Methods Design	**Decisions Needed in Data Collection**	**Recommendations for Designing a Mixed Methods Study**
Convergent design	Will the two samples include different or the same individuals?	If intent is to compare the data sets, use the same individuals.
	Will the samples be of the same size?	Consider what option to use, such as stating that it is not a problem to have different sizes, choosing equal sample sizes, or stating that unequal sizes is a limitation of the study.
	Will the same concept be assessed qualitatively and quantitatively?	Create parallel questions for the qualitative and quantitative data collection.
	Will the data be collected from two independent sources or from a single source?	Collect independent qualitative and quantitative data sets from two sources.
Explanatory design	Will the same or different individuals be used in both samples?	Individuals who participate in the qualitative phase must be individuals who participated in the quantitative phase.
	Will the samples be of the same size?	The qualitative follow-up phase has a smaller size than the quantitative phase.
	What quantitative results will be followed up?	Consider multiple options depending on the follow-up needed (e.g., significant results, significant predictors).
	How will follow-up participants be selected?	Select follow-up participants based on initial, quantitative results.
	How should the emerging follow-up phase be described for IRB approval?	Describe the follow-up phase as tentative and file an addendum as needed.
Exploratory design	Who and how many individuals should be in the quantitative follow-up phase?	For the quantitative phase, use a different sample than the qualitative phase, and obtain a large sample.

(Continued)

Table 6.2 (Continued)

Type of Mixed Methods Design	Decisions Needed in Data Collection	Recommendations for Designing a Mixed Methods Study
	How should the emerging follow-up phase be described for IRB approval?	Describe the follow-up phase as tentative and file an addendum as needed.
	What qualitative results will be used to inform the quantitative data collection?	Use themes, codes, and quotes to help design the instrument (e.g., themes become variables) or the taxonomy (e.g., different groups)
	In instrument design, how do you develop a good instrument?	Use rigorous procedures in scale development.
	How do you convey the rigor of the instrument design?	Use a diagram of the procedures to convey the multiple steps in this process.
Embedded design	Why and when should the embedded data be used in the study?	Give reasons for embedding the data, and consider the timing for embedding.
	Will embedding of a second data set introduce bias?	Collect secondary data unobtrusively (e.g., diaries during an experiment).
	If a design or procedure is used (to tie qualitative and quantitative data together), what will it be?	Consider what designs and procedures have been used in mixed methods research (e.g., case study, social networks).
	What data collection issues might be anticipated?	Examine the literature for the types of issues associated with the chosen design or procedure.
Transformative design	What labels will be used to refer to the participants?	Use labels meaningful to participants in the study.
	How can inclusiveness be promoted in the study?	Design a sampling procedure by collaborating with likely participants.
	How do you collect data that will be credible to the community being studied?	Involve participants as coresearchers (e.g., advisory board).
	What types of instruments can be used that are sensitive to the participants?	Choose measures sensitive to the participants in the study.

Type of Mixed Methods Design	Decisions Needed in Data Collection	Recommendations for Designing a Mixed Methods Study
	How will the data collection be sensitive to the study community?	Create ways to give back to the community (e.g., referrals, sharing findings).
Multiphase design	What multiple sampling strategies will be used in the phases or projects?	Use sampling strategies that fit the phases or projects in the study (e.g., levels, qualitative and quantitative sampling).
	Will both concurrent and sequential sampling occur?	Match the sampling strategy to the needs of the phases or projects.
	How will the project handle measurement and attrition issues?	Consider emergent approaches, recontacting individuals, planning for attrition.
	What overall objective (or theoretical drive) will tie the phases or projects together?	Identify a single objective for the line of inquiry composed of multiple phases or projects.

several care providing institutions to understand immunization practices. When the purpose is to corroborate, directly compare, or relate two sets of findings about a topic, we recommend that the individuals who participate in the qualitative sample be the same individuals who participate in the quantitative sample. In a convergent study by Morell and Tan (2009), 230 elementary students were included in the quantitative sample, and 34 of these same students were included in the qualitative sample. If the mixed methods researcher collects different types of data from different participants, the researcher is introducing extraneous information into the study and potentially influencing the researcher's ability to merge results.

Decide whether the size of the two samples will be the same or different. The mixed methods researcher needs to consider the size of the two samples in relation to the use of the convergent design. A good option is for the two samples to have different sizes, with the size of the qualitative sample much smaller than the quantitative sample. This helps the researcher obtain an in-depth qualitative exploration and a rigorous quantitative examination of the topic. This disparity can raise the question of how to converge or compare the two databases in any meaningful way when the size is so different. There are several options to address this issue for the mixed methods researcher. One option, embraced by some researchers (i.e., qualitative-oriented researchers), is that this size differential is not a problem because the intent of the data gathering is different for the two databases: quantitative data collection aims toward making generalizations to a population while qualitative data collection seeks to develop an in-depth understanding from a few people.

If the difference in sample size is a problem, another option is to use an equal sample size for both the quantitative and qualitative samples. In a convergent study of teacher candidates' multicultural attitudes and knowledge, Capella-Santana (2003) gathered quantitative questionnaire data from 90 undergraduate elementary education teacher candidates. She also invited all 90 participants to be interviewed "to corroborate the information obtained through the questionnaires" (p. 185). This equal sample size approach probably sacrificed some of the richness of the qualitative data. However, this option is recommended for researchers using a data transformation variant where having two large samples of equal size and the same individuals is an important consideration in planning to collect qualitative data that will eventually be quantitized (a process to be discussed further in Chapter 7). When the sizes are equal but small, the researcher may be sacrificing the use of rigorous statistical tests. McVea et al. (1996) evaluated family practices that had adopted prevention materials in their convergent design. They gathered both quantitative data (e.g., using a structured observation checklist) and qualitative data (e.g., through key informant interviews) from the same eight practices. This approach led to the use of descriptive quantitative statistics because of the small size. Some researchers compromise with equal sample sizes that are too small or too large and then discuss the resulting limitations of the study. Bikos et al. (2007a, 2007b) used samples of the same 32 expatriate female spouses who moved to Turkey for the two strands of their study of cultural adaptation. In their conclusions, they noted that their sample size resulted in low statistical power for the quantitative strand, because it was small and may have limited their ability to find individual differences in experiences in the qualitative strand because the sample size was large.

Decide to design parallel data collection questions. We recommend that merging the two databases works best if the researcher designs the study by asking parallel questions in both the qualitative and the quantitative data collection efforts. By parallel questions we mean that the same concepts need to be addressed in both the qualitative and quantitative data collection so that the two databases can be compared or merged. If the concept of "self-esteem" is being addressed on a quantitative survey, then an open-ended question on "self-esteem" needs to be asked during the qualitative one-on-one interviews. In this way, the results can be merged around the concept during data analysis.

Decide if the data will be collected on two, independent sources or a single source. Researchers need to consider whether the two data sets will be collected independently, using different forms (such as gathering quantitative data with a survey questionnaire and qualitative data through focus group interviews), or whether they will both be collected on one form (e.g., a single questionnaire with both closed- and open-ended questions). Although researchers using a data transformation variant often utilize a single format, in general we encourage the use of collecting two independent data sets when using this design. A related issue follows in that the researcher needs to decide the order for collecting the two data sets. Many convergent designs happen to collect one form of data (e.g., surveys) before the other form (e.g., focus groups) simply for logistical reasons. If the researcher is concerned that there may be an interaction between the two forms (e.g., participating in a focus group discussion may change the way that participants respond to the survey items), then one option is to alternate the order for collecting the data. Luzzo (1995) gathered survey packets and individual interviews from students and noted "the order in which these students completed the packet and participated in the interview was counterbalanced" (p. 320).

Explanatory Design

The data collection procedures in the explanatory design involve first collecting quantitative data, analyzing the data, and using the results to inform the follow-up qualitative data collection. Thus, sampling occurs at two points in this design: in the quantitative phase and in the qualitative phase. In this design, the quantitative and qualitative data collections are related to each other and not independent. One builds on the other. The emphasis on data collection may favor either quantitative or qualitative data. Typically, an emphasis is placed on the initial, substantial quantitative data collection with a smaller emphasis on the qualitative follow-up.

The data collection decisions for the explanatory design include who should the participants be in the second phase, what sample sizes to use for both strands, what data to collect from one phase to the other and from whom, and how to secure IRB permissions for the two data collections.

Decide whether to use the same or different individuals in both samples. Since the explanatory design aims to explain initial, quantitative results, the individuals for the qualitative follow-up phase should be individuals who participated in the initial, quantitative data collection. The intent of this design is to use qualitative data to provide more detail about the quantitative results, and the individuals best suited to do so are ones who contributed to the quantitative data set.

Decide on the sizes for the two samples. Although some researchers choose to follow up qualitatively with all participants in the first phase (resulting in equal sample sizes), we recommend that the qualitative data collection come from a much smaller sample than the quantitative data collection. The intent of this design is not to merge or compare the data as in the convergent procedures, so having unequal sizes is not an issue in sequential designs. The important consideration lies in collecting enough qualitative information so that meaningful themes can be developed. In the explanatory design study by Thøgersen-Ntoumani and Fox (2005), the authors collected quantitative data from 312 employees and then followed up in the qualitative phase with a subset of 10 employees who had participated in the quantitative questionnaire.

Decide what quantitative results to follow up. In terms of the follow-up data collection, researchers need to make a decision as to what quantitative results need to be further explored through the qualitative data collection. Several options exist for making this decision. The first step is to conduct the quantitative analysis and examine the results to see which ones are unclear or unexpected and require further information. This will help dictate a strategy. Some results that might be considered for follow-up are statistically significant results, statistically nonsignificant results, key significant predictors, variables that distinguish between groups, outlier or extreme cases, or distinguishing demographic characteristics. The researcher should identify the results that need further information and use these results to guide the design of the qualitative phase research questions, sample selection, and data collection questions.

Decide how to select the best participants for the qualitative follow-up phase. Another decision is how to select the participants to be studied in the qualitative follow-up. Certainly the quantitative results that become the focus of the qualitative strand will suggest who these participants might be. Sometimes the participants will simply be individuals who volunteer to participate in interviews. In an explanatory design study of adoptive fathers and birth fathers, Baumann (1999) asked the fathers completing the quantitative questionnaire in the first phase whether they would volunteer for interviews for the follow-up phase. This approach provides a weaker connection between the phases but may be necessary in studies where identifying information cannot be collected as part of the quantitative data. A more systematic approach is to use the quantitative statistical results to direct the follow-up sampling procedures to select the participants best able to help explain the phenomenon of interest. Way, Stauber, Nakkula, and London (1994) used a systematic approach. These researchers determined that the relationship between depression and substance use differed among surburban and urban high school students, and they decided to use this quantitative result as a basis for studying students in the top 10% of depression scores from the schools in qualitative follow-up interviews.

Decide how to describe the emerging follow-up phase for institutional review board approval. Since this design is implemented in two distinct phases, researchers may consider seeking IRB approval for each phase separately. However, we recommend that the researchers describe the plans for both phases in their initial IRB application materials, noting that the plans for the second, qualitative phase are tentative because they will evolve from the results of the first phase. IRBs require as full a disclosure of data collection procedures as possible. This means that participants should be informed from the start about the potential of being contacted in the future for a second data collection. The researcher may also have to explain to the IRB that identifying information will be collected as part of the quantitative data to facilitate the follow-up process and address the additional ethical concerns associated with this information. We also recommend stating the follow-up phase as "tentative," recognizing that an addendum may need to be filed with the IRB when the follow-up data collection procedures are firmly established.

Exploratory Design

In an exploratory design, the researchers first collect qualitative data, analyze it, and then use the information to develop a follow-up quantitative phase of data collection. The quantitative strand thus builds on the qualitative one. The sampling occurs in two phases, and they are related to each other. However, in some exploratory designs, a three-phase model is in use when the initial, exploratory phase is followed by an instrument design phase, and then a phase testing and administering the instrument. Alternatively, the middle phase might be the development of a typology, the search for an appropriate instrument, or the modification of an instrument. The priority in this design may be placed on any of the phases.

Although both designs are sequential and raise the same types of data collection issues, the considerations for making decisions for an exploratory design differ in many respects from those for the explanatory design. The primary **data collection decisions for the exploratory design** are the determination of samples for each phase, the decisions about results to use from the first phase, and, if a middle phase is used, how to design a rigorous instrument with good psychometric properties.

Decide who and how many individuals to include in the sample for the quantitative phase. Unlike the explanatory design, the individuals who

participate in the quantitative follow-up for the exploratory design are typically not the same individuals who provided the qualitative data in the initial
phase. Because the purpose of the quantitative phase is to generalize the
results to a population, different participants are used in the quantitative follow-up stage than in the initial, qualitative phase. In addition, the second
phase requires a large sample size so that the researcher can conduct statistical tests and potentially make claims about the population in question. In
their study of the factors perceived to be affecting changes in adult education
graduate programs, Milton, Watkins, Studdard, and Burch (2003) conducted
qualitative interviews with 11 faculty members and administrators and then
administered a quantitative instrument to a second population of 131 individuals representing 71 adult education programs.

**Decide how to describe the emerging follow-up phase for institutional
review board approval.** For IRB purposes, only the initial phase of data collection in the exploratory design can be identified with any certainty when
the initial application is prepared because the second phase will evolve from
the first. When filing the initial application with the IRB, the researcher can
provide some tentative details for the second phase and then seek an addendum once the quantitative instrument has been developed. Since the two
phases do not include the same participants, it is also possible to submit the
application as two separate IRB proposals since separate selection, recruitment, and informed consent procedures will need to be described.

**Decide what aspects of the qualitative results to use to inform the quanti-
tative data collection.** In the exploratory design with an intent of developing
and testing an instrument (or taxonomy), one decision is determining what
information from the initial, qualitative phase can be most helpful in designing the instrument for the quantitative phase. During the initial, qualitative
phase, we recommend that a typical qualitative data analysis consist of identifying useful quotes or sentences, coding segments of information, and the
grouping of codes into broad themes (as will be discussed in Chapter 7).
With this configuration of the qualitative data, the mixed methods researcher
can use the central phenomenon as the quantitative construct to be assessed
by the instrument, the broad themes as the scales to be measured, the individual codes within each theme as the variables, and the specific quotes from
individuals as specific items or questions on the instrument.

Decide what steps to take in developing a good quantitative instrument.
Another decision is how to design a good instrument so that it has strong psychometric properties. It takes time and is hard work, and the mixed methods
researcher may use the themes from the initial qualitative phase to locate

published instruments to use that best match the different qualitative themes. Alternatively, mixed methods researchers may decide to develop their own instrument based on the qualitative findings. The best instruments are rigorously developed using good procedures of scale development. A general approach that we recommend has been adapted from DeVellis (1991):

1. Determine what you want to measure, and ground yourself in theory and in the constructs to be addressed (as well as in the qualitative findings).

2. Generate an item pool, using short items, an appropriate reading level, and questions that ask a single question (based on participant language when possible).

3. Determine the scale of measurement for the items and the physical construction of the instrument.

4. Have the item pool reviewed by experts.

5. Consider the inclusion of validated items from other scales or instruments.

6. Administer the instrument to a sample for validation.

7. Evaluate the items (e.g., item-scale correlations, item variance, reliability).

8. Optimize scale length based on item performance and reliability checks.

Another way to learn about mixed methods researchers' procedures for generating an instrument from qualitative findings is to examine published mixed methods studies that use an exploratory design with the intent to develop an instrument. In addition to the study about creating a measure of organizational assimilation in diverse industries (Myers & Oetzel, 2003) found in Appendix C, other examples of this design are an education study about the teaching of reading comprehension (Meijer, Verloop, & Beijaard, 2001), a psychological study of the tendency to see oneself as significant to a romantic partner (Mak & Marshall, 2004), a higher education study about the factors perceived as affecting changes in the size of graduate programs (Milton et al., 2003), and a cross-cultural study of the lifestyle behaviors of Japanese college women (Tashiro, 2002). In Tashiro (2002), the author began by collecting focus group data. She formed a questionnaire using data from the focus groups, as well as from other unpublished sources. The focus group participants were then asked to evaluate the clarity of the questions, and the resulting questionnaire was used in a pilot test with participants similar to those in the study. The questionnaire's content was validated by a number of research experts and checked for inter-item reliability and test–retest reliability. These procedures closely followed those recommended by DeVellis (1991).

Decide how to convey the instrument development component in a procedural diagram. Finally, the connecting phase during which the researcher designs the instrument can be incorporated into a discussion or a diagram of the overall procedures in a mixed methods study. We recommend the use of a diagram to highlight the numerous steps required to design a good instrument. Bulling (2005) designed a mixed methods study addressing how emergency personnel reacted to tornadoes. Figure 6.1 is a figure from her study. This figure indicates the instrument development stages and how they paralleled the qualitative and quantitative procedures in her study.

Embedded Design

In an embedded design, the quantitative and qualitative data can be collected either sequentially or concurrently or both. In this design, one form of data is embedded within another form (e.g., qualitative data embedded within an experimental intervention trial or qualitative interviews embedded within a longitudinal correlational design). A variant on this approach is to embed both qualitative and quantitative data within a traditional design or procedure (e.g., geographic information surveys that include both quantitative and qualitative information). Data collection issues will be discussed in view of these two variants.

What decisions about data collection does a researcher have to face when using this design? When embedding one form of data within the other form, the **data collection decisions for the embedded design** are the rationale for embedding one form of data, the timing of the embedded data, and how to address problems that may arise from the embedding. If using the variant where both forms of data are embedded within a traditional design or procedure (e.g., case study design or social network analysis), the researcher needs to decide what procedure to use and how to work with problems that may result during embedding within the procedure. We begin by discussing the decisions associated with the use of embedding one form of data within another followed by embedding both forms of data within a procedure.

Decide the reason and timing for embedding a second type of data within a larger design. In terms of a rationale and timing for embedding supportive data into another major form of data collection, such as a quantitative experiment or correlational study, reasons need to be advanced for why the supportive data are being used. These reasons might be stated in a purpose statement (see Chapter 5). One approach to an experimental study is to consider introducing the supportive qualitative database in one or more phases of

Figure 6.1 Procedures for an Exploratory Instrument Design Mixed Methods Study

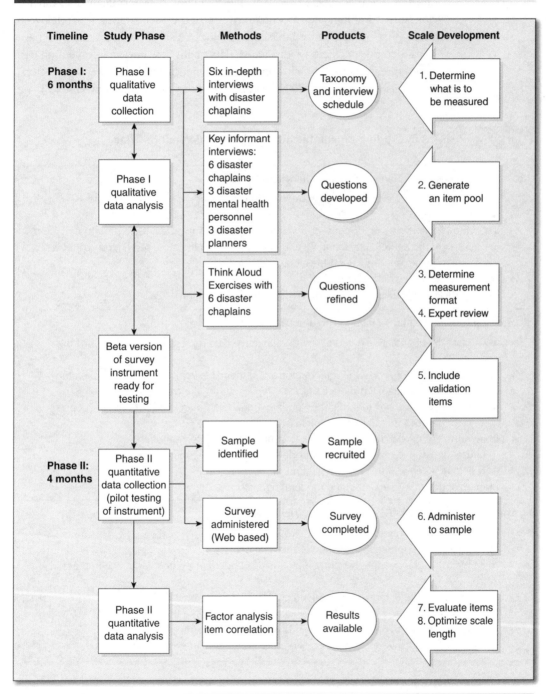

SOURCE: Bulling (2005). Used by permission from the author.

the experiment: before the trial begins, during the trial, or after the trial. Before and after the trial means a sequential introduction of the qualitative data, while during the trial indicates a concurrent use of the two databases. Sandelowski (1996) first conceptualized these possibilities. In subsequent writings, we have expanded on specific ways qualitative data might flow into an intervention trial and the reasons for its inclusion in health science research (Creswell, Fetters, Plano Clark, & Morales, 2009). These reasons are mentioned in Table 6.3. We

Table 6.3 Reasons for Adding Qualitative Research Into Intervention Trials

Reasons for adding qualitative data before the trial begins:

- To develop an instrument for use in an intervention trial (when a suitable instrument is not available)
- To develop good recruiting or consent practices for participants in a intervention trial
- To understand the participants, context, and environment so that an intervention would work (i.e., applying interventions to real-life situations)
- To document a need for the intervention
- To compile a comprehensive assessment of baseline information

Reasons for adding qualitative data during the trial:

- To validate the quantitative outcomes with qualitative data representing the voices of the participants
- To understand the impact of the intervention on participants (e.g., barriers and facilitators)
- To understand unanticipated participant experiences during the trial
- To identify key constructs that might potentially impact the outcomes of the trial, such as changes in the sociocultural environment
- To identify resources that can facilitate the conduct of the intervention
- To understand and depict processes experienced by the experimental groups
- To check the fidelity of the implementation of procedures
- To identify potential mediating and moderating factors

Reasons for adding qualitative data after the trial concludes:

- To understand how participants view the results of the trial
- To receive participant feedback to revise the treatment
- To help explain the quantitative outcomes, such as underrepresented variations in the trial outcomes
- To determine the long-term, sustained effects of an intervention after a trial
- To understand in more depth how the mechanisms worked in a theoretical model
- To determine if the processes in conducting the trial had treatment fidelity
- To assess the context when comparisons of outcomes are made with baseline data

SOURCE: Adapted from Table 9.1 in Creswell et al. (2009).

have also noted that depending on resources and personnel, qualitative data might be added at a single phase, such as before the trial (see Donovan et al., 2002, in which the authors collected qualitative interview data before the randomized controlled trial (RCT) to enhance the rate of consent to participate), or at multiple times during a study (see Rogers, Day, Randall, & Bentall, 2003, in which the qualitative data were collected before the intervention trial to inform the treatment design and after the intervention trial to examine the processes behind patients' adherence to antipsychotic medication). Although the examples here are for variants of embedded experimental designs, the notion of embedding a supplemental data before, during, or after the major data collection is a useful conceptualization regardless of the approach (such as a case study) used when both quantitative and qualitative data are embedded within a major design.

Decide whether the issue of introducing bias within an embedded experiment is a concern. One issue that can occur when a researcher concurrently embeds qualitative data into an intervention experiment is the issue of introducing bias through the qualitative data collection that affects the experiment's internal validity. For example, will gathering focus group data during the trial with the experimental treatment sample affect the outcomes in the experiment? Researchers need to be alert to this possibility and openly discuss it. Steps should be taken to minimize this potential bias. One option for the researcher is to collect unobtrusive qualitative data, such as collecting diaries or making recordings of the activities occurring during intervention sessions. Victor, Ross, and Axford (2004) used diaries in an intervention trial of individuals with osteoarthritis of the knee. They asked individuals in the intervention group to maintain diaries during the intervention to review their symptoms, use of medication, and goals for treatment during the trial. The investigators then collected these diaries after the intervention and reviewed them. Another approach is to equally distribute the qualitative data collection across all treatment and control groups. Finally, investigators might postpone the qualitative data collection until after the intervention is complete by employing a sequential approach to data collection.

Decide what approach will provide the design or procedure for collecting quantitative and qualitative data. Unquestionably, more and more designs and procedures are being used to collect both qualitative and quantitative data (Creswell & Tashakkori, 2007). One way to look at these designs is to consider them as "tools" that disciplines have used that become a framework for collecting both forms of data. For example, ethnography has long been considered a method of data collection (as well as a study of culture-sharing groups) and could be seen as a design for gathering multiple forms of data—both qualitative and quantitative (Morse & Niehaus, 2009). Depending on the research question, design, and purpose, a case study can be considered qualitative, quantitative, or both (Luck, Jackson, & Usher, 2006). A Life History Calendar, geographic information systems (GIS) technology, a Neighborhood History Calendar, or a Microdemographic Community Study can also be seen as procedures for collecting data within social demography (Axinn & Pearce, 2006). Social network analysis becomes a design for incorporating ethnography approaches, such as in the sociological study of the legal handling of rape (Quinlan & Quinlan, 2010).

Whichever design or procedure is chosen, the researcher needs some expertise related to collecting data within this framework in addition to expertise in mixed methods research.

Decide what data collection issues can be anticipated within the chosen design or procedure. Researchers embedding both types of data within a larger design or procedure will face issues in data collection related to the specific design or procedure (e.g., Life History Calendars) being used as a framework as well as the implementation of it within a mixed methods design. Using the examples of social demography mentioned by Axinn and Pearce (2006), we can assemble some of the specific data collection problems involved in embedding both qualitative and quantitative data into "tools" such as Life History Calendars. Life History Calendar methods attempt to strike a balance between more structured and unstructured data collection approaches for retrospective reports. Highly structured survey questions can be used with less structured oral history reports. In combining these sources of data into a Life History Calendar, the researcher builds a mixed methods study. Some of the issues in data collection Axinn and Pearce (2006) mentioned involve standardizing the period of recall for respondents; including multiple timing cues to help limit the recall burden for older members; using calendars among populations who do not employ time records; using flexible recording alternatives; including culturally relevant behaviors and events; confining the data collection to events and experiences and not attitudes, values, or beliefs; limiting the scale of the application of the calendar to larger projects; training for the qualitative interviewers; and employing larger time units than days, such as months, for older respondents.

Transformative Design

The transformative design is a mixed methods design that the researcher frames within a transformative theoretical perspective to help address injustices or bring about change for an underrepresented or marginalized group. Data collection for the qualitative and quantitative strands of the study can proceed concurrently, sequentially, or both. Therefore, this design can include the data-collection decisions already raised for the concurrent and/or sequential designs. Additional data collection decisions for the transformative design are related to sampling, benefits to those participating in the study, and collaboration during the data collection process. Many of these issues surfaced during our study of transformative mixed methods designs and the specific examination of studies with feminist, racial, ethnic, gay and lesbian, and disability lenses (Sweetman, Badiee, & Creswell, 2010).

Decide how best to refer to and interact with participants. Avoid stereotypical labels for participants when collecting data, and use labels that are meaningful to the participants in the study. In a mixed methods study of individuals with disabilities, Boland, Daily, and Staines (2008) mentioned that the interviewers used in the qualitative phase were trained in appropriate language and etiquette related to disability: "Five interviewers were given specific training on the social model of disability, etiquette and language when interviewing clients with disability" (p. 201).

Decide what sampling strategies will promote inclusiveness. Use sampling strategies that improve the inclusiveness of the sample to increase the probability that traditionally marginalized groups are adequately and accurately represented. A collaborative approach to making decisions about sampling might be used. For example, Payne (2008) described how he formed a research team with four street life-oriented Black men, and together this team mapped out street communities of interest, identified "street allies" (p. 11) as gatekeepers, and utilized snowball sampling to identify street life-oriented Black male participants for their mixed methods participatory study of resiliency.

Decide how to actively involve participants in the data collection process. Use methods to ensure that the research findings will be credible to that community, and design data collection to permit effective communication with community members. Use collection methods that are sensitive to the community's cultural contexts and that open up avenues for participation in the social change process. One way that this might occur is by involving the participants as coresearchers or by creating an advisory board that consists of community members. Boland et al. (2008) discussed consulting with an advisory board that consisted of a community member; in addition, the lead interviewer also had a disability. Kumar and colleagues (2000) also utilized an advisory board that consisted of members who varied in religion, caste, politics, gender, and welfare. Shapiro, Setterlund, and Cragg's (2003) project was overseen by members of OWN, a self-help organization dedicated to promoting the rights and dignity of older women.

Decide to use instruments that are sensitive to the cultural context of the group being studied. In addition to making decisions about the entire data collection process, researchers using transformative designs need to carefully select quantitative measures that are sensitive to the constructs and groups under study. Hodgkin (2008) described selecting a particular measure of social capital that was shown to be sensitive to the range of formal and informal activities in which women become involved. McMahon (2007) also discussed selecting a nonstandard measure for use in her study of student–athlete cultures of rape myths. She wrote that the selected measure "represents a departure from the instruments that are typically used to measure students' attitudes toward sexual assault because it was specifically designed to address issues of acquaintance rape on a college campus" (p. 360).

Decide how the data collection process and outcomes will benefit the community being studied. This decision reflects the notion of reciprocity—or giving back to participants. It is not enough to develop and implement a study that may be useful to the community; there has to also be an attempt to disseminate the findings within the community. During the process of conducting a study of Hispanic females, Cartwright, Schow, and Herrera (2006) attempted to share the findings with participants: "Formando was conceptualized with the idea of sharing the findings with the participants as the study progressed, as well as through addressing participants' questions during the process" (p. 100). Referrals can be another source of reciprocity. Filipas and Ullman (2001), studying women sexual assault survivors in a mixed methods study, provided their participants "with a list of medical and mental health resources in the community for dealing with rape and other violence and the cover letter to students gave an additional contact for counseling referrals at the university" (p. 676). In another mixed methods study, Kumar et al. (2000) provided free HIV testing and counseling, medical referrals, food, and the prospect of becoming a peer educator to drug users.

Multiphase Design

The multiphase design combines both sequential and concurrent strands over a period of time in a program of study. Prime examples are large-scale evaluation projects and health science projects. They might involve multiple levels of analysis and typically are conducted over several years. What all of these projects have in common is that (1) they are more complex than the two-phase projects in our basic mixed methods research designs; (2) they typically occur over time; (3) they often involve a team of researchers; (4) they require extensive funding; and (5) they involve collecting multiple quantitative and qualitative databases that build toward an overall objective. These studies often appear as different projects based on the databases and in different publications (with different publication lags) (Morse & Niehaus, 2009), which makes it difficult to discern the specific data collection issues.

Until the procedures of linking various studies and projects over time are discussed in more detail, our discussion can only be a starting point for the conversation about important data collection issues. However, as mixed methods becomes more ingrained in single studies, its use in longitudinal programs will also become more common (Axinn & Pearce, 2006).

The data collection decisions for the multiphase design include sampling, using longitudinal designs, and developing a programmatic objective that binds the multiple projects together.

Decide to use multiple sampling strategies. The multiphase design often involves multiple sampling strategies and may include different sampling procedures for different levels of analysis. In an examination of schools in the Louisiana Effectiveness Study (Teddlie & Stringfield, 1993), the authors used eight different sampling strategies (e.g., types of probability, such as random; types of qualitative purposive, such as typical case sampling; and types of combinations, such as stratified purposive) at five levels of education: state school system, school districts, individual schools, teachers or classrooms, and students within classrooms.

Decide how to sample and collect data for each phase. Multiphase designs may include both concurrent and sequential forms of data collection, and one or both forms may apply to different levels of an organization (Teddlie & Yu, 2007) and/or to different phases of a longitudinal study. This might mean using different samples for different projects to keep from biasing the participants or causing them research fatigue. Bradley et al. (2009) described their different samples and data collection procedures across four phases to improve hospital care for patients with acute myocardial infarction. They first examined quantitative performance data in a database representing all U.S. hospitals. In the next step, they conducted in-depth qualitative studies of 11 hospitals identified as high performers. The sample at each hospital included individuals in various roles, such as cardiologists, emergency medicine physicians, nurses, technicians, ambulance staff, and administrators. From the qualitatively derived practices, the team developed a quantitative Web survey that was administered to a randomly selected sample of 365 hospitals. After disseminating recommendations based on results, the team evaluated the success of the dissemination efforts using a combination of approaches.

Decide how to handle measurement and attrition issues. Specific issues may surface related to the longitudinal aspect of a multiphase design. These include the possible attrition of participants if the data collection methods take place over several years (Axinn & Pearce, 2006). Other issues involve the tension between maintaining close comparability of measures over time and measures that change over time given emerging data collection efforts. In addition, with multiple data collection points, the participants may change (as well as changes in the social context or the subject matter of interest) over the course of a multiphase study that includes longitudinal data collection. Procedures for recontact and cooperation need to be built into the study. If the unit of analysis consists of households, individuals change in households over time. Also, researchers need to keep an emergent approach in mind so that the inquiries build incrementally and not simply develop as separate studies. As pointed out by Axinn and Pearce (2006), "information gleaned from methods used at one point can be used to guide implementation of the next round of alternative methods" (p. 178).

Decide on the programmatic thrust to provide the framework for the multiphase projects. There needs to be consistency among the multiple projects in a multiphase design, which should be provided by a central programmatic thrust. Morse and Niehaus (2009) discussed the importance of a programmatic objective that gives the research program its theoretical thrust. For example, the central programmatic objective of developing complex interventions to improve health was the subject of an article by Campbell et al. (2000). They discussed complex interventions "made up of various interconnecting parts" (p. 694). Examples of their complex intervention studies in the health sciences were for service delivery and organization units (e.g., stroke units), for health professionals' behavior (e.g., strategies for implementing guidelines), for communities (e.g., community-based programs to prevent heart disease), for groups (e.g., school-based interventions to reduce smoking), and for individuals (e.g., cognitive behavioral therapy for depression). The phases of these intervention projects could be sequential or iterative and often were not linear. Their phases consisted of a preclinical or theoretical phase, a phase defining components of the intervention, a phase defining the trial and intervention design, and a phase promoting effective implementation. In these phases, the investigators included qualitative and quantitative data collection methods and discussed method issues of the difficulty of randomization, concealing allocation of treatment, and poor recruitment.

SUMMARY

Qualitative and quantitative data collection involve the key components of sampling, obtaining permissions, selecting types of data, preparing forms for recording data, and administering the data collection. For each component, the approaches differ for quantitative and qualitative approaches to data collection. In mixed methods research, it is helpful to conceptualize the two types of data collection as either concurrent or sequential and to relate the data collection procedures to the specific types of mixed methods designs. General principles for collecting data in mixed methods studies involve gathering information to address the research questions, providing details for the procedures, being familiar with both quantitative and qualitative data collection, and using sampling that draws on the approaches found in both qualitative and quantitative research.

There are also specific decisions related to data collection associated with each of the mixed methods designs. For the convergent design, the decisions relate to what individuals to include in the samples for each strand, the relative size of the two samples, the use of parallel questions across data forms, and the format and order of the collection of the two types of data. For the explanatory design, the decisions relate to who participates in the second phase, the relative size of the two samples, how to design the qualitative data collection based on the quantitative results, how to select the best individuals for the second sample, and how to secure IRB approval for the two-phase approach. For the exploratory design, the decisions are similar due to the two-phase approach, but the considerations are different. These decisions relate to the determination of the samples and sample sizes for each phase, the approach for securing IRB approvals, the decisions about what results to use from the first phase, how to design a rigorous instrument with good psychometric properties, and how to convey this important step within the design. For an embedded design, the data collection issues include creating a rationale for embedding the data, deciding the timing of the embedded data, addressing problems of bias that may arise from the embedding, deciding on a design or procedure as an overarching perspective to use, and working with problems that may result during embedding within a traditional design or procedure. In the transformative design, the decisions relate to identifying and interacting with participants from marginalized groups in sensitive ways, using inclusive sampling strategies, actively collaborating with participants during the data collection process, selecting instruments that are sensitive to the contexts of the participants, and using procedures that lead to benefits to those participating in the study. For a multiphase design, the data collection decisions are emerging, and we do not know much about the specific concerns, but they relate to the need for multiple sampling strategies, designing procedures for multiple data collections that unfold longitudinally over time, addressing measurement and attrition issues, and developing a programmatic objective that binds the multiple projects together.

ACTIVITIES

1. Examine one qualitative and one quantitative journal article. The two studies should display the two forms of sampling: purposeful sampling and random or systematic sampling. Discuss the different approaches used.

2. Find a convergent mixed methods design study published in a journal. Draw a diagram of the data collection and analysis activities. Indicate in the drawing specifics about the sampling strategies, the sample sizes, the participants, and the different forms of data collection.

3. Find an explanatory mixed methods design study published in a journal. Examine how the author selected the participants for the second phase and the reason(s) the author gives for selecting those individuals. List the ways that the author used the quantitative results to guide the collection of data in the qualitative phase.

4. Find an exploratory mixed methods design study in which the intent was to develop an instrument. List the steps the authors used to develop the instrument from the qualitative database. Compare these steps with those in DeVellis (1991) mentioned in this chapter.

5. Find a feminist transformative mixed methods design study. Look closely at the data collection, and determine how the authors collaborated with participants, were respectful of their rights, used data collection procedures that were sensitive to the participants' contexts, and built support to engage in their research.

6. Find a multiphase design in the evaluation or the health sciences. Identify the various projects that were linked together. Draw a diagram of their data collection procedures within each of the projects.

7. Write a description of the data collection procedures that you might use in a mixed methods study of your choice. Specify your sampling strategy, sample size, data collection types, forms for recording information, and administration procedures for the quantitative and qualitative strands of your study.

ADDITIONAL RESOURCES TO EXAMINE

There are numerous books available to develop a good understanding of quantitative and qualitative methods or both. Books are available within specific discipline fields, but here are two resources that include both quantitative and qualitative methods and are broadly aimed at the social sciences and education:

Creswell, J. W. (2008b). *Educational research: Planning, conducting, and evaluating quantitative and qualitative research* (3rd ed.). Upper Saddle River, NJ: Pearson Education.

Plano Clark, V. L., & Creswell, J. W. (2010). *Understanding research: A consumer's guide*. Upper Saddle River, NJ: Pearson Education.

There is also a Web source on research methods that provides an overview of key ideas in both quantitative and qualitative research:

Trochim, W. M. *The research methods knowledge base* (2nd ed.). Retrieved from http://www.socialresearchmethods.net/kb/

For a detailed overview of the steps involved in constructing an instrument and in scale development, see the following resource:

DeVellis, R. F. (1991). *Scale development: Theory and application*. Newbury Park, CA: Sage.

For a detailed discussion about transformative mixed methods studies and specific issues of data collection, see the following resource:

Sweetman, D., Badiee, M., & Creswell, J. W. (2010). Use of the transformative framework in mixed methods studies. *Qualitative Inquiry*. Prepublished April 15, 2010, DOI: 10.1177/1077800410364610.

For a discussion of collecting both qualitative and quantitative data embedded within a design or procedure, see the following resource:

Axinn, W. G., & Pearce, L. D. (2006). *Mixed method data collection strategies*. Cambridge, UK: Cambridge University Press.

CHAPTER 7

ANALYZING AND INTERPRETING DATA IN MIXED METHODS RESEARCH

Data analysis in mixed methods research consists of analyzing separately the quantitative data using quantitative methods and the qualitative data using qualitative methods. It also involves analyzing both sets of information using techniques that "mix" the quantitative and qualitative data and results—the mixed methods analysis. These analyses are in response to the research questions or hypotheses in a study, including the mixed methods questions. Thus, our focus here will be primarily on the types of analyses that will be used to address the mixed methods questions in studies. Data are analyzed to address these questions through distinct steps and through key decisions made by the researcher. The steps and decisions vary among the six different mixed methods research designs that we introduced in Chapter 3. Using these analysis procedures, the mixed methods researcher represents, interprets, and validates the data and results. Computer programs can help in the quantitative, qualitative, and mixed methods analyses.

Our discussion, however, begins with a review of the basics of quantitative and qualitative data analysis and interpretation. Thus, in this chapter, we

- review the basics of quantitative and qualitative data analysis and interpretation,
- summarize key principles in mixed methods data analysis and interpretation,
- discuss steps in mixed methods data analysis for all six mixed methods designs,
- highlight the decisions made in merging data for concurrent designs and for connecting data in sequential designs,
- relate validity issues to mixed methods designs, and
- identify ways computer software programs can be used in mixed methods data analysis.

● THE BASICS OF QUANTITATIVE AND QUALITATIVE DATA ANALYSIS AND INTERPRETATION

For both quantitative and qualitative data analysis, researchers go through a similar set of steps: preparing the data for analysis, exploring the data, analyzing the data, representing the analysis, interpreting the analysis, and validating the data and interpretations. These steps unfold in a linear fashion in quantitative research but are often implemented both simultaneously and iteratively in qualitative research. As shown in Table 7.1, the procedures associated with each step also differ for quantitative and qualitative research.

These steps may be familiar, so the presentation here will review and highlight essential aspects of data analysis (see Creswell, 2008b, for a more detailed presentation).

Preparing the Data for Analysis

In quantitative research, the investigator begins by converting the raw data into a form useful for data analysis, which means scoring the data by assigning numeric values to each response, cleaning data entry errors from the database, and creating special variables that will be needed, such as recoding items on instruments with inverted scores or computing new variables that comprise multiple items that form scales. Recoding and computing are completed with statistical computer programs, such as those made by Statistical Program for the Social Sciences (SPSS) (http://www.spss.com) and the Statistical Analysis System (SAS) (http://www.sas.com). A codebook that lists

Table 7.1	Recommended Quantitative and Qualitative Data Analysis Procedures for Designing Mixed Methods Studies	
Rigorous Quantitative Data Analysis Procedures	**General Procedures in Data Analysis**	**Persuasive Qualitative Data Analysis Procedures**
• Code data by assigning numeric values. • Prepare the data for analysis with a computer program. • Clean the database. • Recode or compute new variables for computer analysis. • Establish codebook.	Preparing the data for analysis	• Organize documents and visual data. • Transcribe text. • Prepare the data for analysis with a computer program.
• Visually inspect data. • Conduct descriptive analyses. • Check for trends and distributions.	Exploring the data	• Read through the data. • Write memos. • Develop qualitative codebook.
• Choose an appropriate statistical test. • Analyze the data to answer the research questions or test hypotheses. • Report inferential tests, effect sizes, and confidence intervals. • Use quantitative statistical software programs.	Analyzing the data	• Code the data. • Assign labels to codes. • Group codes into themes (or categories). • Interrelate themes (or categories) or abstract to smaller set of themes. • Use qualitative data analysis software programs.
• Represent results in statements of results. • Provide results in tables and figures.	Representing the data analysis	• Represent findings in discussions of themes or categories. • Present visual models, figures, and/or tables.
• Explain how the results address the research questions or hypotheses. • Compare the results with past literature, theories, or prior explanations.	Interpreting the results	• Assess how the research questions were answered. • Compare the findings with the literature. • Reflect on the personal meaning of the findings. • State new questions based on the findings.

(Continued)

Table 7.1   (Continued)

Rigorous Quantitative Data Analysis Procedures	General Procedures in Data Analysis	Persuasive Qualitative Data Analysis Procedures
• Use external standards. • Validate and check the reliability of scores from past instrument use. • Establish validity and reliability of current data. • Assess the internal and external validity of results.	Validating the data and results	• Use researcher, participant, and reviewer standards. • Use validation strategies, such as member checking, triangulation, disconfirming evidence, and external reviewers. • Check for the accuracy of the account. • Employ limited procedures for checking reliability.

the variables, their definitions, and the numbers associated with the response options for each also needs to be developed.

For qualitative data analysis, preparing the data means organizing the document or visual data for review or transcribing text from interviews and observations into word processing files for analysis. During the transcription process, the researcher checks the transcription for accuracy and then enters it into a qualitative data analysis software program, such as MAXQDA (http://www.maxqda.com), Atlas.ti (http://www.atlasti.com/), NVivo (http://www.qsrinternational.com), or HyperRESEARCH (http://www.researchware.com).

Exploring the Data

Exploring the data means examining the data with an eye to developing broad trends and the shape of the distribution or reading through the data, making memos, and developing a preliminary understanding of the database.

Exploring the data in quantitative data analysis involves visually inspecting the data and conducting a descriptive analysis (the mean, standard deviation [SD], and variance of responses to each item on instruments or checklists) to determine the general trends in the data. Researchers explore the data to see the distribution of the data and determine whether it is normally or nonnormally distributed so that proper statistics can be chosen for analysis. The quality of the scores from the data collection instruments is also examined using procedures to assess their reliability and validity. Descriptive statistics are generated for all major variables in the study—especially the main ones, such as the independent and dependent variables.

Exploring the data in qualitative data analysis involves reading through all of the data to develop a general understanding of the database. It means recording initial thoughts by writing short memos in the margins of transcripts or field notes. In this general review of the data, all forms of data are reviewed, such as observational field notes, journals, minutes from meetings, pictures, and transcripts of interviews. Making these memos becomes an important first step in forming broader categories of information, such as codes or themes. The memos are typically short phrases or ideas written in the margins of transcripts or field notes. At this time as well, a qualitative codebook can be developed. The codebook is a statement of the codes for a database. It is generated during a project and may rely on codes from past literature, as well as codes that emerge during an analysis. The process of generating this codebook helps organize the data, and it facilitates agreement (if several individuals code the data) on the contents of the transcripts as new codes are added and other codes removed during the coding process. Not all qualitative researchers use such a systematic procedure, but this process helps to organize large databases.

Analyzing the Data

Analyzing the data consists of examining the database to address the research questions or hypotheses. In both quantitative and qualitative analysis, we see multiple levels of analysis. In **quantitative data analysis**, the researcher analyzes the data based on the type of questions or hypotheses and uses the appropriate statistical test to address the questions or hypotheses. The choice of a statistical test is based on the type of questions being asked (e.g., a description of trends, a comparison of groups, or the relationship among variables), the number of independent and dependent variables, the types of scales used to measure those variables, and whether the variable scores are normally or nonnormally distributed. Information in research methods texts discusses each of these considerations (e.g., Creswell, 2008b). Researchers should also look for evidence of practical results, reported as effect sizes and confidence intervals. The quantitative data analysis proceeds from descriptive analysis to inferential analysis, and multiple steps in the inferential analysis build a greater refined analysis (e.g., from interaction effects to main effects to post hoc group comparisons).

Qualitative data analysis involves coding the data, dividing the text into small units (phrases, sentences, or paragraphs), assigning a label to each unit, and then grouping the codes into themes. The coding label can come from the exact words of the participants (i.e., in vivo coding), phrases composed by the researcher, or concepts used in the social or human sciences. If the researcher codes directly on the printed transcript, the transcript pages need to be typed with extra-large margins so that codes and memos can be placed in the margins. In this hand-coding process, researchers assign code words to text segments in one margin (e.g., the left side) and record broader themes in the other margin (e.g., the right side).

A more practical approach today is to use one of the many qualitative data analysis software programs (see Creswell & Maietta, 2002). These programs all contain some combination of the following features. **Qualitative computer software programs** can store text documents for analysis; enable the researcher to block and label text segments with codes so that they can be easily retrieved; organize codes into a visual, making it possible to diagram and see the relationship among them; and search for segments of text that contain multiple codes. The programs vary in how and the extent to which they carry out these functions.

The core feature of qualitative data analysis is the coding process. Coding is the process of grouping evidence and labeling ideas so that they reflect increasingly broader perspectives. Evidence from a database is grouped into codes, and codes are grouped into broader themes. Themes then can be grouped into even larger dimensions or perspectives, related, or compared. A typical example of relating themes can be seen in grounded theory, in which researchers form themes or codes (called categories) and then relate them in a theoretical model. Another example can be seen in narrative research, in which a chronology of an individual's life is composed using a sequence of codes, or themes, from the data. In this process, the themes, interrelated themes, or larger perspectives are the findings, or results, that provide answers to the qualitative research questions.

Representing the Data Analysis

The next step in the analysis process is to represent the results of the analysis in summary form in statements, tables, or figures. These summaries may be statements summarizing the results. In quantitative research, this involves representing the findings in statements summarizing the statistical results: "The scores varied for the four groups in the experiment. The analysis indicated a statistically significant difference ($p < .05$) among the groups, $F(4,10) = 9.98$, $p = .023$, effect size = .93 SD."

Tables in quantitative research can report results related to descriptive questions or inferential questions. If hypotheses are tested, tables report whether the results of the test were statistically significant (as well as the effect size and confidence intervals). Researchers usually present only one statistical test in each table. Tables need to be well organized, with a clear, detailed title and with the rows and columns labeled. There is standard information that should be reported for each type of statistical procedure, and various statistics books provide sample tables as models.

Researchers use figures to present quantitative results in a visual form, such as in bar charts, scatterplots, line graphs, or charts. These visual forms depict the trends and distributions of the data. The information needs to augment rather than duplicate information provided in the text, be easy to read and understand, and omit visually distracting details. Some statistical programs permit figures to be copied directly into word processing documents.

In qualitative research, representing the results may involve a discussion of the evidence for the themes or categories; the presentation of figures that depict the physical setting of the study; or diagrams presenting frameworks, models, or theories. When discussing the evidence for a theme, or category, the basic idea is to build a discussion that convinces the reader that the theme, or category, emerges from the data. Writing strategies for providing this evidence include conveying subthemes, or subcategories; citing specific quotes; using different sources of data to cite multiple items of evidence; and providing multiple perspectives from individuals in a study to show the divergent views (see Creswell, 2008b, for specific examples of these strategies). Apart from these discussions, researchers may represent their findings through visuals, such as figures, maps, or tables that present the different themes. The interrelated themes may comprise a model (as in grounded theory), a chronology (as in narrative research), or comparison tables (as in ethnography). A map may show the physical layout of the setting in which the research took place.

Interpreting the Results

After presenting the results or findings, the researcher next makes an interpretation of the meaning of the results. This often comes in a discussion section of a report. Basically, an **interpretation of results** involves stepping back from the detailed results and advancing their larger meaning in view of the research problems, questions in a study, the existing literature, and perhaps personal experiences. For quantitative research, this means comparing the results with the initial research questions asked to determine how the question or hypotheses were answered in the study. It also means comparing the results with prior predictions or explanations drawn from past research studies or theories, which provide explanations for what the researcher has found.

In qualitative research, the interpretation provides similar explanations about the results but with a few differences. The qualitative researcher needs to address how the research questions were answered by the qualitative findings. Also, comparisons can be made of the findings with past research studies in the literature. But, in addition to these approaches, qualitative researchers also may bring in their personal experiences and draw personal assessments of the meanings of the findings. This last feature sets qualitative research apart from quantitative approaches, and it reflects the role of the qualitative researcher, who believes that research (and its interpretations) can never be separated from the researcher's personal views and characterizations.

Validating the Data and Results

Another component of all good research is to utilize procedures to ensure the validity of the data, results, and their interpretation. Validity differs in quantitative and qualitative research, but in both approaches, it serves the purpose of checking on the quality of the data, the results, and the interpretation.

In quantitative research, the researcher is concerned about issues of validity at two levels: the quality of the scores from the instruments used and the quality of the conclusions that can be drawn from the results of the quantitative analyses. **Quantitative validity** means that the scores received from participants are meaningful indicators of the construct being measured. The standards are drawn from a source external to the researcher and the participants: statistical procedures or external experts. Researchers look for evidence of content validity (how judges assess whether the items or questions are representative of possible items), criterion-related validity (whether the scores relate to some external standard, such as scores on a similar instrument), or construct validity (whether they measure what they intend to measure). The *Standards for Educational and Psychological Testing* (American Educational Research Association, 1999) uses terminology based on the type of evidence collected (test content, response processes, internal structure, other variables, and consequences testing) and focuses on the interpretation of test scores in relation to the proposed use of the test. To assess validity for a study, investigators establish the validity of their instruments through content validity and of their scores through criterion-related and construct validity procedures. Quantitative researchers also consider the validity of the conclusions that they are able to draw from their results. This means that quantitative researchers need to design their studies to reduce the threats to internal validity and external validity. Internal validity is the extent to which the investigator can conclude that there is a cause and effect relationship among variables. The investigator can only draw correct cause and effect inferences if threats, such as participant attrition, selection bias, and maturation of participants, are accounted for in the design (see Creswell, 2008b). Internal validity is of highest concern in experimental studies. External validity is the extent to which the investigator can conclude that the results apply to a larger population, which is usually of highest concern in survey designs. This means that correct inferences can only be drawn to other persons, settings, and past and future situations if the investigator has used procedures such as selecting a representative sample.

Quantitative researchers also consider issues of reliability. **Quantitative reliability** means that scores received from participants are consistent and stable over time. Reliability of scores from past uses, assessed in terms of reliability coefficients, and instrument test–retest results need to be addressed. In a study researchers need to check for the reliability of scores (through statistical procedures of internal consistency) and any test–retest comparisons while exploring the data. The reliability of scores needs to be established before assessments of their validity can be addressed.

In qualitative research, there is more of a focus on validity than reliability to determine whether the account provided by the researcher and the participants is accurate, can be trusted, and is credible (Lincoln & Guba, 1985). Qualitative validity comes from the analysis procedures of the researcher, based on information gleaned while visiting with participants, and from external reviewers. Reliability plays a minor role in qualitative research and relates primarily to the reliability of multiple coders on a team to reach agreement on codes for passages in text.

Qualitative validation is important to establish, but there are so many commentaries and types of qualitative validity that it is difficult to know which approach to adopt. We will work from standards we have set in prior publications (Creswell, 2007; Creswell & Miller, 2000). Overall, checking for qualitative validity means assessing whether the information obtained through the qualitative data collection is accurate. There are strategies available to determine this validity, and qualitative researchers typically use more than one procedure. Member-checking is a frequently used approach, in which the investigator takes summaries of the findings (e.g., case studies, major themes, theoretical model) back to key participants in the study and asks them whether the findings are an accurate reflection of their experiences. Another validity approach is triangulation of the data drawn from several sources (e.g., transcripts and pictures) or from several individuals. This procedure is a common data analysis practice: The inquirer builds evidence for a code or theme from several sources or from several individuals. Another approach consists of reporting disconfirming evidence. Disconfirming evidence is information that presents a perspective that is contrary to the one indicated by the established evidence. The report of disconfirming evidence in fact confirms the accuracy of the data analysis, because in real life, we expect the evidence for themes to diverge and include more than just positive information. A final approach is to ask others to examine the data. These others may be in the form of peers (e.g., graduate students or faculty), who are familiar with qualitative research as well as the content area of the specific research, or external auditors, individuals not affiliated with the project who review the database and the qualitative results using their own criteria (Creswell, 2007).

Reliability has limited meaning in qualitative research, but it is popular in qualitative research when there is interest in comparing coding among several coders. Called **intercoder agreement in qualitative research**, the basic procedure involves having several individuals code a transcript and then compare their work to determine whether they arrived at the same codes and themes or different ones (Miles & Huberman, 1994). Typically, coders look for text passages that they have all coded and, using a predetermined coding scheme, identify whether they assigned the same or different codes to the text passage. Rates are developed for the percentage of codes that are similar, and reliability statistics (kappas) can be computed for systematic data comparisons.

● DATA ANALYSIS AND INTERPRETATION WITHIN MIXED METHODS DESIGNS

Mixed methods data analysis consists of analytic techniques applied to both the quantitative and the qualitative data as well as to the mixing of the two forms of data concurrently and sequentially in a single project or a multiphase project (see a similar definition in Onwuegbuzie & Teddlie, 2003). Data analysis can occur at a single point in the process of mixed methods research or at multiple points. It also involves certain steps undertaken by the researcher and key decisions made at different steps. Once analyses are complete, **mixed methods interpretation** involves looking across the quantitative results and the qualitative findings and making an assessment of how the information addresses the mixed methods question in a study. Teddlie and Tashakkori (2009) call this interpretation drawing "inferences" and "meta-inferences" (p. 300). Inferences in **mixed methods research** are conclusions or interpretations drawn from the separate quantitative and qualitative strands of a study as well as across the quantitative and qualitative strands, called "meta-inferences." Teddlie and Tashakkori (2009) see mixed methods as a vehicle for improving the quality of inferences that are drawn from both the quantitative and qualitative methods. Before discussing details of data analysis within different mixed methods approaches, it is useful at the outset, we believe, to review how our under-standing of mixed methods data analysis has evolved.

Insight into mixed methods data analysis has emerged slowly over the years. The first discussions about data analysis in mixed methods identified several of the general procedures that could be used. These procedures were not related to specific designs but viewed as generic approaches to analyzing data. A case in point is the discussion of four analytic strategies by Caracelli and Greene in 1993. Their four strategies were

- data transformation—the conversion or transformation of one data type into the other so that both can be analyzed together;
- typology development—the analysis of one data type so that it yields a typology (or set of categories) that is then used as a framework applied in analyzing the other data type;
- extreme case analysis—the identification of "extreme cases" from the analysis of one data type, which are examined with data of the other type to test and refine the initial explanation for the extreme cases; and
- data consolidation or merging—the joint review of both data types to cre-ate new or consolidated variables or data sets used in further analyses.

By 2003, a more substantive conversation was taking place around data analysis that was linked more to the process of conducting research. Onwuegbuzie and Teddlie (2003) discussed a model for mixed methods data analysis around seven stages in the data analysis process:

1. Data reduction—reducing data collected through statistical analysis of quantitative data or writing summaries of qualitative data

2. Data display—reducing the quantitative data to, for example, tables and the qualitative data to, for example, charts and rubrics

3. Data transformation—transforming qualitative data into quantitative data (i.e., quantitizing qualitative data) or vice versa (i.e., qualitizing quantitative data)

4. Data correlation—correlating the quantitative data with quantitized qualitative data

   5. Data consolidation—combining both data types to create new or con-
      solidated variables or data sets

   6. Data comparison—comparing data from different sources

   7. Data integration—integrating all data into a coherent whole

   The first two steps in this process of analysis follow logical steps in data
analysis, but the last five steps (from transformation to integration) in this list
of procedures appear to be alternative options for analysis rather than steps
that follow one after the other. Further, these steps do not specifically relate
to mixed methods designs.
   A more recent editorial begins to bring the discussion about mixed
methods data analysis into research designs. Bazeley (2009) discussed
emerging ways to consider mixed methods data analysis: through a substantive common purpose for a study (e.g., intensive case analysis, extreme or
negative cases, or inherently mixed analysis, such as social network analysis);
through employment of the results in one analysis in approaching the analysis of another form of data (e.g., typology development); through synthesis
of data from several sources for joint interpretation (e.g., comparing theme
data with categorical or scaled variables using matrixes); through conversion
of one form of data into the other (e.g., data transformation); through the
creation of blended variables; and through multiple, sequenced phases of
iterative analyses.
   Bazeley's (2009) list, we believe, foreshadows many of the mixed methods data analysis procedures that are central to mixed methods designs.
Looking across the six major designs, we find that the mixed methods data
analysis involves considering the steps in analysis typically undertaken for
each of the designs as well as key decisions that the researcher makes when
implementing these steps. **Steps in mixed methods data analysis** refers to
the procedures taken in a logical order by the researcher when conducting
data analysis for a mixed methods design. **Decisions in mixed methods data
analysis** refers to those critical points in data analysis when the researcher
needs to decide what option to select for analysis. An overview of the steps
and key decisions is found in Table 7.2.

Steps and Key Decisions in
Data Analysis for Each Mixed Methods Design

As shown in Table 7.2, in the convergent design, after collecting both quantitative and qualitative data concurrently, the researcher analyzes the information

Table 7.2 Steps and Decisions in Mixed Methods Data Analysis by Design

Type of Mixed Methods Design	Type of Mixed Methods Data Analysis	Data Analysis Steps in the Design	Data Analysis Decisions
Convergent design	Merging data analysis to compare results	1. Collect the quantitative and qualitative data concurrently.	
		2. Independently analyze the quantitative data quantitatively and the qualitative data qualitatively using analytic approaches best suited to the quantitative and qualitative research questions.	
		3. Specify the dimensions by which to compare the results from the two databases.	Decide how the two data sets will be compared (e.g., dimensions, information).
		4. Specify what information will be compared across the dimensions.	
		5. Complete refined quantitative and/or qualitative analyses to produce the needed comparison information.	

(Continued)

Table 7.2 (Continued)

Type of Mixed Methods Design	Type of Mixed Methods Data Analysis	Data Analysis Steps in the Design	Data Analysis Decisions
		6. Represent the comparisons. 7. Interpret how the combined results answer the quantitative, qualitative, and mixed methods questions.	Decide how to represent or present the combined analysis. Decide if further analysis is needed.
Convergent design	Merging data analysis through data transformation (example of quantitizing qualitative data)	1. Collect the quantitative data and the qualitative data concurrently. 2. Independently analyze the quantitative data quantitatively and the qualitative data qualitatively using analytic approaches best suited to the quantitative and qualitative research questions.	

Type of Mixed Methods Design	Type of Mixed Methods Data Analysis	Data Analysis Steps in the Design	Data Analysis Decisions
		3. Define a quantitized variable based on the qualitative results, and develop a rubric for scoring the qualitative results.	Decide how to quantify the qualitative data (i.e., scoring rubric).
		4. Systematically score the qualitative results to determine the quantitized variable.	
		5. Analyze the quantitative data, including the quantitized variable, quantitatively using analytic approaches best suited to the mixed methods research question.	Decide on the statistics to use in relating the two data sets.
		6. Interpret how the merged results answer the qualitative, quantitative, and mixed methods questions.	
Explanatory design	Connected data analysis to explain results	1. Collect the quantitative data.	
		2. Analyze the quantitative data quantitatively using analytic approaches best suited to the quantitative research question.	

(Continued)

Table 7.2 (Continued)

Type of Mixed Methods Design	Type of Mixed Methods Data Analysis	Data Analysis Steps in the Design	Data Analysis Decisions
	Connected data analysis to explain results	3. Design the qualitative strand based on the quantitative results. 4. Collect the qualitative data. 5. Analyze the qualitative data qualitatively using analytic approaches best suited to the qualitative and mixed methods research questions. 6. Interpret how the connected results answer the quantitative, qualitative, and mixed methods questions.	Decide what participants to follow up with and what results need to be explained. Decide how the qualitative results explain the quantitative results.
Exploratory design	Connected data analysis to generalize findings	1. Collect the qualitative data. 2. Analyze the qualitative data qualitatively using analytic approaches best suited to the qualitative research question. 3. Design the quantitative strand based on the qualitative results.	Decide what data can be used in the quantitative follow-up.

Type of Mixed Methods Design	Type of Mixed Methods Data Analysis	Data Analysis Steps in the Design	Data Analysis Decisions
		4. Develop and pilot test the new instrument (or the new intervention treatment).	Decide how best to assess the psychometric quality of the instrument.
		5. Collect the quantitative data.	
		6. Analyze the quantitative data quantitatively using analytic approaches best suited to the quantitative and mixed methods research questions.	
		7. Interpret how the connected results answer the qualitative, quantitative, and mixed methods questions.	Decide how the quantitative results build or expand on the qualitative findings.
Embedded design	Merged or connected data analysis depending on whether the design is concurrent or sequential	1. Analyze the primary data set to answer the primary research questions.	Decide how to use the secondary data results.
		2. Analyze the secondary data (qualitative or quantitative) where it is embedded within the primary design by merging or connecting using the steps involved in the convergent, explanatory, or exploratory designs.	Decide when the secondary data should be incorporated into the primary data set.

(Continued)

Table 7.2 (Continued)

Type of Mixed Methods Design	Type of Mixed Methods Data Analysis	Data Analysis Steps in the Design	Data Analysis Decisions
		3. Interpret how the primary and secondary results answer the qualitative, quantitative, and mixed methods questions.	Decide how the secondary data support or augment the primary data.
Transformative design	Merged or connected data analysis depending on whether the design is concurrent or sequential	1. Analyze the quantitative and qualitative data by merging or connecting using the steps involved in the convergent, explanatory, or exploratory designs. 2. Interpret how the results answer the quantitative, qualitative, and mixed methods questions	Decide on the analyses that will best provide evidence for the transformative lens. Decide the data analysis decisions identified for the corresponding merging or connecting data analysis procedures outlined for the convergent, explanatory, or exploratory designs. Decide to what extent the results uncover inequities and call for change.
Multiphase design	Merged or connected data analysis for each phase or project in the multiphase design	1. Analyze the data for each project in the overall program. 2. Employ strategies for merged and connected analysis as the timing of the project dictates. 3. Interpret how the results answer the project's research questions and contribute to the overall objective.	Decide on the applicability of merged and connected data analysis or some combination for each phase in the project. Decide on how to best combine the data analyses from all projects in the study to address the common research objective. Decide to what extent the results advance the program objective.

separately and then merges the two databases. The analysis is conducted in order to merge the results by comparing the two data sets or to merge the data after the researcher transforms one of the data sets. Data analysis in this design occurs at three distinct points in one phase of the research: with each data set independently, when the comparison or transformation of the data occurs, and after the comparison or transformation is completed. Interim steps may occur between these points, such as identifying the dimensions on which the data will be compared, defining what variable will be transformed, and representing the comparisons in data displays or in discussions. In the end, the researcher compares the merged results with the research questions. Key data analysis decisions in this design relate to deciding how to compare the two data sets (e.g., dimensions, information), how to present the combined analyses, and what further analysis to conduct if the results diverge.

The data analysis procedures in the explanatory design involve first collecting quantitative data, analyzing the data, and using the results to inform the follow-up qualitative data collection. The data analysis occurs in three phases: the analysis of the initial quantitative data, an analysis of the follow-up qualitative data, and an analysis of the mixed methods question as to how the qualitative data help to explain the quantitative data. In this design, the data analysis of the initial quantitative phase connects into the data collection of the follow-up qualitative phase. At the interpretation stage in this design, the analysis is used to address the mixed methods question about whether and how the qualitative data help to explain the quantitative results. Key data analysis decisions relate to how to use the quantitative analysis to identify participants to determine what results will be explained qualitatively, and to decide how the qualitative results explain the quantitative results.

In an exploratory design, the researchers first collect qualitative data, analyze it, and then use the information to develop a follow up quantitative phase of data collection. The quantitative strand thus connects to the initial qualitative strand. Like the explanatory design, three analyses are conducted: after the initial qualitative data collection, after the follow-up quantitative data collection, and at the interpretation phase when the researcher connects the two databases to address how the follow up analysis helps to generalize or extend the initial qualitative exploratory findings. Key data analysis decisions in this design relate to the point of interface when the initial qualitative findings are used for the data collection in the follow-up quantitative phase. Other decisions need to be made about the psychometric quality of the instrument, how to analyze data from it, and how the quantitative results build or expand on the initial qualitative findings.

In an embedded design, the quantitative and qualitative data can be collected sequentially, concurrently, or both. In this design, one form of data is embedded within another form (e.g., qualitative data embedded within an experimental intervention trial or qualitative interviews embedded within a longitudinal correlational design). A variant on this approach is to embed both qualitative and quantitative data within a design or procedure (e.g., the procedure of using geographic information surveys that include both quantitative and qualitative information). Data analysis steps thus depend on when and how the embedded data is used in the study. Three major steps are used in data analysis: analysis of the primary data, analysis of the secondary data, and further mixed methods analysis to determine how and in what way the secondary data support or augment the primary data. Key data analysis decisions in this design involve how to use the secondary analysis and when it should be incorporated into the primary data design.

In the transformative design, the researcher frames the study within a transformative theoretical perspective to help address injustices or bring about change for an underrepresented or marginalized group. Data collection for the qualitative and quantitative strands of the study can proceed concurrently, sequentially, or both. Analysis steps may reflect concurrent data analysis procedures (e.g., as in the convergent design) or sequential data analysis procedures (e.g., as in the explanatory or exploratory designs). Data analysis decisions thus occur within each set of data, in merging or connecting the two sets of data, and in the interpretation phase. Similar combinations of analysis may be used in the multiphase design. This design combines both sequential and concurrent strands over a period of time in a program of study. Prime examples of this design are large-scale evaluation and health science projects. They might involve multiple levels of analysis and typically are conducted over several years. In these projects, both quantitative and qualitative databases build toward an overall objective. Data analysis decisions relate primarily to obtaining results to address the common research objective and advance the overall research program.

As suggested in Table 7.2, many detailed decisions are involved in analyzing the data for each of the six designs. As we look across the designs, two analytic procedures—merging when concurrent data are collected and connecting when sequential data are collected—are used by mixed methods researchers. These procedures involve a more detailed discussion of specific techniques of mixed methods analysis to appear in recent years.

Decisions for Merged Data Analysis in a Concurrent Approach

In four types of designs that may utilize concurrent procedures—the convergent design, the embedded design, the transformative design, and the multiphase design—the researcher employs mixed methods data analysis and interpretation approaches for merging quantitative and qualitative data. As mentioned in Chapter 5, the prototypical mixed methods question to be answered when merging data is as follows: To what extent do the quantitative and qualitative results converge (convergent design)? Examples of other mixed methods research questions that call for merged data analysis procedures include the following: Are the qualitative findings significantly related to the quantitative results (data transformation variant of the convergent design)? To what extent do the qualitative process findings enhance the understanding of the experimental outcomes (during-experiment variant of the embedded design)? In what ways do the qualitative themes and the quantitative results converge and diverge to uncover injustice and suggest change (transformative design)? These questions all call for merged data analysis procedures and techniques.

The mixed methods data analysis is conducted to answer the mixed methods research question as to whether the results from both analyses converge and how they converge. **Merged data analysis strategies** involve using analytic techniques for merging the results, assessing whether the results from the two databases are congruent or divergent, and, if they are divergent, then analyzing the data further to reconcile the divergent findings. Several options are available for implementing these strategies.

Strategies for comparing results. In data analysis, when comparisons are made after the initial quantitative and qualitative analyses, what options exist for comparing the results? Three options exist for **merged data analysis comparisons**, stated largely in the order in which they are popularly found today in mixed methods studies: side-by-side comparisons in a discussion or summary table, joint display comparisons in the results or interpretation, or data transformation in the results.

The first option for merging, a **side-by-side comparison for merged data analysis**, involves presenting the quantitative results and the qualitative findings together in a discussion or in a summary table so that they can be easily compared. The presentation then becomes the means for conveying the merged results. For example, when it is presented in a discussion within a study, the discussion becomes the vehicle for merging the results. One popular approach is to first present the quantitative results followed by qualitative results in the form of quotes (or vice versa) in a results or discussion section. A comment then follows specifying how the qualitative quotes either confirm or disconfirm the quantitative results. An example can be seen in the results section in a mixed methods social work study addressing the success of coalitions (Mizrahi & Rosenthal, 2001) as shown in Figure 7.1.

Figure 7.1	Excerpt From a Results Section Showing a Side-By-Side Comparison of Quantitative and Qualitative Data Results

Present
QUAN result

> Overall, certain elements were consistently considered to have a great or considerable impact on coalition success, regardless of how success was defined. "Commitment to goal/cause/issue" (95.0 percent) and "competent leadership" (92.5 percent) were the top two elements regardless of definitions of success, followed by "commitment to coalition unity/work" (87.5 percent), "equitable decision-making structure/process" (80.0 percent), and "mutual respect/tolerance" (77.5 percent). Additional important elements of success were having "a broad-based constituency" (75.0 percent), "achieving interim victories" (72.5 percent), "members' continued contributing resources" (67.5 percent), and "shared responsibility and ownership" (65.0 percent). Note that the tangible elements relating to resources (staffing and funding) were given much less import overall. Only three external factors were deemed important by most coalition leaders: "the right timing" and selecting a "critical issue" (at 87.5 percent each), and "appropriate target" (71.5 percent). Whereas coalition leaders cannot control these factors as much, it is clear that these factor into the decision-making processes with respect to the framing of goals and strategies:

Present
corresponding QUAL
result and relate
to QUAN result

> The resources amassed by our coalition are valued and respected. They [the members] all possessed tremendous knowledge about their subject areas and about the political process. Being recognized as experts gives the coalition leverage and clout with the target.

SOURCE: Mizrahi and Rosenthal (2001, p. 70).

In this example, the authors present a passage in which they use a qualitative quote (at the bottom of Figure 7.1) to support the quantitative descriptive findings (presented at the top of the figure). This comparison could easily be reversed in another study with the quantitative used to support the qualitative quotes (see McAuley, McCurry, Knapp, Beecham, & Sleed, 2006, for an example). Also, in the Figure 7.1 example, there is no attempt by the authors to directly merge, or integrate, the data; instead, the discussion highlights a comparison of the results from the two data sets. In effect, the merging of the data occurs through the results and discussion sections.

Another form of a side-by-side comparison can be made using a summary table that merges the quantitative and qualitative findings, such as found in a mixed methods study of preschool inclusion (Li, Marquart, & Zercher, 2000). As shown in Figure 7.2, the authors compared their interview data with their survey data on four major themes (similar information found

Figure 7.2 Comparison of Information From Interviews and Survey Data

Comparison of Information from Interview and Survey Data: Examples of Four of the Eight Themes

Theme	Face-to-face Interviews (QUAL results)	Telephone Survey (QUAN results)
1. How and why child was placed in program	Two aspects of decision: (1) Community-based "inclusive" option (2) Specific child care center Factors affecting choice: • Visited and liked classroom & teacher • Convenience of location • Flexibility in hours • Good reputation of center • Concern if center would not accept child because of behavior	Parents' most important reasons for using program: • Offers special education services or therapies • Provides opportunities for child to learn • Provides opportunities to play with other children
2. Program's appropriateness for child	In successful placement, there is a "match or fit" between child's and family's needs & program. Factors affecting match or fit: • Acceptance by staff & children • Likes activities and routines for child • Child likes program • Sees benefits or specific improvements	• 90% said very important for child to be in inclusive program • 80% indicated child usually or always receives special services needed • 86% were satisfied with way in which child's educational goals were made
3. Helpful and unhelpful players	Characteristics of helpful players: • Consistent presence over time & settings • Personal investment in child • Provides different types of support • Dependable source of information about child Characteristics of unhelpful players: • Minimize or disregard family concerns • Inadequate communication	The most helpful supports were: • Other family members at home • Child's teachers • Other professionals in community and at child's program
4. Child's participation in family and community activities	Factors that affect participation: • Parent's safety concerns about child • Parent's perception of what is expected of child's behavior • Lack of other young children in immediate neighborhood • Family's own style, schedule, and how it participates in the community • An extended family system was so strong a part of family's culture that family did not need or choose to participate much in the community • Young age of children	Limitations on participation: • Child's language skills • Family's schedule and time constraints • Attitudes of others towards child's disability • Child's behavior • Lack of other children to play with in neighborhood

Major Topics

SOURCE: Li et al., 2000, Table 2, pp. 124–125. Reprinted with permission of SAGE Publications, Inc.

in both sources of data). They presented this information in a table so that a reader could see how both sources of data—side-by-side—provided evidence for each topic.

The second merging strategy option is the use of a joint display. A **joint display** is a figure or table in which the researcher arrays both quantitative and qualitative data so that the two sources of data can be directly compared. In effect, the display merges the two forms of data. Researchers using a joint display need to decide on the dimensions to be considered and the specific information to be compared across the dimensions. Several options exist for creating this display, and researchers are continually inventing new options. The most straightforward form is to create a **category/theme display in merged data analysis** that arrays the qualitative themes derived from the qualitative analysis with quantitative categorical or continuous data from items or variables from the quantitative statistical results. An example of such a display is shown in Figure 7.3. This figure portrays data generated in a mixed methods study that explored the building of positive relationships among 16 undergraduates in teacher education in a mentoring program (McEntarffer, 2003).

In this study, pairs of junior or senior mentors and freshman mentees completed the quantitative StrengthsFinder instrument, developed by the Gallup Organization and consisting of 180 paired items about individual strengths or talents (Clifton & Anderson, 2002). The analysis of this instrument led to the identification of the five top strengths of each individual. For all of the mentors and mentees, the top three strengths were input, relator, and achiever, which are shown along the vertical dimension of the chart in Figure 7.3. These strengths represent different levels of a categorical quantitative variable in the display. For the qualitative data on the horizontal dimension of the chart in Figure 7.3, the researcher collected documents, conducted interviews, and took field notes observing the interactions between the mentors and mentees. An analysis of this qualitative data led to three themes: relationship-building strategies, strengths awareness, and relationship outcomes. As shown in Figure 7.3, the investigator developed a display to portray the quantitative (strengths) data and the qualitative (theme) data. The data were merged within the cells with information indicating the number of text units associated with each theme for the different individuals in each category as well as subthemes that convey what the individuals said. A similar type of display could also be used where the researcher develops a typology. A **typology and statistics merged data analysis display** combines in merged analysis qualitative theme data and quantitative data based on a typology or classification. An example of this display can be found in Table A.3 of Appendix A in the study by Wittink, Barg, and Gallo (2006).

| Figure 7.3 | Example of Joint Display Arraying Categories by Themes |

Dimension: QUAL themes

Top Three Strengths from the Gallup StrengthsFinder	Qualitative Themes		
	Relationship-Building Strategies	Strengths Awareness	Relationship Outcomes
Input (*n* = 8)	24 Chilling out. Talked a little bit.	15 Talked about results. Talked about the awkwardness of strengths terminology.	55 We saw an increase in comfort. Conversations got noticeably easier.
Relator (*n* = 6)	32 "How is your week going" conversations. Hot-button conversations.	13 Talked about strengths in a casual manner. Discussed being positive, in a good mood.	13 A special relationship developed between us. We went through an early period of discomfort. Early conversations were superficial.
Achiever (*n* = 5)	22 Talked about our lives. Trusted me with personal information.	3 It was cool to hear about other people's strengths. I notice my strengths in everyday life. Watching a movie helped us to reflect on strengths.	3 The early project jitters are going away. We're not hanging out because we have to. We learned new things about ourselves.

Dimension: QUAN categories

SOURCE: Adapted from the database in McEntarffer (2003). Used by permission.

Another joint display for merging data works differently than arraying the categorical data by themes. To highlight **convergent and divergent findings in a merged data analysis display**, a researcher analyzes both the quantitative and qualitative data, compares the results, and creates a table displaying congruent or incongruent (or discrepant) findings along the horizontal dimension. Along the vertical dimension, the researcher may indicate different topics and/or participant types as indicated by their numeric scores. Within the cells of this display could be quotes, numbers, or both. A joint display provided by Lee and Greene (2007) from their mixed methods study of the predictive validity of students' English as a second language placement test scores illustrates this type of display. These researchers collected and analyzed quantitative and qualitative indicators of student performance, including student grade point average (GPA) and test scores along with questionnaires and interviews with students and faculty. As shown in Figure 7.4, different relationships between placement test scores and GPAs (e.g., low score and low GPA; low score and high GPA) were arrayed with quotes (and associated demographic information) that illustrate selected quotes that were congruent and discrepant with the hypothesis that high test scores would be associated with high academic performance and low scores with low performance.

A final example of a **case-oriented merged analysis display** positions cases on a quantitative scale along with qualitative text data about the individual cases. This example of a display indicates how researchers can be creative as they develop matrixes to fit their needs. An example of this display is found in a study of women's health behaviors among mother–daughter dyads from six ethnic groups (Israeli, European, North African, former Soviet Union, American/Canadian, and Ethiopian) (Mendlinger & Cwikel, 2008). As shown in Figure 7.5, four individual cases are arrayed on a scale for health assessment ranging from poor to excellent. Quotes are provided for daughters and mothers to indicate what they said about their health, which formed the basis for the quantitative ratings. Country categories are also assigned to the daughters and mothers. In this way, the resulting figure illustrates the combination of numeric rating scores as well as textual qualitative data in a single display.

The third strategy option for the concurrent use of data is through **data transformation merged analysis**. In this form of merging, the researcher transforms one type of data into the other type so that both databases can be easily compared and further analyzed. Data transformation is a topic that has been addressed in the mixed methods literature (Caracelli & Greene, 1993; Onwuegbuzie & Teddlie, 2003; Sandelowski, Voils, & Knafl, 2009). Hesse-Biber and Leavy (2006), for example, asked researchers, Do you want to use quantitative data

Figure 7.4 Excerpt From a Joint Display for Presenting Congruent and Discrepant Findings

Dimension: Congruent and Discrepant Examples

Linking Computerized Enhanced ESL Placement Test (CEEPT) Scores and Student and Faculty Member Quotes Regarding the Role of English Proficiency in Course Performance

CEEPT Score	GPA	Faculty Members' Responses		Students' Responses	
		Congruent	Discrepant	Congruent	Discrepant
2	Above 3.18	"He has problems with listening, speaking, and reading. In my course, he got an A–; and the lowest grade I assigned was B+ . *He is the second to last.*" (ID 0624, I) (3.80) (technology)	"*Her submitted assignments are well considered and prepared.*" (ID 2005, Q) (3.75) (humanities) "*He is the second highest among 12 students.* He asked very good questions based on lectures. *His notes were quite good.*" (ID 0620, I) (4.00) (science)	"*Lack of knowledge about idiomatic expressions prevents me from understanding questions on the homework assignment.*" (ID 2037, I) (3.22) (business) "Because of my poor listening, I am struggling with catching up with my content courses." (ID 0605, I) (3.5) (humanities) "*I understand only 60–70% of the lectures. It has made my scores less than my expectation.*" (ID 0620, Q) (4.00) (science)	"*It is easy to understand the lectures and participate in class discussions.* The instructor speaks slowly." (ID 0624, I) (3.80) (technology)
3	Below 3.18	No data available	No data available	No data available	No data available
	Above 3.25	"The student is *quiet in class, but did well in* her oral presentation. Her written work on homework is comparable to other international students (overall, very good)." (ID 2036, Q) (3.57) (humanities)	"She is in the *top 10%.* She got a grade of A. I do not think she has problems because of her language." (ID 2032, I) (4.00) (science)	"I still have some problem in speaking. This difficulty doesn't affect my ability to do well in all the courses I take." (ID 0603, Q) (3.39) (technology) "*Listening is a problem.* Lack of cultural knowledge interferes with understanding the concept." (ID 0610, I) (3.53) (technology)	No data available

Dimension: QUAN relationships between test scores and GPA

SOURCE: Adapted from Lee and Greene (2007, Table 5, p. 383). Used with permission of SAGE Publications, Inc.

Figure 7.5 Example of a Joint Display Using the Case Approach to Position Individual Cases on a Scale and Provide Text

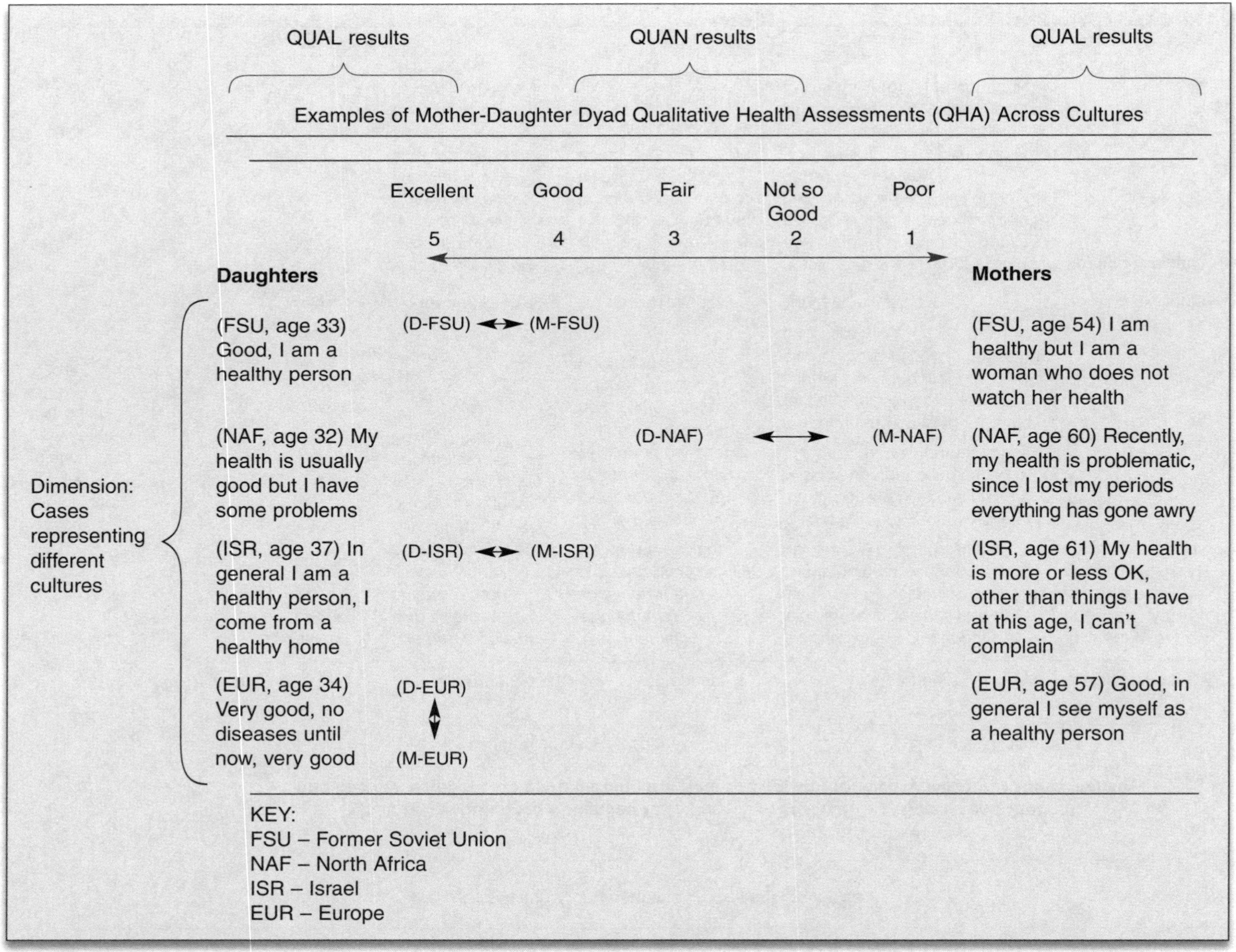

QUAL results   QUAN results   QUAL results

Examples of Mother-Daughter Dyad Qualitative Health Assessments (QHA) Across Cultures

	Excellent	Good	Fair	Not so Good	Poor
	5	4	3	2	1

Daughters

(FSU, age 33) Good, I am a healthy person — (D-FSU) ⟷ (M-FSU)

(NAF, age 32) My health is usually good but I have some problems — (D-NAF) ⟷ (M-NAF)

(ISR, age 37) In general I am a healthy person, I come from a healthy home — (D-ISR) ⟷ (M-ISR)

(EUR, age 34) Very good, no diseases until now, very good — (D-EUR) ⟷ (M-EUR)

Dimension: Cases representing different cultures

Mothers

(FSU, age 54) I am healthy but I am a woman who does not watch her health

(NAF, age 60) Recently, my health is problematic, since I lost my periods everything has gone awry

(ISR, age 61) My health is more or less OK, other than things I have at this age, I can't complain

(EUR, age 57) Good, in general I see myself as a healthy person

KEY:
FSU – Former Soviet Union
NAF – North Africa
ISR – Israel
EUR – Europe

SOURCE: Mendlinger and Cwikel (2008, Figure 3, p. 288). Reprinted with permission of SAGE Publications, Inc.

to inform your qualitative data, or do you want to use qualitative data to inform your quantitative data? Unquestionably, it is easier to transform qualitative data into numeric counts (quantitative data) than vice versa.

Transforming qualitative data into quantitative data involves reducing themes or codes to numeric information, such as dichotomous categories. Some of the most specific information about the procedures are based on writings from Onwuegbuzie and Teddlie (2003). A key issue in this procedure is deciding what aspect of qualitative data to quantify and how to quantify it. Perhaps the simplest approach is to define a new dichotomous variable that indicates whether a theme or code is present (scored a one) or not present (scored a zero) for each participant. Other approaches can involve counting, such as the number of times a theme or code appears in the data. Onwuegbuzie and Teddlie (2003) provided detailed approaches to counting, such as counting the

- frequency of a theme within a sample by converting it to percentages,
- number of units for each theme by converting it to a percentage,
- percentage of total themes associated with a phenomenon,
- percentage of people selecting or endorsing multiple themes,
- observations, interviews, text—the time sequence and length of the unit,
- number of times behavior was observed per hour,
- number of times a significant statement appears per page, and
- amount of time that elapses before a unit of analysis is observed.

In an article by Daley and Onwuegbuzie (2010) the process of data transformation is discussed in a study of violence attribution of male juvenile delinquents. The authors sought to correlate closed-ended items with open-ended items using a convergent design. From the open-ended responses, seven themes emerged. The researchers binarized each theme by assigning a score of one or zero for each individual in the sample, depending on whether the theme was represented by that individual. Then they developed an interrespondent display for comparing for each individual and correlated scores from the open-ended themes with scores on the closed-ended items. In another study, Sandelowski (2003) discussed the "quantitizing" (p. 327) of qualitative data in her study of transition to parenthood. She and her colleagues transformed interview data into a display that compared the number of couples having and not having amniocentesis with the number of physicians encouraging or not encouraging them to have the procedure. They then used a statistical test to report nonsignificant findings. Another approach to quantitizing qualitative data includes assigning scores based on a theoretical model (for example, see Idler, Hudson, & Leventhal, 1999, who developed a theoretical model to develop a rubric for scoring qualitative responses on a six-point health assessment scale).

Far fewer examples exist of the transformation of quantitative data into qualitative data. Punch (1998), however, provided one example for this procedure. He cited a case in which quantitative data could be loaded into factors in a factor analysis and the factors viewed as aggregated units similar to themes, so that the factors (derived quantitatively) could be directly compared with the themes developed qualitatively. As another example, Teno, Stevens, Spernak, and Lynn (1998) reported transforming quantitative data (i.e., medical records, closed-ended interviews, survival predictions) into qualitative narrative summaries as part of their study on the use of written advance directives.

Strategies for interpreting merged results and reconciling differences. Researchers must interpret their combined results (whether merged in a side-by-side presentation, in a joint display, or through data transformation) to assess how the analysis answers the mixed methods research question. These interpretations therefore relate specifically to the design being used. For example, a researcher using data transformation might interpret whether a significant relationship is found among the transformed data with other data and what meaning can be drawn from the relationship and what limitations must be considered. As another example, a researcher using an embedded during-intervention design might synthesize the findings about the process with the experimental outcomes to enhance the understanding of the experimental conditions. Most commonly, however, researchers using these approaches interpret the extent to which the two databases converge, whether differences or similarities are found, and what conclusions can be drawn from the differences and similarities.

In merging the two data sets for the purpose of convergence, what differences should the mixed methods researcher look for when they make an interpretation of findings? And, if differences occur, how will the researcher address the inconsistencies? In terms of the first question, what the mixed methods researcher looks for in comparing the two data sets is not fixed and rigid. As discussed earlier, Lee and Greene (2007) looked for congruent and discrepant evidence between the databases. Other ways of comparing the two data sets is to look for consistencies or inconsistencies, conflicts (which can be made sense of), and contradictions (that present an either/or situation) (see Slonim-Nevo & Nevo, 2009). Slonim-Nevo and Nevo (2009) cited the illustration of an assessment of family functioning using a quantitative standardized scale and in-depth interviews with members of the family. In this situation, the in-depth interviews told " . . . a different story. . . ." (p. 112), leading the authors to ask, "Which method, then, is right—the quantitative or the qualitative?" (p. 112).

The authors then went on to discuss potential discrepancies between their quantitative and qualitative results in a study of immigrant adolescents and their parents and teachers in Israel. In view of the different ways that results might be interpreted, we favor the idea of noting "discrepant" and "congruent" results as suggested in the convergent studies of Lee and Greene (2007) and Slonim-Nevo and Nevo (2009). In this sense, we believe that the mixed methods researcher should look for how the quantitative and qualitative databases tell different stories and to assess whether the statistical results and the qualitative themes are more congruent than discrepant.

What if the data tell the story of discrepant findings? Several options exist for handling this situation. The discrepancy may well be a result of methodological problems in the quantitative or qualitative aspects of the study, such as quantitative sampling problems or qualitative theme development issues. In this case, the mixed methods researcher would need to state limitations of the study. Alternatively, the researcher could collect additional data to help resolve the discrepancies or cite that they had more trust in the results of one form of data than the other. Researchers could also view the problem as a springboard for new directions of inquiry (Bryman, 1988).

The best and least costly alternative, however, is to reexamine the existing databases. This was the approach taken by Padgett (2004) in social work. Padgett's (2004) study recounted how a team of researchers returned to their initial database to gain additional insight. Her project was called the Harlem Mammogram Study—a project funded by the United States National Cancer Institute. It examined factors that influenced delay in response to an abnormal mammogram among African American women living in New York City. Padgett's research team had collected both structured quantitative data and open-ended interview data. After data analyses, the team concluded that the women's decisions to delay were not driven by factors in their quantitative model. The researchers then turned to their qualitative data, highlighted two qualitative themes, and reexamined their quantitative database for support for the themes. To their surprise, the quantitative data confirmed what the participants had said. This new information, in turn, led to a further exploration of the literature, in which they found some confirmation for the new findings.

Decisions for Connected
Data Analysis in a Sequential Approach

When two databases are implemented sequentially and connected—as in the explanatory and exploratory designs and some embedded, transformative, and multiphase designs—the mixed methods data analysis is much simpler than in the concurrent data designs. This is so because the quantitative and qualitative data are analyzed in different phases and not merged. This does mean, however, that special decisions are needed in analysis for sequential approaches, and these decisions relate primarily to how to best analyze data sets to support the follow-up actions.

In addition, the analysis needs to be responsive to answering the mixed methods research questions that arise when the researcher is using a sequential approach as discussed in Chapter 5 and Table 5.2. When quantitative data are collected before qualitative data (in explanatory designs and in some embedded, transformative, and multiphase designs), the mixed methods question may be the following: In what ways do the qualitative data help to explain the quantitative results? Variations of this question appear in other sequential designs, such as the following: To what extent do the qualitative follow-up findings explain the participants' reactions to the different treatment conditions (after-experiment embedded design)? When the qualitative data are collected before the quantitative data (in exploratory designs or some embedded, transformative, or multiphase designs) the typical mixed methods question is the following: In what ways do the quantitative results generalize the qualitative findings? A variation of this question might be this: In what ways did the qualitatively informed procedures enhance the experiment (before-experiment embedded design)?

In these sequential situations, the issues related to data analysis do not focus in on the analysis per se but on the procedures for using the analysis from the separate quantitative and qualitative analyses. Specific steps and decisions for the sequential approaches were identified earlier in Table 7.2. Here, we discuss the specific strategies for mixed methods data analysis.

Strategies for connected data analysis. As researchers think about data analysis within sequential approaches, they engage in **connected mixed methods data analysis** in which the analysis of the first data set is connected to data collection in the second data set. In addition, since the second data set is dependent on the results of the first phase, researchers should also consider how the analysis of the second data set can build on what was learned in the first phase. We consider each of these aspects for the two basic sequential approaches, starting with explanatory approaches.

When the intent is to explain quantitative results, the researcher may need to include analytic procedures during the first phase to help guide participant selection in the second phase. Although participants from the first phase may simply volunteer to participate in a second, qualitative phase, a stronger connection can be made when participants are determined through information arising from the quantitative data analysis. Researchers need to assess available options, such as the following:

- Select participants who are typical or representative of different groups for the follow-up to understand how groups differ. This may entail conducting quantitative analyses to describe typical scores or trends within groups of interest in the quantitative sample. From these analyses, purposefully select individuals who are typical of the groups for the second phase. For example, in Ivankova and Stick's (2007) study of students' persistence in a distributed doctoral program, the researchers identified typical scores for each of four groups (a beginning group, a matriculated group, a graduated group, and a withdrawn/inactive group) and then selected one individual per group whose scores were similar to the typical scores for the corresponding group.

- Select participants who scored at extreme levels outside the norm to understand why they might have scored as they did. This may entail graphically displaying scores for the participants in the first phase to identify outliers or using procedures such as calculating z scores to identify scores that are extreme (e.g., by setting a level as a specific number of SDs from the sample mean). Then sample these individuals based on their scores and ask them questions about why their scores were so extreme.

- Select participants from groups that might have differed in their statistical results. This will permit an analysis as to why groups might have differed. Ask the same questions to all of the groups to see why they might have differed. For example, Weine et al. (2005) studied Bosnian refugees engaged in multiple-family support and education groups in Chicago. They compared two groups—those who engaged and those who did not engage through statistical analysis in the first phase of the study. The factors that distinguished between the two groups then became key issues explored in follow-up qualitative interviews.

- Select participants who differed in their scores on significant predictors (positive scores, neutral scores, and negative scores) so that reasons behind different results might be further examined. This involves analyzing the data to identify significant predictors but then also examining responses to identify participants whose scores matched the patterns of interest. Once participants are selected, focus the follow-up questions on significant predictors, and ask participants to explain their thoughts about the predictors. In the mixed methods study of student note taking, Igo, Kiewra, and Bruning (2008) found puzzling results in the quantitative dependent measures of student learning that were inconsistent with previous research. The follow-up qualitative phase was then aimed at explaining these results. Participants were selected based on several of the previous criteria (demographics, statistical results, and so forth) and the data collection questions related to predictors and group factors.

Researchers using a sequential approach to explain results (e.g., explanatory design or after-trial embedded design) may also want to reflect on how best to analyze the second, qualitative data set. This qualitative analysis should use persuasive procedures to address the qualitative research question (as summarized in Table 7.1), but it should also ensure that the researcher will be able to answer the mixed methods research question (as to how the qualitative data help to explain the quantitative results). Therefore, the researcher may utilize the initial quantitative results to inform aspects of the qualitative data analysis. For example, the researcher may include some predetermined topic codes in the qualitative analysis that are based on the important factors identified in the quantitative results. As another example, if the researcher plans on explaining group differences with the qualitative follow-up data, then the strategy may link the demographic group variables to qualitative themes in the mixed methods analysis.

When building from qualitative data collection and analysis to a quantitative phase (as in the exploratory design or a before-trial embedded design), the analysis focuses on using data analysis results from the qualitative analysis in the follow-up phase. The researcher will begin with typical qualitative data analysis procedures and may include some techniques to facilitate the utility of the results for developing an instrument (or a typology or an intervention treatment) for the second phase:

• Analyze the qualitative data to best design a typology or an instrument. For example, look for natural differences in responses so that categories can be formed for a typology or pay attention to participants' language to develop good terms to use when writing items. This analytic process may also include analyzing the qualitative data to identify useful quotes, codes, and themes that can be used in designing the items, variables, and scales on an instrument. Developing a table of themes, codes, and quotes is particularly useful for specifying the content to be included when developing an instrument based on the qualitative results. In an exploratory mixed methods study, Meijer, Verloop, & Beijaard (2001) examined language teachers' knowledge about teaching reading comprehension to 16- to 18-year-old students. They first conducted a qualitative study consisting of interviews and concept mapping, and used the data to develop a questionnaire. They described in some detail the procedure of designing this questionnaire: using the qualitative categories to organize the questionnaire drawing on specific teacher expressions, formulating the teacher expressions into items, creating Likert-type scales for the items, and then translating each questionnaire into languages of participants in the study.

- Analyze the qualitative data to best enhance a larger design. Analyze the qualitative data for information that can be used in designing an intervention or planning the next phase in quantitative data collection in a longitudinal correlational design or a multiphase design. To determine how to use the data analysis results in an embedded design, consider the suggestions offered in Table 6.2. For example, in a randomized controlled trial (RCT) study testing different treatment options for prostate cancer, the researchers were not successfully recruiting adequate numbers of participants. Donovan et al. (2002) added a qualitative strand to learn how prospective participants understood the recruitment information. Based on their qualitative thematic analysis and results, they described four specific types of changes made to the recruitment procedures, which resulted in a major increase of individuals willing to enroll in the quantitative study.

- As with the explanatory design, researchers using an exploratory design can also consider how best to analyze the quantitative data in the second phase to not only answer the quantitative research question but also to provide the results needed to answer the mixed methods question. This quantitative analysis should use rigorous procedures, such as those listed in Table 7.1. This means that if the researcher is developing a new instrument, procedures should be included to assess the reliability and validity of the scores from this instrument for the population under study. The researcher should also consider how best to analyze the quantitative data so that they generalize or test the initial qualitative results. These analyses may include descriptive statistics to determine the relative prevalence, or importance, of different dimensions and/or inferential statistics to test relationships among variables as suggested by the qualitative findings.

Strategies for interpreting connected results. When data are connected across phases or projects, how does the mixed methods researcher interpret the findings? This interpretation could also be called "drawing conclusions" or "drawing inferences." Here are some basic ideas about drawing "inferences" and "meta-inferences" in a mixed methods study in which the analysis is based on connected data:

- Although inferences can be drawn after each phase (quantitative or qualitative), the meta-inferences are drawn at the end of the study and included in the larger interpretation being made in the conclusion or discussion section of a study.

- For an explanatory design, the meta-inferences relate to whether the follow-up qualitative data provide a better understanding of the problem than simply the quantitative results. This inference addresses the mixed methods question in an explanatory design. For an exploratory design, the meta-inferences relate to whether the follow-up quantitative strand provides a more generalized understanding of the problem than the qualitative database alone. Again, this interpretation relates to answering the mixed methods question. For an embedded design with sequential data collection before or after an experiment or a correlational study, the inferences attach to when the supporting data is being used. If qualitative data is used before an experiment, for example, then one meta-inference to be drawn is whether the before-experiment qualitative data leads to a better design of the intervention or better recruitment of participants to the trial than a trial without a before-design qualitative component (e.g., see the qualitative data analysis results that enhanced the rate of consent to participate in the trial by Donovan et al., 2002). In a transformative design, the inferences are drawn from the combination of the two phases and interpreted as to how the results help to uncover injustices and suggest change. In a multiphase design, the inferences can be drawn from the collection of analysis results from several projects or from the entire research program that has unfolded through multiple phases.

● VALIDATION AND MIXED METHODS DESIGNS

Discussions about validity in mixed methods have been described as being in their infancy (Onwuegbuzie & Johnson, 2006). The topic has been identified as one of the major six issues in mixed methods research and as the most important aspect of a research project (Tashakkori & Teddlie, 2003a). As Maxwell and Mittapalli (in press) pointed out, validity has been rejected by some mixed methods scholars, either because of its overuse, its meaninglessness, or because it is routinely used in quantitative research and is therefore disliked by qualitative researchers. For those who have addressed validity, early discussions focused on identifying both quantitative and qualitative approaches to it (see, for example, Tashakkori & Teddlie, 1998 and more recently Onwuegbuzie & Johnson, 2006). Also, a recent discussion has arrayed traditional quantitative, traditional qualitative, and mixed methods types of validation under a general framework of construct validation, and has incorporated several discussions of mixed methods validity under a common rubric (Dellinger & Leech, 2007). In addition, authors have discussed how it relates to the research design and data collection, to data analysis, and to interpretation of findings (Onwuegbuzie & Johnson, 2006).

The perspective of Onwuegbuzie and Johnson (2006) illustrates how validity in mixed methods can be related to stages in the process of research. We have seen this focus in the orientation of other mixed methods researchers. For example, Teddlie and Tashakkori (2009) addressed validity in mixed methods as it related to the design and the interpretation stage of research. They discussed design quality (suitability given the questions, fidelity of the quality and rigor of procedures, consistency across all aspects of the study, and analytic implementation of procedures) and interpretive rigor (consistency with findings, consistency with theory, interpretations given to participants and scholars, and distinctiveness in terms of credible or plausible conclusions). Onwuegbuzie and Johnson (2006), on the other hand, focused on data analysis, calling validity "legitimization" (p. 57), and specified a typology of forms of mixed methods validity. In addition, they addressed how to conceptualize legitimization (e.g., the design, the data analysis, or the interpretation) and specific procedures that mixed methods researchers might employ in the data analysis phase of their research.

Consequently, when discussing validity in mixed methods, our focus will be on the strategies that researchers might use in all three of the phases of data collection, data analysis, and interpretation of research. However, we will take the discussion one step further and relate validity and phases of research to specific analysis techniques used in mixed methods to merge data or to connect the data. We feel that validity cannot be adequately addressed (or made specific) as a procedure unless the researcher conceptualizes it within a research design. The very act of combining qualitative and quantitative approaches raises additional potential validity issues that extend well beyond the validity concerns that arise in the separate quantitative or qualitative methods procedures.

In general, we recommend the following for mixed methods researchers writing about or conducting validity procedures:

- Since mixed methods research involves both quantitative and qualitative strands of data, there is a need to address the specific types of validity checks that will be done for both strands. Discussions about the specific forms of validity for both quantitative and qualitative research are outlined in Onwuegbuzie and Johnson (2006) as well as many research methods books.

- Even though different terms are available in the mixed methods literature, we believe that the best term to use is *validity* because of its acceptance by both quantitative and qualitative researchers today and that such use presents a common language understandable to many researchers.

- We define validity in mixed methods research as employing strategies that address potential issues in data collection, data analysis, and the interpretations that might compromise the merging or connecting of the quantitative and qualitative strands of the study and the conclusions drawn from the combination. These compromises and strategies that mixed methods researchers might use to address them are detailed in Tables 7.3 and 7.4.

Table 7.3 Potential Validity Threats and Strategies When Merging Data in Concurrent Convergent, Embedded, Transformative, and Multiphase Designs

Potential Validity Threats When Merging Data	Strategies for Minimizing the Threat
Data collection issues	
• Selecting inappropriate individuals for the qualitative and quantitative data collection	• Draw quantitative and qualitative samples from the same population to make data comparable.
• Obtaining unequal sample sizes for the qualitative and quantitative data collection	• Use large qualitative samples or small quantitative samples so that the same number of cases can be selected.
• Introducing potential bias through one data collection on the other data collection (adding qualitative data into a trial while the trial is going on)	• Use separate data collection procedures, and collect data at the end of an experiment.
• Collecting two types of data that do not address the same topics	• Address the same question (parallel) in both quantitative and qualitative data collection.
Data analysis issues	
• Using inadequate approaches to converge the data (e.g., uninterpretable display)	• Develop a joint display with quantitative categorical data and qualitative themes or use other display configurations.
• Making illogical comparisons of the two results of analysis	• Find quotes that match the statistical results.
• Utilizing inadequate data transformation approaches	• Keep the transformation straightforward (e.g., count codes or themes), and use procedures to enhance reliability and validity of transformed scores.
• Using inappropriate statistics to analyze quantitized qualitative results	• Examine the distribution of scores, and consider use of nonparametric statistics, if needed.
Interpretation issues	
• Not resolving divergent findings	• Use strategies such as gathering more data, reanalyzing the current data, and evaluating the procedures.

Potential Validity Threats When Merging Data	Strategies for Minimizing the Threat
• Not discussing the mixed methods research questions	• Address each mixed methods question.
• Giving more weight to one form of data than the other	• Use procedures to present both sets of results in an equal way (e.g., a joint display) or provide a rationale for why one form of data provided a better understanding of the problem.
• Not interpreting the mixed methods results in light of the advocacy or social science lens	• Return in the interpretation of a transformative study to the lens used in the beginning of the study, and advance a call for action based on the results.
• Not relating the stages or projects in a multiphase study to each other	• Consider how a problem, a theory, or a lens might be an overarching way to connect the stages or projects.
• Irreconcilable differences among different researchers on a team	• Have researchers on a team evaluate the overall project objectives, and negotiate philosophical and methodological differences.

As shown in these tables, designs involved in the merging of data have different validity considerations than designs that connect the data. We have raised many of these issues in our discussions about data collection and analysis and in our recommendations for procedures in using the designs. Issues that might compromise data collection involve the selection of the sample, the size of the sample, and the recording and use of data that may not lend itself to comparison, may bias results, or may lead to faulty follow up procedures. Issues that may compromise data analysis include inadequate representations of the data and unclear or inappropriate analysis that becomes the focus for follow-up procedures. In interpretation, compromises occur when results related to research questions are not stated, contradictions are left unattended, the researchers favor one set of results over the other, merging approaches are used when the data sets are not meant to be compared, the order of discussing the data sets is reversed, and full advantage of both sets of data

Table 7.4 Potential Validity Threats and Strategies When Connecting Data in Sequential Explanatory, Exploratory, Embedded, Transformative, and Multiphase Designs

Potential Validity Threats for Connecting Data	Strategies for Minimizing the Threat
Data collection issues	
• Selecting inappropriate individuals for the qualitative and quantitative data collection	• Select the same individuals to follow up on findings; select different individuals when building and testing new components, such as an instrument, typology, or intervention.
• Using inappropriate sample sizes for the qualitative and quantitative data collection	• Use a large sample size for quantitative and a small sample size for qualitative.
• Choosing inadequate participants for the follow-up who cannot help explain significant results	• Choose individuals for the qualitative follow-up that participated in the quantitative first phase.
• Not designing an instrument with sound psychometric (i.e., validity and reliability) properties	• Use rigorous procedures for developing and validating the new instrument.
Data analysis issues	
• Choosing weak quantitative results to follow up on qualitatively	• Weigh the options for follow-up, and choose the results to follow-up that need further explanation.
• Choosing weak qualitative findings to follow up on quantitatively	• Use major themes as the basis for the quantitative follow-up.
• Including qualitative data in an intervention trial without a clear intent of its use	• Specify how each form of qualitative data will be used in the study.
Interpretation issues	
• Comparing the two data sets when they are intended to build rather than merge	• Interpret the quantitative and qualitative data sets to answer the mixed methods research question.
• Interpreting the two databases in reverse sequence	• Order the interpretation to fit the design (e.g., quantitative then qualitative or vice versa).
• Not taking full advantage of the potential of "before" or "after" qualitative data findings for an intervention trial	• Consider the reasons for using qualitative data in an intervention trial.

Potential Validity Threats for Connecting Data	Strategies for Minimizing the Threat
• Not interpreting the mixed methods results in light of the advocacy or social science lens	• Return in the interpretation of a transformative study to the lens used in the beginning of the study, and advance a call for action based on the results.
• Not relating the stages or projects in a multiphase study to each other	• Consider how a problem, a theory, or a lens might be an overarching way to connect the stages or projects.
• Irreconcilable differences among different researchers on a team	• Researchers on a team need to agree to the overall project objectives and to negotiate philosophical and methodological differences.

is not taken. Researchers should actively use strategies to minimize the validity threats in their studies and also discuss the limitations of the study's design as part of the larger interpretation of the study in the discussion section of a report.

SOFTWARE APPLICATIONS AND ● MIXED METHODS DATA ANALYSIS

Quantitative and qualitative software packages have been available for years to assist researchers in the analysis of both quantitative and qualitative data. Only recently have attention and discussion developed around the topic of computer software applications and mixed methods. In our opinion, the conversation has been lodged mainly at the more general level, but the commentary is becoming increasingly specific. The following examples help to illustrate this point. Two writers in particular—Bazeley (2009) and Kuckartz (2009)—have begun substantive conversations about mixed methods and specific software products. Bazeley (2009) reviewed the range of software packages that might be used in mixed methods research. She cited Excel for mixed methods tasks and then highlighted two software programs primarily designed for qualitative analysis that might be used for mixed methods analysis—NVivo (http://www.qsrinternational.com) and

MAXQDA (http://www.maxqda.com). She mentioned another program from Provalis (http://www.provalisresearch.com) that has subprograms (i.e., SimStat and QDA Miner) for both quantitative and qualitative data analysis capabilities. Bazeley described mixed methods applications using these software packages, such as to

- compare how cases with different characteristics discuss an issue,
- review changes in individual experiences over time on a case-by-case or group basis,
- consider the impact of changing settings on the evolution of an experience,
- examine the interrelationship of exported codes, and
- conduct quantitative comparative analysis of cases.

Kuckartz (2009) goes into more detail about the relationship of mixed methods to qualitative computer analysis—specifically the use of mixed methods with MAXQDA. He feels that the strongest argument for using MAXQDA in mixed methods research lies in linking, coding, and memoing; transforming the qualitative data into quantitative data; and creating visual representations of code distributions for exporting to statistical software. Some specific mixed methods applications of MAXQDA he mentioned are

- quantifying qualitative data—counting the number of times that a code occurs;
- linking text and variables using text codes and the "attributes" features (demographic or other quantitative variables);
- exporting and importing data into a statistical program—a researcher can create a data display of demographic variable names on the horizontal axis and themes on the vertical axis with counts in the cells and export this display into a statistical computer program; and
- using word counts—analyzing the qualitative data for the frequency of words used and linking the word counts to the codes or to the variables.

The suggestions by Bazeley (2009) and by Kuckartz (2009) provide useful starting points for conceptualizing the use of software in mixed methods data analysis. Also, we know that quantitative and qualitative software can help to separately analyze the strands of data (such as in the convergent design or in the explanatory design). We also know that qualitative software enables us to compare categorical variables with qualitative themes (for developing joint displays). We know that qualitative codes can be output to Excel or SPSS spreadsheets (for developing joint displays). We can also derive quantified counts of words from the qualitative software program, a procedure useful in data transformation designs.

Our recommendation is that mixed methods researchers consider how the software might be used to perform the data analyses needed within types of mixed methods designs and to specifically address the mixed methods questions. The reason why this link has not been forged before may be due to the need to think conceptually about research designs and understand the steps in data analysis undertaken within each of the designs.

How might the current software aid in these analyses to address research questions associated with merging or with the sequential analysis of data? We feel that one key is to consider the idea of the joint display as a technique for mixed methods analysis. The application of this technique moves beyond simply merging data; it can also be used in the sequential analysis of data. Our conceptualization as to how joint displays might be used is illustrated in Table 7.5. This table arrays the different designs and merging and connecting data analysis with types of mixed methods research questions and types of displays that might be developed. The types of mixed methods questions were identified earlier in Table 5.2. The displays proposed in this table can be easily designed using analysis from existing quantitative and qualitative software packages. They flow into a mixed methods study at different points—some displays can be inserted during the data analysis stage (e.g., linking qualitative codes/themes with variable data) and some can be implemented in a stage between data analysis and data collection (e.g., display relating qualitative data such as quotes, codes, and themes to scale development), and some can be used in the interpretation stage (e.g., comparing qualitative themes as explanations of quantitative results). Other mixed methods questions might yield additional displays that can be used. By using computer programs to analyze data and develop displays, the mixed methods question data can be represented and then answered. Undoubtedly, other applications for computer use in analyzing data for mixed methods studies will emerge in the future, and these are but a few possibilities for the mixed methods researcher to consider.

Table 7.5 Mixed Methods Designs, Research Questions, Mixed Methods Data Analysis, and Joint Displays

Type of Design	Mixed Methods Questions (Illustrative Questions)	Merging or Connected Data Analysis	Types of Joint Displays That Can be Developed That Link Quantitative and Qualitative Data
Convergent design	To what extent do the quantitative and qualitative results converge?	Merging analysis	• A display that places quantitative results side by side with qualitative themes (see Figure 7.2) • A display that combines qualitative codes or themes with quantitative categorical or continuous variable data (see Figure 7.3) • A display relating qualitative themes to quantitative ratings for transforming qualitative data into quantitative scores
Explanatory design	In what ways do the qualitative data help to explain the quantitative results?	Sequential analysis	• A display that links quantitative results and demographic characteristics to participants purposefully selected for the follow-up sample • A display at the end of the study that links qualitative themes to quantitative results for the purpose of explanation
Exploratory design	In what ways do the quantitative results generalize the qualitative findings?	Sequential analysis	• A display of quotes, codes, and themes that match proposed items, variables, and scales for instrument development • A display at the end of the study to show how the quantitative results generalize the qualitative themes and codes

Type of Design	Mixed Methods Questions (Illustrative Questions)	Merging or Connected Data Analysis	Types of Joint Displays That Can be Developed That Link Quantitative and Qualitative Data
Embedded design	How do the qualitative findings provide enhanced understanding of the quantitative results?	Merging or sequential analysis	• A display that links the qualitative themes to recruitment strategies for an intervention trial • A display that links qualitative themes to specific intervention activities • A display that compares themes about the processes individuals have experienced with outcome data • A display that compares the statistical results with qualitative follow-up themes • A display that compares qualitative themes with significant correlations at each stage in the research study
Transformative design	Add to the previous questions, "In order to explore inequities"	Merging or sequential analysis	• A display that compares the strategies in a call for action with the quantitative statistical results or with the qualitative theme results or both
Multiphase design	Multiply the previous questions in different phases	Merging or sequential analysis	• A display of the themes and quantitative results across studies and how these results have changed over time

SUMMARY

In mixed methods data analysis, the researcher needs to incorporate good procedures of data analysis for both quantitative and qualitative strands of the study. This involves preparing the data for analysis, exploring the data, analyzing the data to answer the research questions or test the research hypotheses, representing the data, interpreting the results, and validating the data, results, and interpretation. In addition, in mixed methods research, the additional process of mixed methods data analysis occurs. Mixed methods data analysis consists of analytic techniques applied to both the quantitative and the qualitative data as well as to the mixing of the two forms of data concurrently and sequentially in a single project or a multiphase project. Mixed methods data analysis has evolved through several writings in the field. It basically relates to analyzing the data to address the mixed methods questions. Mixed methods data analysis consists of steps that are conducted within each of the mixed methods designs. It also involves making key decisions about analysis within the steps. Two specific analytic techniques that cross the different designs are strategies for merging the data and for connecting the data. In merging the data, several procedures may be used, such as the side-by-side comparison, the joint display, or data transformation. Connecting data analysis focuses on the use of the data analysis results, and this use involves examining the analysis to select follow-up participants, specify follow-up research questions, or to design follow-up phases, such as the construction of a typology, an instrument, or an experimental intervention. After analyzing the quantitative and qualitative data, the researcher makes an interpretation by drawing inferences from both strands of analysis as well as the overall mixed methods analysis. Whatever interpretations are made, they need to be validated first in terms of inferences drawn from both the quantitative and qualitative strands as well as from the overall issues likely to arise in the mixed methods design. Finally, computer applications can be used in mixed methods research, although they are only being discussed recently in the literature. Using a framework of thinking about the data analysis in each of the mixed methods designs, computer-based analysis can lead to displays that provide useful results for each of the mixed methods designs.

ACTIVITIES

1. Develop sections on quantitative and qualitative data analysis for your study that include how you will prepare the data for analysis, how you will explore the data, how you will analyze the data for results to your questions

or hypotheses, how you will represent the data, how you will interpret the results, and how you will validate the data.

2. Assume that your data analysis will be to merge the quantitative and qualitative data. Discuss the three options for merging data, and discuss which one you would use and why.

3. Assume that your data analysis will consist of connecting quantitative and qualitative data. Indicate your option for using results from the initial phase, and discuss why you would select the option.

4. Given either a concurrent or a sequential design, identify which validity issues may arise in your study and how you will resolve them.

5. For a proposed mixed methods study that you would like to conduct, design a joint display that you would use to represent data and answer your mixed methods research question.

ADDITIONAL RESOURCES TO EXAMINE

For an overview of quantitative and qualitative data analysis procedures, see the following resource:

Creswell, J. W. (2008). *Educational research: Planning, conducting, and evaluating quantitative and qualitative research* (3rd ed.). Upper Saddle River, NJ: Pearson Education.

For options in mixed methods data analysis, examine the following resources:

Bazeley, P. (2009). Integrating data analyses in mixed methods research [Editorial]. *Journal of Mixed Methods Research, 3*(3), 203–207.

Caracelli, V. J., & Greene, J. C. (1993). Data anlaysis strategies for mixed-method evaluation designs. *Educational Evaluation and Policy Analysis, 15*(2), 195–207.

Onwuegbuzie, A. J., & Teddlie, C. (2003). A framework for analyzing data in mixed methods research. In A. Tashakkori & C. Teddlie (Eds.), *Handbook of mixed methods in social & behavioral research*. Thousand Oaks, CA: Sage.

For examples of applications of mixed methods analysis and joint displays, see the following resources:

Logan, T. K., Cole, J., & Shannon, L. (2007). A mixed-methods examination of sexual coercion and degradation among women in violent relationships who do and do not report forced sex. *Violence and Victims, 22*(1), 76.

Plano Clark, V. L., Garrett, A. L., & Leslie-Pelecky, D. L. (2009). Applying three strategies for integrating quantitative and qualitative databases in a mixed methods study of a nontraditional graduate education program. *Field Methods*. Prepublished December 29, 2009, DOI:10.1177/1525822X09357174

For recent validity discussions, consult the following resources:

Dellinger, A. B., & Leech, N. L. (2007). Toward a unified validation framework in mixed methods research. *Journal of Mixed Methods Research, 1*(4), 309–332.

Onwuegbuzie, A. J., & Johnson, R. B. (2006). The validity issue in mixed research. *Research in the Schools, 13*(1), 48–63.

Teddlie, C., & Tashakkori, A. (2009). *Foundations of mixed methods research: Integrating quantitative and qualitative approaches in the social and behavioral sciences.* Thousand Oaks, CA: Sage.

For discussions about computer applications and mixed methods research, see the following resources:

Bazeley, P. (2009). Integrating data analyses in mixed methods research [Editorial]. *Journal of Mixed Methods Research, 3*(3), 203–207.

Kuckartz, U. (2009). Realizing mixed-methods approaches with MAXQDA. Unpublished manuscript, Philipps-Universitaet Marburg, Marburg, Germany.

CHAPTER 8

WRITING AND EVALUATING MIXED METHODS RESEARCH

We focus now on the stage of composing and writing a mixed methods study. Part of this stage is thinking about how to structure and organize the written report. With multiple forms of data collection and analysis, it is easy for a reader to get lost in the complex forms of data and multiple layers of analysis. In addition, the length of an article on a mixed methods study may be a problem for journals because of the inclusion of both qualitative and quantitative approaches. Care needs to be taken so that both structure and writing are lean and concise. Some readers may not know what a mixed methods study "looks" like, and a well-designed structure for presentation will educate individuals new to this approach. Ways of evaluating a mixed methods study may also be unfamiliar to readers. Attention has not been given to the criteria for assessment of a "good" study, and such criteria would seem invaluable to graduate student advisors, editors of journals, proposal reviewers, and individuals designing and conducting this form of research. Standards help reviewers know what to look for in studies, and they help researchers locate exemplars and good models of research.

This chapter will address

- general guidelines for writing a mixed methods study;
- the structure to use in writing a mixed methods graduate student mixed methods proposal, a doctoral dissertation or thesis, a proposal for federal funding, and a mixed methods journal article; and
- criteria for evaluating a mixed methods study.

● GENERAL GUIDELINES FOR WRITING

Excellent books are available on grammar, syntax, and scholarly writing, so this discussion will focus on the structural aspects of writing a mixed methods study (see additional readings at the end of this chapter). Some of the ideas apply to all scholarly writing, but we will discuss them with a special focus on mixed methods projects. As with all writing, the audience must be kept in mind when the writer is organizing and structuring material.

The choice of a design may need to be partly based on what type of writing structure the anticipated audiences will accept and embrace. For example, an exploratory design will probably appeal to the qualitative research community, especially if priority is given to the qualitative component and the article begins with a clear exploration of the subject matter. Alternatively, an explanatory design will probably appeal to quantitative researchers for similar reasons.

The writing can serve the function of educating a reader about mixed methods research. Complete methods discussions can be placed into proposals, dissertations, and journal articles. Writers can use mixed methods terms (e.g., the term *convergent parallel design*), provide a definition of mixed methods research, include references to the mixed methods literature and specific mixed methods studies, and embed in the writing the parts of research with mixed methods components (e.g., a mixed methods research question, discussion of the mixed methods data analysis). Graduate students can encourage this educational process by selecting a published mixed methods study in their field and sharing it with committee members prior to presenting their proposal or dissertation/thesis.

Because of the complexity of mixed methods research, readers will need aids to help them understand a mixed methods study. These aids include diagrams of procedures, well-designed purpose statements following the scripts advanced in this text, and clear headings that separate the quantitative and the qualitative elements of data collection and analysis.

Scholarly research writing involves telling a good story. It aids the reader when the writer takes care to tell a coherent and cohesive story throughout the qualitative and quantitative aspects of the research study. The reasons for including more than one type of data (e.g., qualitative and quantitative) become clear if the writing builds from one to the other and helps provide this transition. The two databases need to link in some way, and the more carefully this link is established, the more coherent and cohesive the overall story will be.

Consider what point of view fits the mixed methods design being used. The point of view—who is telling the story—can be crafted from the first person (I, we), the second person (you), or the third person (he, she, they). It can also be described from how the story is told—from the subjective to the objective (Bailey, 2000). The first person subjective approach is typically found in qualitative research. We see the use of first person pronouns, such as "I" or "we," used throughout the study. The subjective stories of individuals are presented through quotes. In quantitative research, the subjective first and second person are not typically used. Instead, the objective third person is the norm by describing factually the results or using impersonal referrals, such as "the investigator" or "the researcher." The researcher is largely in the background, objectively reporting the results and having an unseen and unheard personal voice. How is one to proceed, then, in mixed methods with both qualitative and quantitative points of view to prevail? Consider one of two possibilities: to write the mixed methods report using one consistent voice throughout or to write the report by varying the voice, with the objective approach used in the quantitative sections and the subjective voice in the qualitative sections. The decision to choose one or the other may be based on the type of design (explanatory design starts out strong in the quantitative approach), personal writing style, and, of course, the audience for whom you are writing.

Adapt the structure of the written study to mixed methods research and the type of mixed methods design being used. Because the components of research (e.g., purpose statement, data collection) differ for each of the major types of mixed methods designs, it should not come as a surprise that the structure used in writing about them will differ. In fact, the structure of the written study can help the reader better understand the type of design.

RELATE THE STRUCTURE TO THE MIXED METHODS DESIGN ●

We have long embraced the idea of thoughtful organization of the research report before the study begins. At the same time, we believe in allowing the design to emerge, and, in many cases of sequential types of designs, the steps down the road are unknown. But before the study begins, it is helpful to have an image in mind as to what the final mixed methods study might look like.

This image may be a vague picture at the outset, but it becomes clearer as the study proceeds.

Thus, because a good plan is central to sound mixed methods research, we provide several outlines that may be useful: the structure of a graduate student proposal for a dissertation or thesis, the structure of a final mixed methods dissertation, the structure of an outline of topics addressed in a proposal for federal funding, and the structure of topics to include in a mixed methods empirical journal article. Looking at these various structures, we see some common features of mixed methods research being introduced that we have discussed in earlier chapters in this book. However, the features differ depending on the type of mixed methods design chosen and whether the writing is a plan for a study or the report of a completed study.

Structure of a Proposal for a Mixed Methods Dissertation or Thesis

A proposal for a dissertation or thesis needs to convince graduate committees and advisors that the topic is worth pursuing and that it will be studied in a rigorous and insightful way, and that it is feasible for the student. The proposal needs to be convincing, and when the design is mixed methods, special components need to be included in the overall plan that relate to mixed methods and the type of design. Proposal formats will differ from campus to campus, and students need to obtain copies of past proposals to review, to see how they are structured. Our first recommendation, then, is that graduate students visit with their faculty and ask for examples of proposals from prior dissertations or theses that have been completed. A search through Dissertation and Theses Abstracts using an academic library search or a search engine will also yield mixed methods dissertations to examine. It is also helpful to locate several mixed methods studies and important to design a proposal that contains major elements of mixed methods, as well as information about a specific design.

Examine Table 8.1, which provides a sample outline of the topics generally found in mixed methods dissertation and thesis proposals.

Our discussion here will focus on the sections of the proposal with mixed methods components, and we have simply placed mixed methods components into a traditional format for a proposal:

- The title should be stated so that it foreshadows the mixed method study and type of design. This title should focus on the topic of the study, mention that it is a mixed methods study, and mention the participants and the research site.

Table 8.1	Outline of the Structure of a Proposal for a Mixed Methods Dissertation or Thesis

Title

- Foreshadows mixed methods study and type of design

Introduction

- The research problem
- Past research on the problem
- Deficiencies in past research and one deficiency related to a need to collect both quantitative and qualitative data
- The audiences that will profit from the study

Purpose

- The purpose of the project (use the scripts in Chapter 5) and reasons for design type
- The research questions and hypotheses (ordered to match the design)
    - ○ Quantitative research questions or hypotheses
    - ○ Qualitative research questions
    - ○ Mixed methods research question(s)

Philosophical and Theoretical Foundations

- Worldview
- Theoretical lens (social science or advocacy)

Literature Review (include quantitative, qualitative, and mixed methods studies, if they are available)

Methods

- A definition of mixed methods research
- The type of design used and its definition
- Challenges in using this design and how they will be addressed
- Examples of use of the type of design (in your field, if possible)
- Reference to and inclusion of a procedural diagram in an appendix
- Quantitative data collection and analysis
- Qualitative data collection and analysis and qualitative data transformation, if used (in exploratory design, place qualitative before quantitative)
- Mixed methods data analysis procedures
- Validity approaches in quantitative, qualitative, and mixed methods research

Potential Ethical Issues

Researcher's Resources and Skills

Timeline for Completing the Study

References

Appendixes With Instruments and Protocols and Procedural Diagram

- The introduction is rather standard for scholarly research in that it basically sets forth the research problem and why it is important to be studied. However, one of the deficiencies in past research is that there is a need for what mixed methods research has to offer, such as a more comprehensive analysis, multiple viewpoints, a chance to explore or confirm, and others. These rationales for mixed methods were delineated in Chapters 1 and 5.

- The purpose statement needs to convey a mixed methods approach, and the research questions can convey the qualitative and quantitative strands of the study as well as the mixed methods question, framed using one of the scripts in Chapter 5. It is important to include the rationale for mixed methods in the purpose statement, as mentioned earlier. The order of the research questions depends on which design is being used in sequential studies, with the order mirroring the procedures proposed for the study (e.g., qualitative followed by quantitative for an exploratory design).

- The philosophical foundation for the use of mixed methods needs to be described and a rationale presented for the use of one or more worldviews. Further, if a theoretical lens is used (e.g., a social science or an advocacy lens), this lens needs to be mentioned, and the proposal needs to detail how the lens will flow into the final study. If no theoretical lens is used in the study, then this section would only address the worldview perspective being taken by the researcher.

- In a mixed methods study, we would recommend including a literature review section. It should cover the literature (divided into subtopics) for studies examining the research problem in the study, and it should include qualitative, quantitative, and mixed methods studies. The end of this literature review should point out how the proposed study significantly adds to the literature.

- The methods section typically begins with information about mixed methods research and the specific type of design being used. In Chapter 3, we provide an example of this opening paragraph that might be adapted for a proposed study.

- The methods section needs to be carefully shaped to convey the details of the procedures of the mixed methods design. Since proposal reviewers might not be familiar with mixed methods, providing a definition for it is important. Then, the specific mixed methods design needs to be mentioned and described. Here would be a place as well to describe some of the challenges of using this design, and we refer you to Tables 7.3 and 7.4 on validity issues. A diagram of the procedures would be useful (as discussed in Chapter 4). The detailed data collection for both qualitative and quantitative research using the elements of good procedures are found in Table 6.1. The mixed methods data analysis steps, as found in Chapter 7, for merging or connecting can be discussed, followed by specifying qualitative, quantitative, and mixed methods validity considerations.

- Also important is the need to identify potential ethical issues that are likely to present challenges in the dissertation or thesis and the strategies that will be used to address them.

- The researcher's skills in conducting mixed methods research and the time involved in gathering both forms of data need to be mentioned. The researcher needs to know both quantitative and qualitative research and the forms of data collection and analysis used in both approaches. Providing a timeline is useful in mixed methods research, given the extensive time involved in collecting and analyzing two forms of data.

Structure of a Mixed Methods Dissertation or Thesis

The ideal structure of a mixed methods dissertation or thesis mirrors the proposal but adds the results or findings and the conclusions. An example of the table of contents for a mixed methods dissertation can serve to illustrate the structure of a final study. The content will differ for different types of mixed methods designs and program requirements, and the example we provide from the field of communication studies on hurtful communication from college teachers to students illustrates an exploratory design with the intent of developing an instrument.

The dissertation structure as illustrated in Table 8.2 by Maresh (2009) was an exploratory mixed methods design with the intent to develop and test an instrument. It began with collecting and analyzing qualitative interview data from students. From this, Maresh then analyzed the results to obtain nine themes of hurtful messages that teachers communicate to students. An instrument was then developed from these themes, and the instrument was administered to a large sample of students. As shown in the structure of this dissertation, the study consisted of six chapters. The first three chapters conveyed the introduction, the relevant literature, and the methodology. In the methodology chapter, the author advanced philosophical assumptions, stated the mixed methods research design, and provided a figure to illustrate the procedures. Then, the methodology discussion conveyed the phases of the research from the initial qualitative beginnings to the interim phase of instrument development and on to the quantitative data collection. Separate chapters were then included for first the qualitative results (including the instrument), the quantitative results, and then the final discussion. In summary, this dissertation table of contents showed more chapters than typically found in a quantitative five-chapter dissertation, and chapters were shaped around the specific results, presented in the order of the design from qualitative to quantitative.

Table 8.2　Example Structure for a Mixed Methods Dissertation or Thesis

Chapter One: Introduction

Defining Teacher Misbehaviors and Recognizing Their Impact (Establishing the Importance of the Problem)
Face in the Teacher–Student Relationship (Description and References of a Key Idea)
Defining and Rationalizing the Study of Hurt (Description and References of a Key Idea)
Purpose of the Present Study
Summary

Chapter Two: Overview of Relevant Literature

Hurtful Messages in Human Relationships
Individuals' Responses to Hurtful Messages
Theoretical Rationale
Face Theory
Relational Consequences of Hurtful Messages
Content-Oriented Consequences of Teacher Misbehaviors
Summary

Chapter Three: Methodology

Epistemological Assumptions
Research Design
Communication Studies and Mixed Methods Research
Limitations of Mixed Methods Inquiry
Phase One: Qualitative/Interpretive
　　Participants
　　Data Collection
　　Focused Interviews
　　Data Analysis
　　Data Validation
Interim Phase: Instrument Development
　　Mixed Methods Validity
Phase Two: Quantitative
　　Participants
　　Data Collection
　　Instrument
Summary

Chapter Four: Qualitative Results

Types of Hurtful Messages
Hurtful Messages and Teacher Misbehaviors
Students' Responses to Hurtful Messages
Face and Hurtful Messages
Advice for Teachers
Perceived Impact of Hurtful Messages
Affective Learning
Relational Satisfaction
Instrument Development
Summary

Chapter Five: Quantitative Results

Research Questions
Hypotheses
Summary

Chapter Six: Discussion

Hurtful Messages in the Teacher–Student Relationship
Face Theory and Students' Attributions of Hurtful Messages
The Impact of Hurtful Messages in the College Classroom
Significance of the Study
Implications of Conclusions
Practical Application
Limitations
Directions for Future Research
Summary

References

Appendixes

SOURCE: Adapted from Maresh (2009). Used with permission from the author.

Structure for a National Institutes of Health Proposal

It is helpful to discuss how a proposal for federal funding might be adapted to fit a mixed methods study, as funding agencies are increasingly interested in funding mixed methods research. We have chosen the guidelines for the National Institutes of Health (NIH) for our illustration, but we might just as easily have chosen the National Science Foundation (NSF) or a private foundation as an illustration.

NIH has issued some guidelines (National Institutes of Health, 1999) for designing an NIH proposal that includes a combination of qualitative and quantitative approaches. NIH has also held workshops focused on mixed methods research, such as the one in the summer of 2004 for social work and health professionals. NIH's 1999 guidelines mention the challenge of conducting "combined" research. They recommend that the combined components relate to the research questions and hypotheses, that the authors mention how the data will be integrated (i.e., mixed), and that the authors explain how the results will be interpreted, taking into account data from two different research paradigms. The guidelines call for expertise in both approaches and a complete description of both methods and their contributions, rather than a superficial approach to either of the two methods. They discuss the integrated (i.e., convergent) and sequential models and point out that integrated approaches are challenging and require extensive explanation. They also imply that qualitative methods may not be enough in a study when factors are known that must be controlled and when prior studies characterize a field of study. They finally recommend that adequate time be made available for this form of research and that the investigators be sure they have sufficient expertise.

Guidelines for the grants process for the U.S. Department of Health & Human Services, Office of Extramural Research at NIH can be found at their Web site: http://grants.nih.gov/grants/oer.htm. The proposal that investigators write for NIH research grants follows the SF424 (R&R) standard form. This form indicates that the research plan needs to include a 1-page statement of specific aims along with a 12-page narrative for a RO1 proposal or a 6-page narrative for a RO3 and R21 proposal that describes the significance, innovation, and approach. As shown in Table 8.3, we have extracted the major elements of this narrative section and modified them to include components of mixed methods research and designs.

The proposal narrative should include the following mixed methods considerations:

- The specific aim section of the proposal can include the mixed methods purpose statement, as well as the mixed methods research question(s).
- In the significance section, include the reasons for collecting both quantitative and qualitative data as a deficiency of past literature.
- In the innovation section, discuss the innovative use of mixed methods within the proposed project.
- In the approach section, begin with an overview of mixed methods research and the specific type of design being used. Examine Figure 3.9 as a model for how to design this paragraph.

Table 8.3	National Institutes of Health Guidelines for a Proposal Narrative for a Mixed Methods Study

A. Specific Aims (Limit to one page)

- Goals of the proposed research (include a mixed methods purpose statement) with quantitative questions or hypotheses, qualitative questions, and mixed methods questions (include mixed methods research questions). Mention the impact the proposed research will have on the content field as well as the mixed methods literature. Discuss reasons for using mixed methods research and how the study will advance understanding of mixed methods research.
- Specify objectives of the proposed research (include objectives that will be accomplished through the mixed methods design).

B. Research Strategy (Match page length to proposal type)

(a) Significance (add statements about how mixed methods will enhance the study of the research problem)

- Explain the importance of the problem or critical barrier to progress in the field that the proposed project addresses.
- Explain how the proposed project will improve scientific knowledge, technical capability, and/or clinical practice in one or more broad fields.
- Describe how the concepts, methods, technologies, treatments, services, or preventative interventions that drive this field will be changed if the proposed aims are achieved.

(b) Innovation (add information about the innovative use of mixed methods as an approach and research methodology)

- Explain how the application challenges and seeks to shift current research or clinical practice paradigms.
- Describe any novel theoretical concepts, approaches or methodologies, instrumentation or intervention(s) to be developed or used, and any advantage over existing methodologies, instrumentation or intervention(s).
- Explain any refinements, improvements, or new applications of theoretical concepts, approaches or methodologies, instrumentation, or interventions.

(c) Approach

- Describe the overall strategy, methodology, and analyses to be used to accomplish the specific aims of the project. Include how the data will be collected, analyzed, and interpreted as well as any resource sharing plans as appropriate. Include
  - Overall approach (mixed methods) and definition of mixed methods research
  - Type of mixed methods design used (define, give reasons for using design, cite studies using design in field)

(Continued)

Table 8.3	(Continued)

- Diagram of procedures (may include as appendix)
- Data collection (order quantitative and qualitative collection methods according to design)
- Data analysis (order quantitative and qualitative analyses according to design)
- Validity (quantitative, qualitative, and mixed methods)
- Discuss potential difficulties and limitations of the proposed procedures (include the challenges to the design and how they will be addressed)
- Provide a tentative sequence or timetable for the project (include this timetable in the mixed methods diagram)

- Discuss potential problems, alternative strategies, and benchmarks for success anticipated to achieve the aims.
- If the project is in the early stages of development, describe any strategy to establish feasibility and address the management of any high risk aspects of the proposed work.
- Point out any procedures, situations, or materials that may be hazardous to personnel and precautions to be exercised.
- Add preliminary studies for new applications (these studies can include research that uses quantitative, qualitative, and mixed methods approaches). Add in experiences and competence of the researchers (include individual and team skills in quantitative, qualitative, and mixed methods research).

SOURCE: Application Guide for NIH and Other PHS Agencies (pp. I–108 - I–110) found at http://grants .nih.gov/grants/funding/424/SF424_RR_Guide_General_Adobe_VerB.pdf

- Give the reasons that mixed methods research is appropriate to address the research problem. The general reasons in Chapter 1 and the discussion of the evolution of this form of research provided in Chapter 2 can provide useful guidance.
- Include, as an insert or as an appendix, a diagram of the procedures in the mixed methods design. Include a timeline for each phase in the process in this diagram.
- Report the quantitative and qualitative data collection and the quantitative and qualitative data analysis as separate sections in the approach section to keep them distinct, clear, and easy to review.
- Include a separate section on mixed methods data analysis that specifies the procedures for "mixing" the data that relate to the type of design.
- Write about the challenges that relate to the specific type of design. These challenges were introduced in Chapter 3 and were advanced as validity issues in Chapter 7.

- Discuss ethical issues that may arise in the particular type of design being used.
- In the preliminary studies section, include references to quantitative, qualitative, and mixed methods studies to illustrate the forms of research that may relate to the problem being studied.

Structure of a Mixed Methods Journal Article

Our examples thus far relate to planning a mixed methods study in a proposal or reporting a full dissertation project. After completing a mixed methods study, many authors submit their work to scholarly journals. We have included numerous examples in this book of mixed methods studies published in journals, and, although they vary in writing structure, there are some common elements of mixed methods research that do (or should) flow through all these studies, as well as features that earmark each study as having used one of the six major types of design. In Table 8.4, we present a general structure for a mixed methods journal article.

The specific mixed methods components and the design components of a mixed methods journal article are as follows:

- The title needs to reflect the fact that this is a mixed methods study, with words such as *mixed methods research* incorporated into the title. As discussed in Chapter 5, the title can also foreshadow the type of design, with neutral language or with language that gives priority to quantitative or qualitative research.
- The introduction can note a deficiency in previous studies that relates to a need for collecting both quantitative and qualitative research. It can also include a purpose statement, written using the scripts in Chapter 5, and quantitative, qualitative, and mixed methods research questions.
- The methods section can open with a statement about mixed methods research and the type of design. The reasons for using the type of design and examples of studies can also be incorporated into this section. A procedural diagram, found with increasing frequency in mixed methods studies today, should be provided, perhaps as an appendix to the journal article. Both quantitative and qualitative data collection and analysis procedures should also be mentioned.
- It is in the results section that mixed methods journal articles vary in structure, but knowing the types of designs helps the writer to understand the different structures. In a convergent design, the results

Table 8.4   Outline of the Structure for a Mixed Methods Journal Article

Title (foreshadows the mixed methods research and design)

Introduction

- Statement of the problem
- Issue
- Literature on the research problem or issue (focused on establishing the need for studying the research problem or issue)
- Deficiencies in previous studies (incorporate need for collecting both quantitative and qualitative data)
- Audiences for the study
- Purpose statement (written using script that applies to type of design)
- Research questions (order quantitative and qualitative questions according to timing and priority in study)
  - Qualitative research questions
  - Quantitative research questions or hypotheses
  - Mixed methods research questions

Related Literature Review (optional, depending on priority and use of theory; a broad literature review about the topic of the study that narrows the focus to the specific issue or problem of the study)

Methods

- Overall approach (mixed methods) and definition of mixed methods research
- Type of mixed methods design used (define, give reasons for using design, cite studies using design in field)
- Diagram of procedures (may include as appendix)
- Data collection (order quantitative and qualitative collection methods according to design)
- Data analysis (order quantitative and qualitative analyses according to design)
- Validity

Results

- Merge results in convergent, embedded, transformative or multiphase designs (sometimes we see separate reports of quantitative and qualitative results, with merging in the discussion section)
- Connect results in sequential, embedded, transformative, or multiphase designs (present the results in the sequence in which they are used—e.g., qualitative results followed by quantitative results)

Discussion

- Summarize results (merged or connected)
- Explain results
- State limitations
- State future research
- Reiterate the unique contribution of the study

References

Appendixes (tables, figures, instruments, protocols)

section might report the separate results from both the quantitative and qualitative data analysis, or it might report the results of both types of data analysis plus the results of the mixed methods merged analysis. When the latter is presented, the researcher might present joint displays that relate the themes to the quantitative variables. Alternatively, the merged analysis may be saved for the discussion section and be seen more as a side-by-side comparison of results from the two databases. In an explanatory design, the quantitative results are presented first followed by the qualitative results. The reverse is true for an exploratory design with the qualitative findings discussed first followed by the quantitative results. An interim phase, such as found in the sample outline of a mixed methods dissertation in Table 8.2, illustrates how the results might have a qualitative phase, an instrument development interim phase, and a quantitative phase. In an embedded design, the results section typically focuses either on the quantitative data or the qualitative data, depending on which is the major data set in the study. Because many authors who use the embedded design report their quantitative and qualitative studies separately, the results may not focus on both data sets but on either the primary or the secondary data set. In many of the embedded design experimental trials we have reviewed, the authors have reported the quantitative trial in an article separate from that in which they report the qualitative study. Stange, Crabtree, and Miller (2006) have discussed writing forms in the health sciences, such as publishing from a mixed methods study in separate quantitative and qualitative papers, staging their papers as separate articles in a single issue of a journal, or integrating their methods into a single article. In transformative as well as multiphase designs, the authors may also report the qualitative and quantitative results separately in different reports. If the researcher is publishing results in more than one journal article, we recommend that both articles make reference to the use of mixed methods research and that the studies be cross-referenced so they both can be identified and located.

- In the discussion section, we find the interpretation of the results, as well as a discussion relating this interpretation to the literature, the limitations of the study, as well as future research. How should the interpretations be reported in a mixed methods study? In a convergent design, the interpretation of the results may reflect the merging of the data, and the authors will compare the findings from the quantitative and qualitative analysis in order to answer the mixed methods research question. In explanatory and exploratory designs, the interpretation often mirrors the sequence of the data collection and analysis (i.e., first quantitative results explained, then qualitative results explained in an explanatory design). Then the researcher reports on the conclusions drawn from answering the mixed methods question. In embedded designs, the focus in the interpretation of the major findings related to the primary data set, but the author needs as well to comment on how the mixed methods question was answered. In transformative designs, the researcher will interpret how the merged or connected findings address the mixed methods research question and suggest a plan of action for social change. In multiphase designs, some combination of a concurrent, merged summary of findings and a sequential, connected summary of findings will be interpreted in terms of how the findings advance the overall objective of the program of inquiry.

● EVALUATING A MIXED METHODS STUDY

A writing structure that conveys the elements of mixed methods research and is organized to reflect the type of design used adds to the sophistication and credibility of a completed study. For those conducting mixed methods research, it is important to consider how to evaluate the quality of their study and to reflect on the criteria that others, such as graduate committee members, funding agencies, journal editors, and readers in general might use to assess the quality of a mixed methods study.

There are several ways to think about the evaluation of a mixed methods study. Assuming that the study is persuasive and rigorous in both the qualitative and quantitative strands, we can use the standards of both approaches that are available in the literature. In addition, mixed methods studies by themselves should be subject to standards of quality, and we will review several standards that are emerging in the mixed methods area.

Quantitative and Qualitative Evaluation Criteria

The standards for evaluating a quantitative study often reflect the type of research design and the methods of data collection and analysis (Hall, Ward, & Comer, 1988). A rigorous quantitative study phase in mixed methods research must use a type of design that matches the research question, a theory that frames the study, and data collection that will lead to reliable and valid scores. The statistical test must be appropriate and robust. The overall study needs to have accurate measures and be generalizable, valid and reliable, and replicable.

The standards for evaluating a qualitative study depend on how the researcher positions herself or himself in the study. Qualitative researchers differ in the criteria they would use, such as philosophical criteria, participatory and advocacy criteria, or procedural, methodological criteria (see Creswell, 2008b). We stress the importance of procedural or methodological criteria, such as emphasizing rigorous data collection, framing the study within philosophical assumptions of qualitative research, using an accepted approach to inquiry (e.g., ethnography, case study), focusing on a single phenomenon, using validity strategies to confirm the accuracy of the account, conducting multiple levels of data analysis, and writing a study that is persuasive and engages the reader (see Creswell, 2007). To this list we could add that researchers need to disclose their role (i.e., reflexivity) and its impact on the interpretations they make in a study.

Mixed Methods Evaluation Criteria

Our stance is that while mixed methods research must be responsive to both qualitative and quantitative criteria, there is a separate set of expectations for a mixed methods study beyond those needed for quantitative and qualitative research. Bryman (2006) calls this the "bespoke" approach where criteria are developed especially for mixed methods studies. Also, we see mixed methods evaluation criteria reflecting trends that seem to exist within qualitative research. As we discussed earlier, in qualitative research several perspectives exist about evaluation and that one's viewpoint depends on their orientation. In mixed methods, this orientation may be as a methods person, a methodologist, a philosopher, or as a theoretically oriented scholar. Policy makers who fund research want to know whether the research questions are adequately answered, researchers who engage in mixed methods studies want to know if they can trust the findings and take action on them, research participants want to know if they have had a good experience, and teachers of research need to convey standards by which studies will be judged (O'Cathain, in press). For all of these stakeholders, we need to establish criteria for assessing mixed methods studies. As we introduced in Chapter 1, our selection of a good mixed methods study reflects a methods orientation. To evaluate a mixed methods study, the researcher

- collects both quantitative and qualitative data,
- employs persuasive and rigorous procedures in the methods of data collection and analysis,

- integrates or "mixes" (merges, embeds, or connects) the two sources of data so that their combined use provides a better understanding of the research problem than one source or the other,
- includes the use of a mixed methods research design and integrates all features of the study consistent with the design,
- frames the study within philosophical assumptions, and
- conveys the research using terms that are consistent with those being used in the mixed methods field today.

We use these methods-oriented criteria with our students completing mixed methods studies and in reviewing manuscripts submitted for publication. They are a set of criteria consistent with ideas set forth in this book. At a more applied level, an approach we used with articles submitted to the *Journal of Mixed Methods Research* (JMMR) might be helpful to start thinking about how to apply criteria to a mixed methods article. Bear in mind that these procedures addressed the "research" approach used in the study, not the content or topic of focus of the study. Although our procedure did not always follow the lockstep guide identified next, we tended to use these steps:

1. *We looked in the methods section first.* We examined the methods section to see if the researcher collected both quantitative and qualitative data in response to research questions or hypotheses. We looked for the typical qualitative approaches of open-ended interviews, observations, documents, or audiovisual materials and the closed-ended data forms of quantitative research consisting of instruments, observational checklists, and documents reporting numeric data. Sometimes this division was unclear because one form of data (e.g., patient records) could be viewed as having both qualitative data (notes from provider) and quantitative data (values reported on screening tests).

2. *Next we examined the method section in detail.* We looked at the methods to determine whether they were pursued thoroughly. This meant that we examined the qualitative methods to see if they were detailed persuasively and the quantitative methods to determine if they were developed rigorously (see Chapters 6 and 7 on data collection and data analysis).

3. *Next we looked at the results and discussion for evidence of mixing of data.* We were interested in whether the researcher actually "mixed" the two methods as opposed to collecting data for both strands and keeping them separate throughout the study. This was sometimes difficult to pinpoint. Helpful in assessing whether mixing

occurred was evidence that the author mentioned a rationale for why the two strands were being collected (e.g., qualitative was being collected in order to explain the quantitative results). This rationale could be found anywhere in the study from the beginning, to the methods, or to the end. Other signs for mixing consisted of tables or figures that contained both databases, separate phases of the study with one devoted to quantitative data and the other qualitative (or vice versa), or results or interpretation sections in which the authors explicitly brought together the two databases.

4. *Finally, we looked for mixed methods terms.* The use of mixed methods terms in a study denoted that the authors made a conscious attempt to use mixed methods procedures, were familiar with the literature on mixed methods, and sought to have their study understood and evaluated by readers as a mixed methods study. We looked for mixed methods terms in such places as the title (did they include the words *mixed methods*), in their method discussion and a specification of a type of mixed methods design, in their rationale for their choice of research approach, and in the advantages they cited for mixed methods in the conclusions of the article.

Another approach for considering quality in mixed methods research is to study researchers' perceptions. Bryman, Becker, and Sempik (2008) asked specifically about quality criteria for mixed methods research in a mixed methods study of researchers' perceptions. The quantitative results found that over two thirds of the surveyed researchers felt that different criteria should be used to judge the quality of the quantitative and qualitative components of a mixed methods study. The analysis of the interview data identified four themes concerning criteria that can be applied to mixed methods studies:

- The use of mixed methods needs to be relevant to the research questions
- There needs to be transparency about the mixed methods procedures
- The findings need to be integrated or mixed
- A rationale for the use of mixed methods needs to be provided

An alternative approach to evaluating mixed methods studies is to consider mixed methods within the larger process of research. In 2008, O'Cathain, Murphy, and Nicholl (2008) developed a set of criteria for Good Reporting of a Mixed Methods Study (GRAMMS). Building on this work, O'Cathain (in press) provided a recent set of evaluation criteria in which she pointed out that evaluation discussions in mixed methods have been derived from literature reviews, from researchers' expertise and from interviewing researchers, and from mapping exercises with researchers. O'Cathain's framework included

- the planning quality of the mixed methods study (e.g., feasibility and transparency),
- design quality (e.g., detailed description of the design, suitability of the design, strength, and rigor),
- data quality (e.g., detailed description, rigor, and adequacy of sampling and analysis),
- interpretive rigor (e.g., relationship of findings to methods, inconsistencies found, credibility, and the likelihood that others reach the same conclusion),
- inference transferability (e.g., conclusions applied to other settings),
- reporting quality (e.g., successful completion of a study, reporting transparency, and yield of the study),
- synthesisability (e.g., whether the study is worthy of inclusion in a synthesis of evidence), and
- utility (e.g., whether results are usable).

O'Cathain ends by stating that there may be too many criteria. We agree that a parsimonious set of criteria will be most useful for those designing a mixed methods study—especially those with limited experience who are beginning their first mixed methods study. We are also mindful that the best criteria perhaps lie within a specific study application and its design. Because mixed methods researchers use specific research designs in their studies, the best criteria would refer to the essential characteristics of the six designs advanced in Chapter 3 and use these characteristics in assessing the quality of a mixed methods study. A contextualized set of criteria would best serve the researcher who wants to assess the quality of a particular study; it may be less valuable, however, for the broader community of mixed methods researchers who are policy makers or editors who need a more general set of criteria.

SUMMARY

General guidelines can help researchers write a mixed methods study. Writers need to consider the writing structure most accommodating to anticipated audiences, how their report and its composition will educate audiences, how their mixed methods study will be understood by audiences because of its complexity, and how it tells a coherent story in a consistent point of view or in a point of view natural to a specific type of design.

Because planning in advance is helpful in all forms of research, we provide examples of structures for designing mixed methods studies. We suggest outlines for writing a dissertation or thesis proposal, a final dissertation, an NIH proposal for funding, and a mixed methods journal article. Using a structure for the type of writing that is consistent with mixed methods research adds to the sophistication and credibility of a study. Most important to recognize with these structures is how the reporting approach changes based on different types of mixed methods designs.

We also suggest several sets of criteria that might be used for evaluating the quality of a mixed methods study, recognizing that various stakeholders, such as graduate committee members, funding agencies, journal editors, and readers, all need some criteria for determining the quality of a mixed methods study. Quality might be assessed for the qualitative and quantitative strands separately, and research methods books detail well these criteria. However, we feel that mixed methods research deserves its own set of criteria, recognizing that no one set of criteria currently exists. We suggest, however, using our "methods" criteria for assessing a good quality study published in journals as a starting point. Another set of criteria draws on recent writings about quality in mixed methods research, and it reflects across the spectrum of planning a study, using a research design, gathering high-quality data, making rigorous interpretations, providing quality reports, and using mixed methods studies for literature syntheses and practice. A final recommendation is to consider the characteristics of the research designs we have advanced in this book and look at the key characteristics of these studies to see whether a particular mixed methods study incorporates those characteristics.

ACTIVITIES

1. Develop an outline for the structure of a graduate student dissertation or thesis proposal that is sensitive to the type of design you plan to use.

2. Locate a published mixed methods journal article in your field. Use the points made in this chapter for selecting and evaluating a good published mixed methods journal article to critique the selected study.

3. Obtain the guidelines for a research proposal from a private foundation or federal agency other than NIH. Take the outline of topics for the NIH proposal as found in Table 8.3. Adapt them to fit the guidelines for the funding agency.

4. For a mixed methods project that you are designing, use the criteria mentioned by O'Cathain (in press) to critique your project.

ADDITIONAL RESOURCES TO EXAMINE

For discussions about writing a mixed methods study, see the following resources:

Sandelowsi, M. (2003). Tables or tableaux? The challenges of writing and reading mixed methods studies. In A. Tashakkori & C. Teddlie (Eds.), *Handbook of mixed methods in social & behavioral research*. Thousand Oaks, CA: Sage.

Stange, K. C., Crabtree, B. F., & Miller, W. L. (2006). Publishing multimethod research. *Annals of Family Medicine, 4*, 292–294.

For criteria for evaluating a quantitative study and a qualitative study, see the following resource:

Creswell, J. W. (2008). *Educational research: Planning, conducting, and evaluating quantitative and qualitative research* (3rd ed.). Upper Saddle River, NJ: Pearson Education.

For a discussion about the criteria to use in evaluating a mixed methods study and in scholarly writing, see the following resources:

Creswell, J. W. (2009). *Research design: Qualitative, quantitative, and mixed methods approaches* (3rd ed.). Thousand Oaks, CA: Sage.

O'Cathain, A. (in press). Assessing the quality of mixed methods research: Towards a comprehensive framework. In A. Tashakkori & C. Teddlie (Eds.), *SAGE Handbook of mixed methods in social & behavioral research* (2nd ed.). Thousand Oaks, CA: Sage.

O'Cathain, A., Murphy, E., & Nicholl, J. (2008). The quality of mixed methods studies in health services research. *Journal of Health Services Research and Policy, 13*(2), 92–98.

See the following for a Web resource:

The NIH SF424 (R&R) form, which details the narrative format for an NIH proposal, can be downloaded from the NIH Web site: http://grants.nih.gov/ grants/forms.htm.

CHAPTER 9

SUMMARY AND RECOMMENDATIONS

In the approximate 20-year history of mixed methods (Greene, 2008), the landscape of this field has developed dramatically. Consequently, in this book, we have attempted to update our ideas, incorporate new trends, and cite many articles published in the mixed methods literature. The circle of scholars and fields embracing mixed methods continues to expand (Tashakkori, 2009). In this chapter, we provide overall recommendations for the conduct and design of a mixed methods study by touching on key topical areas that are important to consider and that have been touched on throughout this book. To this end, we shall suggest in this chapter that those designing and conducting a study do as follows:

- Prepare a methodological paper that advances the mixed methods literature in addition to a content paper about the study.
- Define mixed methods.
- Use mixed methods terms in the study report.
- Take a philosophical stance, and discuss it.
- Be explicit about the mixed methods design, and convey issues in making it rigorous and persuasive.
- Advance the value added by the use of a mixed methods approach.

● ON WRITING A METHODOLOGICAL PAPER

Earlier, we discussed the current state of the field of mixed methods as expressed in the writings by Creswell (2008a, 2009b), Greene (2008), and Tashakkori and Teddlie (2003b). In conducting a mixed methods study, many authors first consider the importance of developing a study that contributes to the specific content area within the field. We urge you to not only complete this study but also to consider writing a methodological article that discusses how your study advances the field of mixed methods research. The steps involved in crafting this article might be as follows:

- First examine the topics that have been discussed in the mixed methods literature by consulting Table 2.2, which maps many of the current issues under discussion.
- Consider what topic your paper might contribute to or extend. Review related writings in the mixed methods field so that you can adequately position your study within that body of literature.
- Write an article that briefly explains the empirical study and then advances the unique mixed methods features of the article. Start the article with the methods orientation in which you review the authors who have studied the methods orientation and mention how your project makes a unique contribution to that orientation. Then advance your empirical study, providing a good but brief analysis of the major components of the article (e.g., the problem, the questions, the methods, the results, and the significance). End the article with a recap of the significant methodological aspects of your article.

One of the challenges in writing a methodological article is to decide how to blend the discussion of the empirical study on a topic with the discussion of the study's unique methodological characteristics. A good article to examine to see how this might be done is a recent *Journal of Mixed Methods Research* (JMMR) article by Woolley (2009). Woolley conducted a sociological study of the interplay of structural factors (i.e., gender, education, and employment) and personal agency (i.e., confidence, independence, and proactivity in pursuing personal interests and plans) in the lives of young people aged 18 to 25 in Derby, England. This article was based on her dissertation, and it showed how to craft an article that would report both the empirical study and the methodological innovations in the right proportions. She began her article by discussing a topic in the mixed methods literature— the integration of quantitative and qualitative data—and how authors in the mixed methods field had a range of difficulties that hindered progress toward more frequent and successful integration of data sets. She then described her sociological study and a rationale for why mixed methods was chosen as her preferred approach. She also detailed her methods and provided a diagram of the mixed methods procedures and discussed her results from combining both qualitative and quantitative data. The conclusion of this article then focused on the methodological issue of how the study illustrated successful integration. In summary, an outline of this article would show a beginning point of lodging the study within a problem area or topic of mixed methods research, the description of the empirical study, and a return, at the end of the article, to mixed methods implications. It balanced just enough information about the empirical study with a discussion about its contribution to understanding integration in the mixed methods literature.

ON DEFINING MIXED METHODS ●

As you design your project, how do you plan on defining mixed methods research? The answer to this question depends on your orientation to mixed methods—whether you have more of a methods orientation, a methodological orientation (i.e., mixing quantitative and qualitative research throughout the study), a philosophical orientation, or more of a phenomenological orientation, wanting to make sense of the world through multiple perspectives and means. You have several definitions from which to choose (as introduced in Chapter 1). As is evident throughout this book, we have chosen a "methods" orientation because we are applied research methodologists and believe that it is a clean way to talk about mixed methods. We also identified core characteristics of mixed methods research, a small set of key principles that can easily be conveyed to others. Mixed methods researchers need today to be able to convey their research approach in a simple, straightforward way to audiences, and the key characteristics and "methods" approach, we believe, accomplishes this message.

One challenge in defining mixed methods research may be determining the "boundaries" of what constitutes this form of inquiry. Here are some recommendations related to issues in defining mixed methods that researchers often raise:

● *Separate mixed methods research from qualitative research.* Researchers who are new to mixed methods research and to qualitative research may think that mixed methods and qualitative research are the same. However, we view mixed methods research as a method approach that is separate and distinct from qualitative research. Quantitative, qualitative, and mixed methods research represent the three main methodological approaches used in the social and behavioral sciences (Tashakkori & Teddlie, 2003a). This question often arises out of an incomplete understanding of mixed methods research and qualitative research.

● *Recognize the difference between "mixed methods" and "mixed model" studies in quantitative research.* The two names are similar, even though they represent different research procedures. Mixed model research is the name given to a category of quantitative statistical techniques that take into account both fixed and random effects during quantitative data analysis and parameter estimation (Cobb, 1998). Therefore, this approach does "mix" models (fixed and random) during analysis, but it does not mix quantitative and qualitative data.

● *Realize that surveys with both open-ended and closed-ended questions provide a minimal qualitative database because of the short open-ended responses.* Consider a survey study that includes a few open-ended questions as part of the survey. The researcher analyzes the qualitative responses to validate the quantitative findings. Is this a mixed methods study? The qualitative data may consist of short sentences and brief comments—hardly the type of qualitative data that involves rich context and detailed information from participants (Morse & Richards, 2002). Although it may not include a rich collection of qualitative data, this approach does include the collection of both quantitative and qualitative data and we consider it an example of mixed methods research.

● *Consider a mixed methods study as one in which the researcher collects both quantitative and qualitative data.* In a content analysis study, only one type of data (qualitative) is collected, and this approach falls short of collecting *both* qualitative and quantitative data. For example, a researcher would collect only qualitative data but would analyze the data both qualitatively (developing themes) and quantitatively (counting words or rating responses on predetermined scales). A more typical content analysis study would be one in which the researcher collects only qualitative data and transforms it into quantitative data by counting the number of codes or themes. Are either of these examples mixed methods research? Certainly they use "mixed methods data analyses" (Onwuegbuzie & Teddlie, 2003) consisting of both qualitative and quantitative data analysis, but the data collection procedure involves collecting only qualitative data (and not quantitative data). Under a strict definition of mixed methods that includes collecting both qualitative and quantitative data, this type of study would not be mixed methods. Under a "methodological" definition—combining at any stage in the process of research—the study would be considered mixed methods because both qualitative and quantitative data analysis is going on.

- *Separate mixed methods research from multimethod research.* In multimethod research (Morse & Niehaus, 2009), the researcher collects, analyzes, and mixes multiple forms of either qualitative or quantitative data. For example, a researcher could collect multiple forms of qualitative data, such as community documents for a participatory action research study and interviews during grounded theory research. A researcher could collect, analyze, and mix different types of quantitative data (e.g., quantitative surveys with structured observations). These forms of research are generally referred to as multimethod research instead of mixed methods research, because they are based on multiple qualitative *or* quantitative methods and data sets.

ON USING TERMS ●

In describing the nature of mixed methods, certain terms are apt to be used. Right now in the mixed methods field, glossaries are being compiled (and this book is no exception) that convey the language of mixed methods research (see also Morse & Niehaus, 2009; Teddlie & Tashakkori, 2009). However, the terms to use in designing and conducting a mixed methods study are far from settled. The issue being raised is what is the language of mixed methods? The issue raised by Tashakkori and Teddlie (2003b) is whether we need a new language for mixed methods research, a language for mixed methods that is separate from the language of either quantitative or qualitative research. Or should the terms be drawn from qualitative and quantitative research? We are reminded of the language to emerge in qualitative research in the early 1980s around the topic of qualitative validity and how terms such as *trustworthiness* and *authenticity* were used to create a distinct, new language for qualitative inquiry (Lincoln & Guba, 1985).

In writing about validity, Onwuegbuzie and Johnson (2006) intentionally called validity, "legitimation," to create a separate, distinct language for mixed methods. In our work on research designs, we referred to one of our designs as an exploratory sequential design to not only advance a new, distinct name for a design but also to signify that the research would first explore qualitatively and then follow up sequentially, quantitatively (Creswell, Plano Clark, et al., 2003). These examples illustrate the creation of a new language for mixed methods research. Illustrating an example of made-up terms, writers in a recent mixed methods psychology text used the term *qualiquantology* to express their discomfort with mixing qualitative and quantitative methods (Stenner & Rogers, 2004). When Teddlie and Tashakkori (2009) speak of "inference transferability" (p. 311) they have created blended terms with both quantitative (inference) and qualitative (transferability) meanings.

There are counterexamples of terms that give a strong quantitative orientation to mixed methods research. For example, Teddlie and Tashakkori (2009) have used the terms *inferences* and *meta-inferences* to denote when the results are incorporated into a coherent conceptual framework to provide an answer to the research question. These terms seem to lean in the direction of quantitative research in which investigators draw inferences from a sample to a population. Another example of a quantitative leaning term is *construct validity*, which has been used by Leech, Dellinger, Brannagan, and Tanaka (2010) as an umbrella validity concept for mixed methods research. This term is drawn from long-established quantitative research and measurement ideas and is not typically associated with qualitative research. Recently, in discussing the rationales for conducting mixed methods, Collins, Onwuegbuzie, and Sutton (2006) included rationales that have a strong association in quantitative research. Instrument fidelity (e.g., assessing the appropriate and/or utility of existing instruments, creating new instruments, and monitoring performance of human instruments) and treatment integrity (i.e., assessing fidelity of intervention) were included as reasons for conducting mixed methods research.

Terms for mixed methods research have been advanced as well that take on qualitative orientations. For example, Mertens (2009) uses her "transformative-emancipatory" framework (p. v) when talking about mixed methods research, a standpoint perspective often associated with qualitative inquiry. Unquestionably, the language that has emerged is both new and oriented toward quantitative and qualitative inquiry. We are inclined to support the conclusion reached by Tashakkori and Teddlie (2003b) that ultimately a bilingual language will win the day in mixed methods research. Further, the list of bilingual terms continues to grow. If this is the case, our recommendation is that mixed methods researchers develop a vocabulary of distinct mixed methods terms and use them in designing and writing their mixed methods study. Furthermore, citing references for key terms helps clarify your mixed methods approach as well as alert readers to individuals that are leading the way in advancing new terms.

● ON USING PHILOSOPHY

One of the most confusing issues for individuals designing a mixed methods study is whether to discuss the philosophical foundations and assumptions that provide a framework for conducting their studies (as introduced in Chapter 2) in their proposals and reports. On the qualitative side of mixed methods, philosophical foundations are made explicit and are often a necessary part of describing a study; on the quantitative side, the philosophical assumptions are seldom mentioned. The recent concern from qualitative researchers that alleges mixed methods favors postpositivist thinking over more interpretive approaches (Denzin & Lincoln, 2005; Howe, 2004) is a symptom that certain researchers have begun to question the philosophy behind mixed methods research. These varied stances on mixed methods place the researcher in a precarious position of not knowing whether philosophy should be included, what philosophical stance might be taken, and how it would actually be written into a study.

Should a passage on philosophy be included in your mixed methods study? This decision depends on the audience for your study and whether the audience tends to be more qualitative than quantitative. Qualitative researchers have advocated for the use of explicit philosophical discussions about key assumptions that researchers bring to a study, such as their view as to how reality is constructed (ontology), how knowledge is acquired (epistemology), how values need to be honored and stated (axiology), and how the procedures are derived from an inductive, bottom-up approach or more of a deductive, top-down approach (see Guba & Lincoln, 2005, and Chapter 2 for a discussion of these different perspectives). In the structure of topics to be presented in a mixed methods proposal in Chapter 8, we recommended that a section on philosophy be included and that it might be placed between the Purpose and the Literature Review (see Table 8.1). Those familiar with qualitative research will welcome this section. Such a section would take the philosophical assumptions previously mentioned, relate them to a paradigm stance, and discuss specifically how they will be incorporated into your specific study.

How to decide on what paradigm or worldview stance to use is more problematic. In Chapter 2, we discussed several paradigm stances that might be taken. We first discussed that there were several paradigms that could be adopted for mixed methods research: postpositivist, constructivist, advocacy, and pragmatism. Further, as applied to mixed methods, researchers might take one or more "stances" toward using a paradigm or worldview: use one paradigm or worldview, use multiple paradigms, link the type of paradigm to the research design being used, or use the worldview that characterizes their discipline orientation. Further, as we indicated in the research designs discussion in Chapter 3, we began to tie paradigms to types of research designs: For a convergent type of design, the researcher might use pragmatism as an overarching worldview; for a sequential type of design, multiple worldviews might be used that relate to different stages or phases in the research process. Thus, our recommendation is that the mixed methods researcher needs to consider the options for use of worldviews and choose which option makes the most sense given the researcher's beliefs and the audience for their mixed methods study. We personally recommend that the paradigm–design fit is a useful perspective to take and that the worldview can shift during a project.

How worldviews can be applied in mixed methods studies is an area that Greene (2007) has questioned. We need more good models of the writing of worldviews into mixed methods studies. Should it be threaded throughout the study, incorporated into a single section, or included in the literature review? Readers are generally not familiar with the different philosophical assumptions (e.g., ontology, epistemology), nor are they familiar with the different paradigms being discussed by qualitative researchers (e.g., postpositivism, constructivism). Nor are readers acquainted with philosophy being introduced into an empirical study and made explicit. Still, because mixed methods research walks the fine line between quantitative and qualitative research—and since qualitative researchers often include philosophy in their studies—we recommend that a philosophy passage be included which sets forth

- the philosophical assumption behind the mixed methods study;
- the paradigm(s) or worldview(s) being used in the study; and
- how the philosophical assumptions and the worldviews shape the development of the mixed methods study. For example, the use of a deductive theory in the beginning of a study indicates a postpositivist orientation to mixed methods. The discovery of emerging constructs through initial qualitative focus groups indicates a constructivist perspective. The "call for action" at the end of the mixed methods study indicates an advocacy worldview, and the integration of findings by merging them in the discussion shows a pragmatic stance.

These are but a few ways in which philosophy can be integrated into a mixed methods study.

● ON DESIGNING PROCEDURES

In Chapter 3, we reviewed several typologies, or classifications, for types of mixed methods designs. We have provided diagrams of designs and recommended procedures for sketching these diagrams. In the sample studies that we provided, we created diagrams for authors where none had existed. In our perspective, no consensus exists about what designs exist for mixed methods studies, but we have taken the stance of advancing six designs that we feel are most common in the literature on mixed methods today. In terms of design, Kelle (2006) seems on target when he suggested that there is currently no canonization of mixed methods designs.

We know that the basic designs are not complex enough to mirror actual practice, although our thinking is to advance designs for the first-time mixed methods researcher. First, we have highlighted that complex designs are being used and reported in the literature. For example, as shown in Appendix F, Nastasi and colleagues wrote about a complex evaluation design with multiple stages and the combination of both sequential and concurrent phases (Nastasi et al., 2007). Second, we also note that the designs reported in journals and funding proposals have incorporated "unusual blends" of methods, such as combinations of quantitative and qualitative longitudinal data, discourse analysis with survey data, secondary data sets with qualitative follow-ups, and the combination of qualitative themes with survey data to produce new variables (Creswell, in press-b; Plano Clark, 2010). The representation of designs has also advanced joint displays for arraying both quantitative and qualitative data in the same table, an approach encouraged by the matrix feature of qualitative software products (see Kuckartz, 2009).

Our designs and the many classifications bring a typology approach to mixed methods design. As introduced in Chapter 3, there are other approaches to research design. Arguing that we need an alternative to typologies, recall that Maxwell and Loomis (2003) conceptualized a systems approach of five interactive dimensions of the research process, and Hall and Howard (2008) suggested a synergistic approach in which two or more options interacted so that their combined effect was greater than the sum of the individual parts.

One way that designers of mixed methods studies can think about designs is not to try to "type" their design or provide a name for the design (along with accompanying diagram) but to reflect on the actual practice of mixed methods research. As Greene (2008) mentioned, practice will lead the way toward a consensus around designs. Indeed, in the last couple of years, there have been detailed discussions about the procedures used by research teams and individuals as they negotiated issues in completing a mixed methods project. The recent article by Brady and O'Regan (2009), included as Appendix D, discussed a mixed methods study of a youth mentoring program in Ireland and highlighted the journey of the team through adopting a type of design, establishing an epistemological position, and conducting data analysis using various methods and sources. Another recent article from Canada by Vrkljan (2009) discussed the topic of car driving safety by older individuals and their copilots. In this article, she reflected on the decisions involved in constructing a mixed methods study from contextualizing the nature of the study, to the experiences of the researcher, and to the various decisions that were considered to arrive at a final design. Vrkljan detailed key strategies she used to determine how to integrate and interpret qualitative and quantitative data, validating findings and the length of time involved in collecting data and preparing manuscripts for publication. An article by Johnstone (2004), from Australia, tells her story of writing a mixed methods dissertation, from reviewing the literature about paradigm assumptions to a conceptualization of the research process, including 20 steps (or "cells," pp. 265–266) from her personal experiences in health services to a tentative study design of collecting, analyzing, and synthesizing outcomes to asking more questions and modifying the study. She ends the article by discussing how she structured her research report, including introduction, theory and literature, and reporting the findings.

These examples of detailed procedures suggest that mixed methods researchers can now look at examples of the actual decisions and steps taken by other researchers as they conduct mixed methods studies. More and more articles are coming forward that explicitly describe how their mixed methods studies were actually conducted. We would encourage researchers to reflect on the process of conducting their own studies and write methodological articles that describe their procedures, their challenges, and their strategies for resolving design issues. At this point in the mixed methods conversation, it may require looking closely at dissertation projects in which authors typically detail these procedures and, at the same time, reflect on the challenges that arise.

● ON THE VALUE ADDED BY MIXED METHODS

Regardless of the design and its procedures, the utility of mixed methods research—from a pragmatic approach—is tied to whether it is a valuable approach. In our earlier definition (see Chapter 1), we ended with the assumption that the combination of methods provides a better understanding than either quantitative methods or qualitative methods alone. Can this assumption be substantiated? In tracing the recent history of mixed methods, a reference was made to a critical question asked by the president of SAGE Publications, Inc., during a luncheon meeting. He asked, "Does mixed methods provide a better understanding of a research question than either quantitative or qualitative research alone?" (Creswell, 2009a, p. 22). This difficult question is central to justifying mixed methods and giving it legitimacy. Unfortunately, it remains unanswered in the mixed methods community.

We can, however, speculate about how it might be addressed. One approach is to turn to research procedures used in early studies that compared participant observation with survey results (Vidich & Shapiro, 1955) or interviews with surveys (Sieber, 1973) and examine if the two databases converge or diverge in understanding a research problem. A second approach is to proceed with an experiment in which groups of readers examine a study divided into a qualitative, a quantitative, and a mixed methods part. In this experiment, outcomes are specified, such as the quality of interpretation, the inclusion of more evidence, the rigor of the study, or the persuasiveness of the study, and the three groups could be compared experimentally (see Haines, 2010). A third approach is to examine some outcomes suggested by authors of published mixed methods studies. One such outcome might be "yield," such as advanced by O'Cathain, Murphy, and Nicholl (2007), and assessed by a number of publications and whether the authors of a mixed methods study actually integrated the data. Other outcomes could be analyzed using qualitative document analysis approaches, and themes developed from statements of value posed by authors of mixed methods empirical articles and methodological studies. For example, authors from the field of communication studies suggested that the value of mixed methods lies in addressing limitations in the results learned from one method: "to address more thoroughly this question, and account for some of the possible limitations of study—one, a broader based assessment of students' involvement in intercultural communication courses was pursued" (Corrigan, Pennington, & McCroskey, 2006, pp. 15–16).

Pending further studies that shed light on the value question, we can recommend that authors incorporate into their purpose statements a rationale for why they are using mixed methods (as mentioned in Chapter 5). Thus, we encourage "value" statements that can talk about the importance of conducting mixed methods. Sometimes these statements will be "content" focused on how the content of the study is enhanced through mixed methods; at other times, these statements will be "methods" specific and relate to how the methods of data collection, analysis, and interpretation were enhanced by using mixed methods. Discovering the value of mixed methods will unfold over time, and sometimes it will be its value in its own right and sometimes its value as compared with either quantitative or qualitative research or both. Regardless, stronger justification for use of mixed methods will emerge and contribute to its use.

SUMMARY

Throughout this book, we have encouraged planning and executing a mixed methods project that incorporates the latest thinking about mixed methods. We offer several final recommendations for those who are designing and conducting a mixed methods study. Consider developing not only an empirical article from your study but also a methodological article that advances your study as well as indicates its contribution to the mixed methods literature. Further, define mixed methods research for audiences of your report, because it is a reasonably new approach to inquiry. Although methods, methodology, philosophical, and framework definitions all exist in the literature, we suggest that you consider more of a methods definition so it does not become entangled with philosophical ideas, and you can also add a philosophical passage to your study. Moreover, in conveying your study, consider identifying the core characteristics of mixed methods as a good way to summarize the essential components of this mode of inquiry.

It is also important to become familiar with the terms that are emerging to describe mixed methods research. These can be typically found in glossaries in mixed methods books. It is also helpful to realize that these terms are continually being created and that the lexicon of this research approach will change and grow over time. These terms, however, do represent the ideas of authors, and you need to cite the references to terms that you use in a project. We also recommend that your study include a passage on the philosophical foundation that you are using in your mixed methods study. There are several options for where this passage might be presented. Several options as well exist for the stance on philosophy that you can take ranging from the use of one worldview to the use of many and how its discussion might include commentary about philosophical assumptions, worldview possibilities, and how the philosophy informs the stages of your research study.

Many research designs are available to use in mixed methods research. Our recommendation is that you identify the procedures in your mixed methods study, whether this identification includes a specific design, a diagram, notation, or the listing of steps and decisions taken in your study. Our advice also is that you examine closely some of the emerging studies in which authors identify the procedures, challenges, and strategies for addressing these challenges. Then, when you write your mixed methods study, we suggest that you be reflective about the procedures that you have used so they can be conveyed to readers as a separate article or in your research report. Finally, you may need to justify the use of mixed methods for audiences to which you are presenting your report. Consider what value your mixed methods approach adds to understanding either the content or the methods in your study. Be explicit about identifying this value in your final report.

ACTIVITIES

1. Take an empirical mixed methods article that you have developed. Discuss topics in the mixed methods literature to which your study contributes, and rewrite the article to emphasize this contribution at the beginning and at the end of the revised article.

2. In writing your plan or study for mixed methods, examine the glossary of terms at the end of this book. Identify (and list) the terms you will incorporate into your study from the glossary so that you are "speaking" as a mixed methods researcher. Incorporate these terms into your project.

3. At the end of this chapter, we reflect on different ways to begin understanding the "value" of mixed methods research. How would you design a study to address this issue? The project you design could be quantitative, qualitative, or mixed methods.

ADDITIONAL RESOURCES TO EXAMINE

For additional readings on the topics currently under discussion and the controversies about the field of mixed methods research, examine the following resources:

Creswell, J. W. (in press-a). Controversies in mixed methods research. In N. K. Denzin & Y. S. Lincoln (Eds.), *The SAGE handbook of qualitative research* (4th ed.). Thousand Oaks, CA: Sage.

Creswell, J. W. (in press-b). Mapping the developing landscape of mixed methods research. In A. Tashakkori & C. Teddlie (Eds.), *SAGE handbook of mixed methods research in social & behavioral research* (2nd ed.). Thousand Oaks, CA: Sage.

Greene, J. C. (2008). Is mixed methods social inquiry a distinctive methodology? *Journal of Mixed Methods Research, 2*(1), 7–22.

Tashakkori, A. T. (2009). Are we there yet?: The state of the mixed methods community [Editorial]. *Journal of Mixed Methods Research, 3*(4), 287–291.

For a recent discussion about a definition for mixed methods research and the terms to use, see the following resources:

Johnson, R. B., & Onwuegbuzie, A. J. (2004). Mixed methods research: A research paradigm whose time has come. *Educational Researcher, 33*(7), 14–26.

Johnson, R. B., Onwuegbuzie, A. J., & Turner, L. A. (2007). Toward a definition of mixed methods research. *Journal of Mixed Methods Research, 1*(2), 112–133.

On the philosophy behind mixed methods research and recommendations on how to advance philosophy within mixed methods, see this resource:

Morgan, D. L. (2007). Paradigms lost and pragmatism regained: Methodological implications of combining qualitative and quantitative methods. *Journal of Mixed Methods Research, 1*(1), 48–76.

On specific procedures for conducting a mixed methods study, see the varied perspectives introduced by these authors:

Brady, B., & O'Regan, C. (2009). Meeting the challenge of doing an RCT evaluation of youth mentoring in Ireland: A journey in mixed methods. *Journal of Mixed Methods Research, 3*(3), 265–280.

Johnstone, P. L. (2004). Mixed methods, mixed methodology health services research in practice. *Qualitative Health Research, 14*(2), 239–271.

Vrkljan, B. H. (2009). Constructing a mixed methods design to explore the older drive-copilot relationship. *Journal of Mixed Methods Research, 3*(4), 371–385.

Appendix A—An Example of the Convergent Parallel Design

Unwritten Rules of Talking to Doctors About Depression

Integrating Qualitative and Quantitative Methods

Marsha N. Wittink, MD, MBE[1]
Frances K. Barg, PhD[1,2]
Joseph J. Gallo, MD, MPH[1]

[1]*Department of Family Medicine and Community Health, School of Medicine, University of Pennsylvania, Philadelphia, Pa*

[2]*Department of Anthropology, School of Arts and Sciences, University of Pennsylvania, Philadelphia, Pa*

ABSTRACT

PURPOSE: We wanted to understand concordance and discordance between physicians and patients about depression status by assessing older patient's views of interactions with their physicians.

METHODS: We used an integrated mixed methods design that is both hypothesis testing and hypothesis generating. Patients aged 65 years and older, who identified themselves as being depressed, were recruited from the offices of primary care physicians and interviewed in their homes using a semistructured interview format. We compared patients whose physicians rated them as depressed with those whose physicians who did not according to personal characteristics (hypothesis testing). Themes regarding patient perceptions of their encounters with physicians were then used to generate further hypotheses.

RESULTS: Patients whose physician rated them as depressed were younger than those whose physician did not. Standard measures, such as depressive symptoms and functional status, did not differentiate between patients. Four themes emerged in interviews with patients regarding how they interacted with their physicians; namely, "My doctor just picked it up," "I'm a good patient," "They just check out your heart and things," and "They'll just send you to a psychiatrist." All patients who thought the physician would "just pick up" depression and those who thought bringing up emotional content would result in a referral to a psychiatrist were rated as depressed by the physician. Few of the patients who discussed being a "good patient" were rated as depressed by the physician.

CONCLUSIONS: Physicians may signal to patients, wittingly or unwittingly, how emotional problems will be addressed, influencing how patients perceive their interactions with physicians regarding emotional problems.

● INTRODUCTION

The primary health care setting plays a key role for older adults with depression and other psychiatric disturbances, because older persons in the community are unlikely to receive mental health care from a mental health care specialist.[1-3] Nevertheless, evidence on the quality of care for older adults with depression in primary care suggests that often their depression is not diagnosed or actively managed.[4] Although much attention has been focused on understanding physician-based reasons for underdiagnosis of depression, primary care physicians believe that barriers to depression treatment are most often patient centered and related to patient attitudes and beliefs about depression care.[5]

Several previous studies have linked patient-physician communication to important health outcomes and adherence to treatments.[6,7] When patients like the way their physician communicates with them, they are more likely to heed the physician's recommendations and are less likely to sue for medical malpractice in the event of a negative outcome.[8] For depression, how patients perceive the communication between physician and patient

becomes particularly salient, because patients may not readily reveal their feelings or accept the diagnosis, and they may be unwilling to take medicine or seek counseling. Studies of physician communication behaviors have suggested that certain behaviors, such as showing empathy, listening attentively, and asking questions about social and emotional issues, are associated with increased patient willingness to share concerns.[8,9]

Our study focuses on the patient's view of the interactions with their physicians and is based on an integrated mixed methods design that includes elements derived from both quantitative and qualitative traditions,[10,11] alternating hypothesis-testing and hypothesis-generating strategies. This design allowed us to link the themes regarding how patients talk to their physicians with personal characteristics and standard measures of distress. We suspected that patients who identified themselves as being depressed and whose physicians rated them as depressed would report more distress and functional impairment than patients not rated as depressed by their physicians. Our work differs from previous studies of communication and the physician-patient relationship in that most previous work focuses on the interaction of patient and physician at a specific visit and underemphasizes the patient's contribution to and perspective on the active production of the diagnostic process.[9,12,13] In this study, we wanted to understand aspects of the physician-patient relationship (as perceived by the patient) that may influence the way patients communicate about depression. To draw attention to a clinically relevant situation, we focus on older adults who identified themselves as being depressed.

● METHODS

Study Sample

The overarching goal of the Spectrum Study (the parent study from which our sample was derived) was to characterize how older primary care patients report depression. The design of the study was a cross-sectional survey of patients aged 65 and older and their physicians recruited from nonacademic primary care practices in the Baltimore, Md, area (n = 355).[14,15] Subsequently, patients were selected for semistructured interviews using purposive sampling.[10] From the 102 persons who provided semistructured interviews, 48 were selected for this study because they identified themselves as being depressed and had physician ratings of depression (the online-only Supplemental Appendix provides a summary of the sampling methods and is available at http://www.annfammed.rg/cgi/content/full/4/4/302/DC1). The study protocols were approved by the Institutional Review Board of the University of Pennsylvania.

Measurement Strategy

Physician Evaluation of the Patient at the Index Visit

At the index visit, the physician rated the patient's level of depression on a 4-point scale: none at all, mild, moderate, or severe. How well the physician knows the patient was rated as very well, somewhat, or not at all.

Patient Assessment

In addition to obtaining information from the respondents on age, sex, ethnicity, marital status, living arrangements, level of educational attainment, and the number of visits made to the practice for medical care within 6 months of the index visit, we used the following measures to examine selected factors that have been associated with recognition of depression in primary care settings.[16] We used the Center for Epidemiologic Studies Depression (CES-D) scale, which was developed by the National Institute of Mental Health for use in studies of depression in community samples,[17-23] and the Beck Anxiety Inventory (BAI), which was developed to measure the severity of anxiety symptoms.[24,25] Thresholds used to indicate substantial depressive symptoms on the CES-D range from 16 to 21,[19,21] and scores of 14 and above on the BAI typically indicate high levels of anxiety.[24] We used the Beck Hopelessness Scale (BHS) to assess factors (hopefulness about the future, a sense of giving up, and future anticipation or plans)[26] found to be related to suicidal ideation.[27] We measured baseline medical comorbidity with an adaptation of the Charlson index,[28] and we used questions from the Medical Outcomes Study 36-item short-form health survey (SF-36) to assess functional status.[29] Cognition we assessed with a standard measure of global functioning (Mini-Mental State Examination [MMSE]).[30,31]

Semistructured Interviews

Trained professional interviewers carried out semistructured interviews in the patient's home, and these interviews were recorded, transcribed, and entered into N6 software for coding and analysis.[32,33] The interview questions used to examine patient's perceptions of their encounters with physicians are displayed in Table A.1. A multidisciplinary team that included medical anthropologists, family physicians, and older persons from the community

Table A.1	Semistructured Interview Guide Questions

Have you discussed your feelings with your doctor?

    If YES, ask A and B
    If NO, skip to C and D

A. Who brought it up? How do you think the discussion went? Do you think (he/she) would have known if you hadn't brought it up?

B. What does (he/she) say about it?

C. What do you think your doctor thinks about the way you feel emotionally?

D. What words (other than depression) would your doctor use to describe how you feel?

processed each transcript for discussion in weekly team meetings (details are provided elsewhere[10] and at http://www.uphs.upenn/spectrum). Study participants were asked: "Have you ever considered yourself depressed?" In practice, the characterization of the patient as depressed was not based on a single yes-or-no response to this question because the interviewer probed further for whether the patient reported being depressed. In summary, we have captured 3 perspectives about the depression status of each patient: (1) a rating from the physician at the index visit, (2) the patient's responses on a standardized questionnaire (CES-D), and (3) the patient's self-report as depressed.

Analytic Strategy

Our analytic strategy reflects the integration of hypothesis testing and hypothesis generation in a single study that is the hallmark of a mixed methods investigation. In the first phase, we compared the personal characteristics of patients who identified themselves as being depressed while their physicians did not with those who were concordant with their physician's rating of depression (using χ^2 or t tests for comparisons of proportions or means, respectively). We used a level of statistical significance set at $\alpha = .05$, recognizing that tests of statistical significance are approximations that serve as aids to interpretation and inference.

In the second phase, we used the constant comparative method, moving iteratively between codes and text to derive themes related to talking with the physician.[34,35] Originally developed for use in the grounded theory method of Glaser and Strauss,[35] this strategy involves taking 1 piece of data (eg, 1 theme) and comparing it with all others that may be similar or different to develop conceptualizations of the possible relations between various pieces of data. During the process of developing themes, the study team did not have access to the survey data, including whether the patient was rated as depressed by the physician. We focused our attention on responses to interview questions related to discussing feelings and emotional issues with the physician (Table A.1). We then related themes to personal characteristics and whether the patient and physician were concordant about depression status. Data analysis was carried out with the use of SPSS (SPSS Corporation, College Station, Texas) and QSR N6.0 (QSR International, Durham, UK).

● RESULTS

Sample Characteristics

In all, 53 patients from the 102 who participated in semistructured interviews considered themselves to have been depressed. Transcripts of 5 were excluded because of missing data, leaving 48 patients in the sample for this study (Figure A.1). Table A.2 compares the characteristics of patients whom the physician rated as depressed with the patients who were not rated as depressed. Except for age (patients who were identified by their physicians as depressed were younger), no significant differences were found among patients whose physician rated them as depressed at the index visit. There were no significant differences in any SF-36 scale means (data not shown in table).

Figure A.1 Flow diagram. Data from the Spectrum Study (2001–2004)

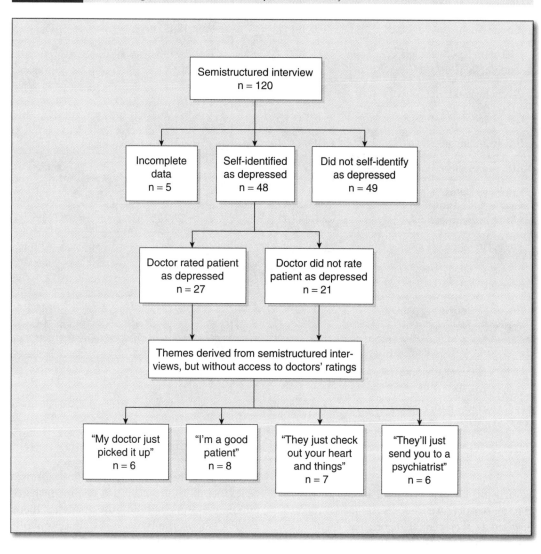

| Table A.2 | Characteristics of Patients Who Identified Themselves as Depressed in Semistructured Interviews (n = 48) | | |

Characteristics	Physician Rated Patient Depressed $n = 27$	Physician Rated Patient Not Depressed $n = 21$	P Value
Sociodemographic characteristics			
Age, mean, No. (SD)	73.0 (5.3)	77.1 (5.3)	.012
Women, No. (%)*	21 (79)	15 (71)	.623
African American, No. (%)*	10 (39)	12 (57)	.173
Education less than high school, No. (%)*	8 (30)	10 (48)	.210
Psychological status			
CES-D score, mean (SD)	18.3 (13.5)	15.6 (10.0)	.450
BAI score, mean (SD)	10.0 (9.2)	11.8 (8.5)	.498
BHS score, mean (SD)	5.5 (4.1)	4.8 (3.7)	.607
Cognitive status			
MMSE score, mean (SD)	27.8 (2.2)	27.1 (3.0)	.371
Physician ratings at index visit			
Physician rates the patient as depressed, No. (%)*	27 (100)	0 (0)	.842
Physician knows the patient very well, No. (%)*	20 (75)	15 (71)	.843

NOTE: Data From the Spectrum Study (2001–2004).

* Column percent.

CES-D = Center for Epidemiologic Studies Depression Scale; BAI = Beck Anxiety Inventory; BHS = Beck Hopelessness Scale; MMSE = Mini-Mental State Examination.

Themes That Emerged in Semistructured Interviews

Several themes emerged from careful review of the transcripts. We describe 4 major themes selected for their clinical importance. The themes relate to the patients' perception of the relationship with their physician.

'My Doctor Just Picked It Up'

In several of the transcripts patients express a belief that their physicians are able to "pick up" on depression without the patient being explicit about their emotions. For example, Mrs K says that her doctor understands how she feels:

"Because she seems to pick up on some things that I don't tell her, and she'll bring it up right now. 'Now you didn't tell me this, let's get down to this. What's going on?' That's the way she is, so I know something is wrong, yes."

This response suggests that the physician has an almost intuitive capability to recognize when something is wrong with a patient, which could reflect the ability of some physicians to recognize nonverbal cues, as is illustrated in the following excerpt from another woman:

"I had one doctor tell me, when I walked into the room, he said, 'Young lady what's your problem?' And um, I was trying to tell him how I was trying to tell him how I was struggling. He said, 'You're depressed.' Yes, he just said, 'You look depressed to me.'"

'I'm a Good Patient'

This theme emerged when patients discussed what the physician thinks of them and often came up specifically in response to the interviewer's question: "What words would your doctor use to describe how you feel?" In this context, patients referred to themselves as "a good patient," suggesting that they perceived themselves as being well-liked by the physician. For example, Mrs S said:

"He thinks I'm a good patient, he thinks I'm doing good. Besides, other people come in there have more pains and that more than I do."

Another patient, Mrs R, said:

"He thinks I'm . . . how does he put it? 'Quite a lady,' and then he told his nurse-practitioner, 'You're going to love her; she's quite a gal.' You know?"

These excerpts illustrate a recognition on the part of the patient that they portray a positive image to the physician. The notion of the good patient is further manifested as a particular role that may be co-constructed by the physician and patient, as seen in the following excerpt from Mr J in response to the interviewer's question: "Did you feel that your doctor understood how you feel?"

"I doubt if I ever discussed it with him. I never felt it important enough to discuss it with him. No,

he wouldn't know, because I go there and cut up and flirt with the girls and kid and everything. He wouldn't know."

Mr J's response illustrates his perception of a role that is perhaps even expected of him during the office visit. For example, when asked, "What do you think your doctor thinks about the way you feel emotionally?" he stated: "He thinks I'm in great physical and mental shape and am very happily married." Nevertheless, this patient considered himself to be depressed and was open to discussing his depression with the interviewer elsewhere in the transcript. Another patient, Mrs R, also discussed how she thought she is a "good patient" in the eyes of her physician. She stated explicitly that her doctor does not care about her feelings:

"No, he don't care. No, in fact . . . he had a substitute come in one time when he wasn't there. . . . This doctor didn't know me. My own doctor does . . . but we don't ever get into my feelings and moods."

Yet when she describes how she thinks her doctor sees her, she evokes the notion of a good patient. When asked, "What do you think your doctor thinks about the way you feel emotionally?" she said:

"He has no idea. He thinks I'm a very, very happy person all the time, wonderful, in excellent health for an old woman, 77 years old. He thinks I'm doing great. He likes me, thinks I'm good. He's always happy to see me, takes enough time to say, 'What are you reading here?' There is only a little bit of small talk."

'They Just Check Out Your Heart and Things'

Several patients mentioned that physicians focus mostly on the physical issues and tend to ignore emotional ones. For instance, Mrs W talks about visits to her physician in the following way:

"[I] just know it's going to be a 3-minute visit, and he'll say, 'Hi, how are you? Good. Need any medicines?' He listens to your chest and back and that's it."

Mr P also portrays his physician as someone who does not focus on emotional issues:

"Well, I don't know—he doesn't bother asking about that. They just check your heart out and things. I'm going to tell you, I don't think they think anything about emotions. I'm just being truthful. I don't think they worry about your mental state, you know, how you feel."

Similarly, Mr R says of his physician:

"He didn't talk about my feelings. All he did, he gave me the numbers that he got from the last blood test, what we're going to do, change the medicine a little bit and that's all."

When asked, "What do you think your doctor thinks about the way you feel emotionally?" he said, "I don't think that it ever occurs to him." Mrs T, another patient, wondered about the reasons that a physician might not want to discuss emotional or mental issues:

"Well it's really not part of, as far as I know, mental exam is not a part of a physical exam at all, you know? So, but even so, doctors, they don't . . . I don't know why they don't address you on it, unless they are afraid that you might not appreciate it, you know? Your mental health is something that is very touchy, something that is very stigmatizing, so people may kind of avoid it if they are not sure how you will react."

'They'll Just Send You to a Psychiatrist'

This theme connotes that patients feel any discussion of emotional issues will lead to a referral to a psychiatrist. We refer to this notion here as *turfing*, a term commonly used among physicians when one passes on difficult issues to another physician with other expertise.

The concept of turfing comes up when patients discuss what their physicians say when the patient brings up emotional issues. For example, in response to the question, "Do you think your doctor is cognizant of your feelings?" Mrs W says, "Oh, I think he knows, yeah, cause he says, 'Well, we'll send you to the psychiatrist.'" And yet when asked whether the physician understands how she feels, she says, "No, no. He just sent me to the psychiatrist." Another patient, Mrs T, also talks about turfing and offers a reason why it may occur when asked, "What do you think you doctor thinks about the way you feel emotionally?"

"I don't know, I think he recommended that I go see a psychiatrist. He's not—obviously, he's not comfortable with trying to treat me—so he never gave me any medicine."

Yet another patient links this notion of turfing to the physicians' focus on the physical aspects of health:

"We never got into emotions that much. They don't get into your emotional health that much. I think if you start complaining about your emotional state, they'll just want to send you to a psychiatrist."

Patient Characteristics and Themes

Table A.3 displays characteristics of patients according to the themes (as indicated in Figure A.1). All of the patients who discussed the theme of "my doctor just picks it up" were women and were concordant with their physicians on the diagnosis of depression. Few of the patients who brought up the "good patient" were rated by their physician as depressed (3 out of 8), and most were women (6 of 8). Among patients who brought up the theme of physicians only focusing on physical illness tended to have more education and to be white; in 4 of 7 cases, the physician rated the patient as depressed. Finally, all of the patients who discussed the notion of being referred when bringing up emotional issues were rated by their physicians as depressed.

● DISCUSSION

Our integrated, mixed methods design allowed us to combine hypothesis testing and hypothesis generation in a single study. Standard measures did not differentiate between patients whose physician rated them as depressed and those whose physician did not (hypothesis testing). When older adults were asked to reflect on how they discuss emotional issues with their physician, however, several themes emerged (hypothesis generating). All the themes represent patients' perceptions of their interaction with their physician regarding feelings and emotional status.

Our study has some potential limitations. First, we relied on the perception of the patient regarding the clinical encounter. Patient perceptions can provide only a partial view of what actually occurs in any given encounter. For the purposes of this study, however, we were specifically interested in the patient's perspective of their interaction with the physician. Because we did not focus on a specific encounter, we considered the narratives in the semistructured interviews to represent the patients' perceptions of encounters over time. We also relied on the patient's self-report of depression because we were interested in the patient's point of view with respect to depression. In doing so, we wanted to recognize that we could not disentangle mild, moderate, and severe depression from somatizing patients, or the worried well. Furthermore, the various assessments were not carried out at the same time. Nevertheless, we attempted to use the quantitative data to sharpen our ability to distinguish themes among participants in a way that can improve our understanding of the role of the physician-patient relationship regarding the identification of depression from both the patient's and physician's points of view. We realize that many system, physician, and patient factors play a role in physician-patient interaction, all of which could not be accounted for in our study. Alternative designs to studying how patient behavior and expectations play a role in identification of depression, such as intensive analysis of physician-patient encounters or interviewing patients immediately following an office visit, would not capture the kind of data we have described here.

"My doctor just picked it up" suggests that these patients might not have known about their depression had the physician not suspected it. The physicians' diagnostic skills, as these patients describe them, appear to include an ability to intuit aspects of the patient's mood without necessarily needing to elicit them directly. This theme emerged only among those patients whose physician rated them as depressed and among patients who reported having discussed their feelings with the physician and who thought their physician understood them. One concern, however, is that for some patients, relying on their physician's ability to " just pick up" on their mood may obviate the need to express mood symptoms at all, leaving depression potentially unaddressed. All the patients who mentioned this theme were women. Perhaps women behave in ways that are stereotypical for depression, leading physicians to pick up on depression without the need for patients to bring it up themselves. It is also possible that physicians, aware that depression is more common among women,[36] are more likely to diagnose depression in women.

"I'm a good patient" may indicate those patients whom physicians do not see as having any negative feelings or being depressed, because the patient and the physician have together created a role that might inhibit any discussion of emotions without happy or positive content. Depression may be seen as a moral failing requiring pulling up oneself by one's bootstraps.[37] The notion of the good patient may be more common among older patients who have grown up in the era of the paternalistic physician. Patients who view themselves as a good patient may operate on the notion that the good patient is one who is respectful of the physician's expertise and recommendations, will be compliant with recommendations, and does not complain or burden their physician. Discussing emotional difficulty with the physician may be seen as unnecessary complaining.

Table A.3	Characteristics of Persons According to Themes Raised in Semistructured Interviews (n = 48)

Characteristics	"My doctor just picked it up" n = 6	"I'm a good patient" n = 8	"They just check out your heart and things" n = 7	"They'll just send you to a psychiatrist" n = 6
Sociodemographic characteristics				
Age, mean y (SD)	73.3 (3.3)	77.5 (4.2)	75.1 (7.8)	71.3 (6.3)
Women, No. (%)*	6 (100)	6 (75)	4 (57)	4 (67)
African American, No. (%)*	2 (33)	3 (38)	2 (28)	3 (50)
Education less than high school, No. (%)*	2 (33)	3 (38)	2 (28)	2 (33)
Psychological status				
CES-D score, mean (SD)	19.0 (11.8)	11.9 (7.4)	15.3 (9.6)	14.0 (10.3)
BAI score, mean (SD)	10.5 (4.9)	10.0 (9.1)	6.4 (4.5)	6.8 (3.8)
BHS score, mean (SD)	4.8 (4.9)	3.8 (3.1)	4.6 (3.7)	5.7 (3.1)
Cognitive status				
MMSE score, mean (SD)	28.7 (1.2)	27.5 (2.2)	28.9 (0.7)	27.8 (1.7)
Physical health				
Physical function score, mean (SD)	64.2 (21.5)	63.6 (31.0)	71.3 (24.8)	56.7 (28.2)
Role physical score, mean (SD)	45.8 (36.8)	65.6 (35.2)	46.4 (44.3)	29.2 (29.2)
Role emotional score, mean (SD)	88.9 (27.2)	72.3 (39.8)	50.0 (50.0)	83.3 (40.8)
Social function score, mean (SD)	75.0 (17.7)	70.3 (34.0)	62.5 (27.0)	72.9 (21.5)
Bodily pain score, mean (SD)	61.3 (17.7)	55.0 (25.8)	50.4 (26.1)	43.8 (24.2)

Appendix A

Characteristics	"My doctor just picked it up" $n = 6$	"I'm a good patient" $n = 8$	"They just check out your heart and things" $n = 7$	"They'll just send you to a psychiatrist" $n = 6$
General health perception score, mean (SD)	41.7 (15.7)	61.3 (17.5)	54.3 (16.4)	42.5 (14.4)
No. of medical conditions, mean (SD)	8.7 (0.8)	6.6 (2.9)	8.0 (3.1)	8.0 (2.3)
No. of visits within 6 months, mean (SD)	2.5 (1.0)	2.8 (1.4)	2.6 (1.5)	2.8 (1.5)
Discussion of depression with physician				
Doctor understood how you feel, No. (%)*	5 (83)	4 (50)	1 (14)	3 (50)
Has discussed feelings with doctor, No. (%)*	5 (83)	3 (38)	1 (14)	2 (33)
Physician ratings at index visit				
Physician rates the patient as depressed, No. (%)*	6 (100)	3 (38)	4 (57)	6 (100)
Physician knows the patient very well, No. (%)*	5 (83)	6 (75)	4 (57)	4 (67)

NOTE: Data From the Spectrum Study (2001–2004).

* Column percents.

BAI = Beck Anxiety Inventory; CES-D = Center for Epidemiologic Studies Depression Scale; MMSE = Mini-Mental State Examination.

"They just check out your heart and things" was mentioned by patients who discuss the tendency of physicians to focus on physical findings and symptoms and who have learned from experience that emotional symptoms are not appropriate for the medical encounter. These patients seem to assume what falls under the purview of physician's expertise is purely physical, namely, patients are clearly not bringing up emotional issues because they may believe their physician will not be interested. Debra Roter and Judith Hall discuss this phenomenon in the following way: "Most patients have particular expectations in mind when they visit the doctor, although they may be reluctant to make these known directly."[12] This expectation appears to lead to a reluctance on the part of the patient to bring up anything that is not viewed as a physical concern.

"They'll just send you to a psychiatrist" was expressed by patients who believe they had been turfed, namely, a sense that the physician will not directly address any emotional issues but will instead send the patient on to a mental health specialist. All the patients who discuss the notion of turfing were rated by the physician as depressed. Thus while these patients tended to discuss turfing in dissatisfied terms, physicians were nonetheless concordant with regard to the depression diagnosis. If patients expect their physician will send them to a psychiatrist when emotional issues are discussed, patients may either avoid discussing emotional issues or they may try to express their emotional issues in physical terms.

We believe our findings have both clinical and methodological implications. Patients come to the physician encounter with experiences and expectations about depression that may have an impact on what patients are willing to tell physicians. The give-and-take between patients and physicians is clearly a dynamic activity, a dance of sorts, with important implications for the ability of physicians to recognize depression and negotiate a treatment plan. From a methodological viewpoint, had we limited the analysis to patient characteristics (a purely quantitative study), we would have missed the patient's perspective. The themes represent patient voices and allowed us to identify possible contributing factors to the dynamic process of physician-patient interaction around depression.

To read or post commentaries in response to this article, see it online at http://www.annfammed.org/cgi/content/full/4/4/302.

Key words: Aged; communication; depression; research methodology; primary health care

Submitted June 8, 2005; submitted, revised, December 9, 2005; accepted December 12, 2005.

Presented in part at the North American Primary Care Research Group (NAPCRG) annual meeting, Banff, Canada, October 28, 2003.

Funding support: The Spectrum Study was supported by grants MH62210-01, MH62210-01S1, and MH67077 from the National Institute of Mental Health. Dr Wittink was supported by a National Research Service Award from the National Institutes of Health (MH019931-08A1). Dr Wittink was supported by a National Research Service Award from the National Institutes of Health (MH019931 08A1) and a Mentored Patient Oriented Research Career Development (K23) award (MH073658).

● REFERENCES

1. Cooper-Patrick L, Gallo JJ, Powe NR, et al. Mental health service utilization by African Americans and Whites: the Baltimore Epidemiologic Catchment Area Follow-Up. *Med Care.* 1999;37:1034–1045.
2. Gallo JJ, Marino S, Ford D, Anthony JC. Filters on the pathway to mental health care, II. Sociodemographic factors. *Psychol Med.* 1995;25:1149–1160.
3. Marino S, Gallo JJ, Ford D, Anthony JC. Filters on the pathway to mental health care, I. Incident mental disorders. *Psychol Med.* 1995;25:1135–1148.
4. Gallo JJ, Bogner HR, Morales KH, Ford DE. Patient ethnicity and the identification and active management of depression in late life. *Arch Intern Med.* 2005;165:1962–1968.
5. Nutting PA, Rost K, Dickinson M, et al. Barriers to initiating depression treatment in primary care practice. *J Gen Intern Med.* 2002;17:103–111.
6. Kaplan SH, Greenfield S, Ware JE, Jr. Assessing the effects of physician-patient interactions on the outcomes of chronic disease. *Med Care.* 1989; 27:S110–127.
7. Stewart MA. Effective physician-patient communication and health outcomes: a review. *CMAJ.* 1995; 152:1423–1433.
8. Roter DL, Stewart M, Putnam SM, et al. Communication patterns of primary care physicians. *JAMA.* 1997;277:350–356.
9. Hall JA, Roter DL, Katz NR. Meta-analysis of correlates of provider behavior in medical encounters. *Med Care.* 1988;26:657–675.
10. Barg FK, Huss-Ashmore R, Wittink MN, Murray GF, Bogner HR, Gallo JJ. A mixed methods approach to understand loneliness and depression in older adults. *J Gerontolo B Psychol Sci Soc Sci.* In press.

Appendix A

11. Tashakkori A, Teddlie C. *Mixed Methodology.* Thousand Oaks, Calif: Sage Publications; 1998.

12. Roter DL, Hall JA. *Doctors Talking With Patients/Patients Talking With Doctors.* Westport, Ct: Auburn House; 1992.

13. Wissow LS, Roter DL, Wilson ME. Pediatrician interview style and mothers' disclosure of psychosocial issues. *Pediatrics.* 1994;93:289–295.

14. Bogner HR, Wittink MN, Merz JF, et al. Personal characteristics of older primary care patients who provide a buccal swab for apolipoprotein E testing and banking of genetic material: the spectrum study. *Community Genet.* 2004;7:202–210.

15. Gallo JJ, Bogner HR, Straton JB, et al. Patient characteristics associated with participation in a practice-based study of depression in late life: the Spectrum study. *Int J Psychiatry Med.* 2005;35:41–57.

16. Klinkman MS. Competing demands in psychosocial care. A model for the identification and treatment of depressive disorders in primary care. *Gen Hosp Psychiatry.* 1997;19:98–111.

17. Radloff LS. The CES-D Scale: A self-report depression scale for research in the general population. *Appl Psychol Measurement.* 1977;1:385–401.

18. Comstock GW, Helsing KJ. Symptoms of depression in two communities. *Psychol Med.* 1976;6:551–563.

19. Eaton WW, Kessler LG. Rates of symptoms of depression in a national sample. *Am J Epidemiol.* 1981;114:528–538.

20. Newmann JP, Engel RJ, Jensen JE. Age differences in depressive symptom experiences. *J Gerontol.* 1991;46: P224–235.

21. Gatz M, Johansson B, Pedersen N, Berg S, Reynolds C. A crossnational self-report measure of depressive symptomatology. *Int Psychogeriatr.* 1993;5:147–156.

22. Miller DK, Malmstrom TK, Joshi S, et al. Clinically relevant levels of depressive symptoms in community-dwelling middle-aged African Americans. *J Am Geriatr Soc.* 2004;52:741–748.

23. Long Foley K, Reed PS, Mutran EJ, DeVellis RF. Measurement adequacy of the CES-D among a sample of older African-Americans. *Psychiatry Res.* 2002;109:61–69.

24. Beck AT, Epstein N, Brown G, Steer RA. An inventory for measuring clinical anxiety: psychometric properties. *J Consult Clin Psychol.* 1988;56:893–897.

25. Steer RA, Willman M, Kay PAJ, Beck AT. Differentiating elderly medical and psychiatric outpatients with the Beck Anxiety Inventory. *Assessment.* 1994; 1:345–351.

26. Beck AT, Weissman A, Lester D, Trexler L. The measurement of pessimism: the hopelessness scale. *J Consult Clin Psychol.* 1974;42:861–865.

27. Hill RD, Gallagher D, Thompson LW, Ishida T. Hopelessness as a measure of suicidal intent in the depressed elderly. *Psychol Aging.* 1988;3:230–232.

28. Charlson ME, Pompei P, Ales KL, MacKenzie CR. A new method of classifying prognostic comorbidity in longitudinal studies: development and validation. *J Chronic Dis.* 1987;40:373–383.

29. McHorney CA. Measuring and monitoring general health status in elderly persons: practical and methodological issues in using the SF-36 Health Survey. *Gerontologist.* 1996;36:571–583.

30. Folstein MF, Folstein SE, McHugh PR. "Mini-mental state". A practical method for grading the cognitive state of patients for the clinician. *J Psychiatr Res.* 1975;12:189–198.

31. Tombaugh TN, McIntyre NJ. The mini-mental state examination: a comprehensive review. *J Am Geriatr Soc.* 1992;40:922–935.

32. DiGregorio S. *Analysis as Cycling: Shifting Between Coding and Memoing in Using Qualitative Software. Strategies in Qualitative Research: Methodological Issues and Using QSR NVIVO and NUD*IST.* London: Institute of Education; 2003.

33. Using N6 in Qualitative Research program. Version N6. QSR International; 2002.

34. Malterud K. Qualitative research: standards, challenges, and guidelines. *Lancet.* 2001;358:483–488.

35. Glaser BG, Strauss AL. *The Discovery of Grounded Theory: Strategies for Qualitative Research.* New York, NY: Aldine Publishing; 1967.

36. Bogner HR, Gallo JJ. Are higher rates of depression in women accounted for by differential symptom reporting? *Soc Psychiatry Psychiatr Epidemiol.* 2004;39:126–132.

37. Switzer J, Wittink M, Karsch BB, Barg FK. "Pull yourself up by your bootstraps": A response to depression in older adults. *Qual Health Res.* In press.

Appendix B—An Example of the Explanatory Sequential Design

Students' Persistence in a Distributed Doctoral Program in Educational Leadership in Higher Education

A Mixed Methods Study

Nataliya V. Ivankova*,† and Sheldon L. Stick**

The purpose of this mixed methods sequential explanatory study was to identify factors contributing to students' persistence in the University of Nebraska-Lincoln Distributed Doctoral Program in Educational Leadership in Higher Education by obtaining quantitative results from surveying 278 current and former students and then following up with four purposefully selected typical respondents to explore those results in more depth. In the first, quantitative, phase, five external and internal to the program factors were found to be predictors to students' persistence in the program: "program", "online learning environment", "student support services", "faculty", and "self-motivation". In the qualitative follow up multiple case study analysis four major themes emerged: (1) quality of academic experiences; (2) online learning

*Assistant Professor, Department of Human Studies, University of Alabama at Birmingham, EB 202, 1530 3rd Ave S, Birmingham, AL, USA.

**Professor, Department of Educational Administration, University of Nebraska-Lincoln, 123 Teachers College Hall, Lincoln, NE 68588-0360, USA.

†Address correspondence to: Nataliya V. Ivankova, Department of Human Studies, University of Alabama at Birmingham, EB 202, 1530 3rd Ave S, Birmingham, AL 35294-1250, USA. E-mail: nivankov@uab.edu

SOURCE: Ivankova, N., & Stick, S. (2007). Students' persistence in a Distributed Doctoral Program in Educational Leadership in Higher Education: A mixed methods study. *Research in Higher Education, 48*(1), 93–135. Reprinted with permission of Springer Science + Business Media, Inc.

environment; (3) support and assistance; and (4) student self-motivation. The quantitative and qualitative findings from the two phases of the study are discussed with reference to prior research. Implications and recommendations for policy makers are provided.

KEY WORDS: *persistence; doctoral students; distributed program; online learning environment.*

● INTRODUCTION

Graduate education is a major part of American higher education, with more than 1850 million students enrolled in graduate programs (NCES, 2002). Approximately one fifth are graduate students pursuing doctoral degrees (NSF, 1998). Out of this number, from 40% to 60% of students who begin their doctoral studies do not persist to graduation (Bowen and Rudenstine, 1992; Geiger, 1997; Nolan, 1999; Tinto, 1993). High failure rate and the ever increasing time to degree are reported as chronic problems in doctoral education (Lovitts and Nelson, 2000; NSF, 1998). In educational majors, attrition from doctoral programs is estimated at approximately 50%. In addition, about 20% give up at the dissertation stage (Bowen and Rudenstine, 1992; Cesari, 1990). Failure to continue in the doctoral program is not only painful and expensive for a student, but is also discouraging for faculty involved, injurious to an institution's reputation, and results in a loss of high-level resources (Bowen and Rudenstine, 1992; Golde, 2000; Johnson, Green, and Kluever, 2000; Tinto, 1993).

Researchers claim a much higher dropout rate among students pursuing their doctoral degrees via distance education (DE) (Carr, 2000; Diaz, 2000; Parker, 1999; Verduin and Clark, 1991). Persistence in DE is a complex phenomenon influenced by a multitude of factors: challenges set by the distance learning environment, personally related internal and external variables, computer literacy, ability to access requisite technology, time management, and absent or questionable support from an employer and/or family (Kember, 1990).

The student population is composed of mainly part-time adult students, who often have numerous and demanding commitments to work, family, and social lives (Finke, 2000; Holmberg, 1995; Thompson, 1998). These students tend to be more vulnerable to factors encroaching on their academic progress because their school-related activities often are not primary life objectives.

Although many studies have been done to understand reasons for persistence of doctoral students in traditional campus-based programs (Bair and Haworth, 1999; Bowen and Rudenstine, 1992; Golde, 2001; Haworth, 1996; Kowalik, 1989), there is much less research on doctoral students' persistence in DE (Tinto, 1998), particularly distributed programs (distributed connotes the material is sent electronically to persons at various locations throughout the world and removes the need for participants to be located at a given site at a given time). Existing studies either focused on DE students' persistence in individual undergraduate and graduate courses, or other than distributed distance learning delivery means (Ivankova and Stick, 2003).

Knowledge and understanding of factors contributing to graduate students' persistence in distributed programs may help academic institutions better meet DE students' needs, improve the quality of their academic experiences, and increase their retention and degree completion rate. This is especially important today when postsecondary institutions have to confront the growing problems of revenue generation and increasing budget cuts and turn to offering graduate programs in distributed environments. Knowledge of the evolving tendencies may serve as a baseline for higher educational administrators in elaborating DE policies, designing and developing graduate distributed programs, and improving distance student support infrastructure.

This article reports on the study conducted to understand students' persistence in the Distributed Doctoral Program in Educational Leadership in Higher Education (ELHE) offered by the University of Nebraska-Lincoln (UNL). The purpose of this mixed methods sequential explanatory study was to identify factors contributing to students' persistence in the ELHE program by obtaining quantitative results from a survey of 278 current and former students and then following up with four purposefully selected individuals to explore those results in more depth through a qualitative case study analysis. In the first, quantitative, phase of the study, the research questions focused on how selected internal and external variables to the ELHE program (program-related, advisor-and faculty-related, institutional-related, student-related factors, and external factors) served as predictors to students' persistence in the program. In the second, qualitative, phase, four case studies from distinct participant groups explored in-depth the results from the statistical tests. In this phase, the research questions addressed seven internal and external factors, found to have differently contributing to the function discriminating the four groups: program, online learning environment, faculty, student support services, self-motivation, virtual community, and academic advisor.

Theoretical Perspective

Three major theories of students' persistence—Tinto's (1975, 1993) Student Integration Theory, Bean's (1980, 1990) Student Attrition Model, and Kember's (1990, 1995) Model of Dropout from Distance Education Courses—served as a theoretical foundation for this study. Tinto's and Bean's models focused primarily on undergraduate campus students and Kember's model was aimed at explaining attrition of distance adult students.

Although these models differed in their approach to persistence, they shared similar core elements and complimented each other. Their principle components helped identify critical internal and external factors presumably impacting students' persistence, such as entry characteristics, goal commitment, academic and social integration, and external forces (family, friends and employers).

Extensive literature review also revealed that graduate students' persistence in a program of study seldom is the result of the influence of one factor. Among those identified were institutional and departmental factors (Austin, 2002; Golde, 1998, 2000; Ferrer de Valero, 2001; Lovitts, 2001; Nerad and Miller, 1996), academic advisors (Ferrer de Valero, 2001; Golde; 2000; Girves and Wemmerus, 1988), support and encouragement (Brien, 1992; Hales, 1998; Nerad and Cerny 1993), motivation and personal goals (Bauer, 1997; Lovitts, 2001; McCabe-Martinez, 1996; Reynolds, 1998), and family and employer relationships (Frasier, 1993; Golde, 1998; McCabe-Martinez, 1996). Based on these factors and the principle components from three theories of students' persistence a set of variables was created to test for the predictive power of internal and external factors on doctoral students' persistence in the ELHE program.

Distributed Doctoral Program in Educational Leadership in Higher Education

The Distributed Doctoral Program in Educational Leadership in Higher Education is offered through the Department of Educational Administration at the University of Nebraska-Lincoln (Stick and Ivankova, 2004). The program was initiated in 1994 and offers students a choice of the PhD or the EdD Degrees in Educational Studies with the emphasis in Educational Leadership in Higher Education. It is possible for students to complete an entire program via distributed means. Innovative teaching methodologies and a distributed learning environment enabled most students to complete their programs of study within a 36- to 60- month period, with minimal disruption to lifestyle, family responsibilities, and employment. Most of the coursework necessary for the degree is provided through distributed learning software, which utilizes the Internet as a connecting link. Most of the program is delivered to students via Lotus Notes and Blackboard groupware, which provides asynchronous and collaborative learning experiences to participants. More than 260 students were enrolled and in varying stages of their programs, with 180–200 active during a given semester. Since 2004 there have been more than 70 students graduated. Some students did partial coursework on campus because either selected courses were not available online, or students wanted the on-campus experience.

● METHODS

Study Design

To answer the study research questions, the researchers used a mixed methods approach (Tashakkori and Teddlie, 2003), which is a procedure for collecting, analyzing and mixing or integrating both quantitative and qualitative data at some stage of the research process within a single study (Creswell, 2005). The rationale for mixing both types of data is that neither quantitative nor qualitative methods are sufficient by themselves to capture the trends and details of situations, such as the complex issue of doctoral students' persistence in the distributed environment. When used in combination, quantitative and qualitative methods complement each other and provide a more complete picture of the research problem (Green, Caracelli, and Graham, 1989; Johnson and Turner, 2003; Tashakkori and Teddlie, 1998).

This study used a sequential explanatory mixed methods design, consisting of two distinct phases (Creswell, Plano Clark, Guttman, and Hanson, 2003; Tashakkori and Teddlie, 1998). In this design, the quantitative, numeric, data is collected and analyzed first, while the qualitative, text, data is collected and analyzed second in sequence, and helps explain, or elaborate on the quantitative results obtained in the first phase. In this study, the quantitative data helped identify a potential predictive power of selected external and internal factors on the distributed doctoral students' persistence and purposefully select the informants for the second phase. Then, a qualitative multiple case study approach was used to explain why certain external and internal factors, tested in the first phase, were significant predictors of students' persistence in the program. Thus, the quantitative data and results provided a general picture of the research problem, while the qualitative data and its analysis refined and explained those statistical results by exploring the participants' views regarding their persistence in more depth.

The priority (Creswell et al., 2003) in the study was given to the qualitative approach, because it focused on in-depth explanations of the results obtained in the first, quantitative, phase, and involved extensive data collection from multiple sources and two-level case analysis. The quantitative and qualitative phases were connected (Hanson, Creswell, Plano Clark, Petska, and Creswell, 2005) when selecting four participants for qualitative case studies and developing the interview protocol based on the results from the statistical tests from the first phase. The results of the quantitative and qualitative phases were integrated (Creswell et al., 2003) during the discussion of the outcomes of the entire study (see Fig. B.1 for a diagram of the mixed methods sequential explanatory design procedures in the study)[1].

Target Population

The target population in this study were active and inactive students, who were admitted to the ELHE program and taking classes during the spring 2003 semester. Also part of the target population were students who had been graduated with an earned

Figure B.1 Visual model for mixed methods sequential explanatory design procedures

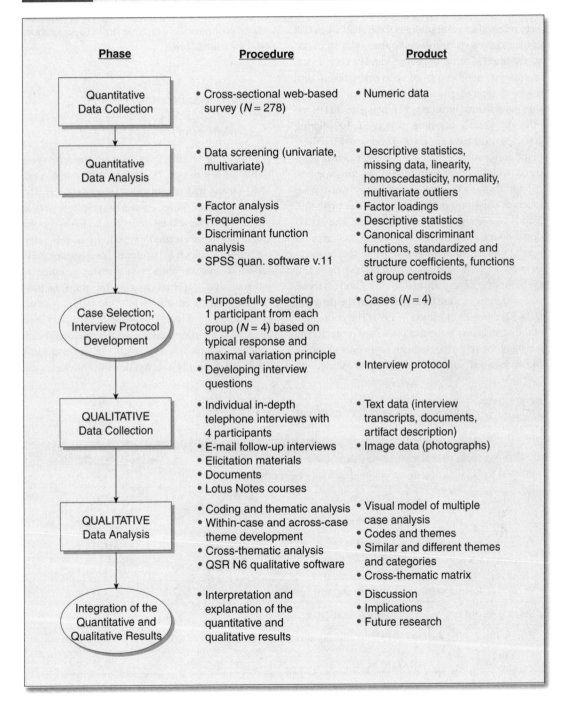

doctoral degree from the program and those who had withdrawn, or had been terminated from the program prior to the spring 2003 semester. Students were referred to as DE students if they had taken half of their classes via distributed means. The students' status varied in terms of progress and/or completion of courses, number of online courses taken, and doctoral degree pursued. Criteria for selecting the participants included: (1) being in ELHE vs. other programs; (2) time period of 1994-Spring 2003; (3) must have done 1/2 of course work online; (4) be either admitted, both active and inactive, graduated, withdrawn, or terminated from the program; (5) for those who just started, they must have taken at least one online course in the program. A total of 278 students met the criteria. The breakdown by their matriculation status in the program was: (1) those admitted and active in the program ($n = 202$); (2) those admitted but inactive ($n = 13$); (3) those who were graduated ($n = 26$), and (4) those who withdrew or were terminated from the program ($n = 37$) since its inception in 1994. The anonymity of the participants in the first phase was protected by assigning them unique numeric passwords to access the web-based survey. In the second phase, the participants selected for case study analysis were assigned fictitious names, thus keeping the responses confidential. In addition, all the names and gender related pronouns were removed from the quotations used for illustrations.

Quantitative Phase

Data Collection

For the first, quantitative, phase, the cross-sectional survey design (McMillan, 2000) was used. The survey instrument was self-developed and pilot tested on 5% of randomly selected participants. The core survey items formed five 7-point Likert type scales related to five internal and external entities affecting students' persistence, and reflected nine variables, representing a range of internal and external to the program factors: "online learning environment", "program", "virtual community", "faculty", "student support services", "academic advisor", "family and significant other", "employment", and "self-motivation". Table B.1 presents the relationship between the

Table B.1 Survey Scales and Predictor Variables in Quantitative Analysis

Survey scales/Factors	Subscales/Predictor variables	Cronbach's alpha	Survey items
Related to ELHE program	Online learning environment	.8503	Q14 a-j
	Program	.8344	Q13 a-g
	Virtual community	.8012	Q13 h-l
Related to faculty and academic advisor	Academic advisor	.9818	Q15 a-m
	Faculty	.9079	Q13 m-r
Related to institution	Student support services	.8243	Q13 s-y
Related to student	Self-motivation	.8948	Q16 a-g
External to ELHE program	Family and significant other	.5829	Q17 a-d
	Employment	.5289	Q17 e-h

survey scales, subscales and variables, and lists the survey items measuring each variable, as well as reliability indexes for each subscale. The survey items and scales were developed based on the analysis of the related literature, three theoretical models of students' persistence (Bean, 1980, 1990; Kember, 1990, 1995; Tinto, 1975, 1993) and an earlier qualitative thematic analysis study of seven ELHE active students (Ivankova and Stick, 2002). A panel of professors teaching in the program was used to secure the content validity of the survey instrument. Based on the pilot testing, some survey items were revised slightly.

The survey was administered online and was accessed through the URL. Active email addresses of the potential participants were obtained through the UNL Department of Educational Administration and identified through other sources. The participants were recruited via e-mail a week before the beginning of the study. The data collection took place between April 1 and July 18, 2003. The procedure was complicated by having to correct 50 inactive email addresses and locate former students, who had withdrawn or graduated from the program. Technological glitches in the system also presented challenges. Twenty-three participants who were willing to complete the questionnaire, could not access the survey, or failed to complete it in full. A hard copy of the survey was mailed, faxed, or sent as a Word document attachment to such participants. Nineteen such participants returned the completed survey.

From 278 potential participants 207 responded, which constituted a response rate of 74.5%. All respondents were organized into four groups based on their matriculation status in the program and similarity of academic experiences: (1) students who had completed 30 or fewer credit hours of course work (Beginning Group) ($n = 78$); (2) students who had completed more than 30 credit hours of course work (Matriculated Group) ($n = 78$); (3) former students who had graduated from the program with the doctoral degree (Graduated Group) ($n = 26$);

and (4) former students who either had withdrawn or had been terminated from the program, or had been inactive during the last three terms (spring, fall, summer) prior to the survey administration (Withdrawn/Inactive Group) ($n = 25$). Reliability and validity of the survey scales and items were established, using descriptive statistics, frequency distributions, internal consistency reliability indexes (Cronbach's alpha, item-total correlation, corrected item-total correlation, and alpha-if-item deleted), as well as inter-item correlations and factor analysis (Ivankova, 2004).

Data Analysis

Both univariate and multivariate statistical procedures were used to analyze the survey data. Survey demographic information and the participants' answers to separate items on each survey sucscale were analyzed using cross tabulation and frequency counts. Discriminant function analysis was used to identify the predictive power of nine selected factors as related to students' persistence in the ELHE program. Prior to the analysis, data screening was conducted at both univariate and multivariate levels, following the procedures outlined by Kline (1998) and Tabachnick and Fidell (2000).

Qualitative Phase

Qualitative Research Design

A multiple case study design (Stake, 1995; Yin, 2003) was used for collecting and analyzing the data in the second, qualitative, phase. The instrumental multiple cases (Stake, 1995) served the purpose of "illuminating a particular issue" (Creswell, 2005, p. 439), such as persistence in the ELHE program. The unit of analysis was a former or current ELHE student. Each case study was bounded by one individual and by the time he or she matriculated in the ELHE program.

Case Selection

A systematic two-stage case selection procedure was developed[2]. During the first stage, typical respondents in each participant group were identified, first, by calculating the summed mean scores and their respective group means for all participants in each of the four groups based on their responses to the survey questions, and then by selecting a few respondents from each group with the mean scores within one standard error of the mean. During the second stage, one "best informant" from each group was selected using a maximal variation strategy (Creswell, 2005). This procedure yielded one male and three females, displaying different dimensions on such demographic characteristics, as age, gender, residency, and family status, which allowed for preserving multiple perspectives on persistence in ELHE program. All four agreed to participate.

Interview Protocol Development

The content of the interview protocol was grounded in the quantitative results from the first phase of the study. Because the goal of the qualitative phase was to explore and elaborate on the results of the statistical tests (Creswell et al., 2003), we wanted to understand why certain predictor variables differently contributed to the function discriminating four participant groups with regards to their persistence. Five open-ended questions explored the role of five factors ("online learning environment", "program", "faculty", "student support services", and "self-motivation"), which demonstrated statistically significant predicting power for this sample of the ELHE students. Two other open-ended questions explored the role of academic advisor and virtual learning community in students' persistence. Although those two factors did not significantly contribute to the function discriminating four participant groups, their important role in students' persistence in traditional

doctoral programs was reported by other researchers (Bowen and Rudenstine, 1992; Brown, 2001; Golde, 2000; Lovitts, 2001). The interview protocol was pilot tested on one participant, purposefully selected from those who had completed the survey in the first phase of the study. As a result, the order of the protocol questions was revised slightly and additional probing questions were developed.

Data Collection

The data was collected from multiple sources to provide the richness and the depth of each case description and included: (1) in-depth semi-structured telephone interviews with four participants; (2) electronic follow-up interviews with each participant to secure additional information on the emerging themes; (3) academic transcripts and students' files to validate the information obtained during the interviews and to get additional details related to the cases; (4) elicitation materials, such as photos, objects, and other personal things, provided by each participant relating to his/her persistence in the program; (5) participants' responses to the open-ended and multiple choice questions on the survey in the quantitative phase; and (6) selected online classes taken by the participants and archived on a Lotus Notes or Blackboard server. The data collection took place during November–December of 2003.

Qualitative Analysis

Each interview was audio taped and transcribed verbatim (Creswell, 2005). The analysis was performed at two levels: within each case and across the cases (Stake, 1995; Yin, 2003), using the QSR N 6, qualitative software for data storage, coding, and theme development. Steps in the qualitative analysis included: (1) preliminary exploration of the data by reading through the transcripts and writing memos; (2) coding the data by segmenting and labeling the text; (3) verifying the codes through inter-coder agreement check; (4) using codes to develop themes by aggregating similar codes together; (5) connecting and interrelating themes; (6) constructing a case study narrative composed of descriptions and themes; and (7) cross-case thematic analysis. Credibility of the findings was secured by triangulating different sources of information, member checking, inter-coder agreement, rich and thick descriptions of the cases, reviewing and resolving disconfirming evidence, and academic advisor's auditing (Creswell, 1998; Creswell and Miller, 2002; Lincoln and Guba, 1985; Miles and Huberman, 1994; Stake, 1995).

● RESULTS

Quantitative Phase

Demographic Information

The study participants were compared on the following demographic characteristics: age, gender, and employment while in the ELHE program, Nebraska (NE) residency status, and family status. The typical participants were: between 36 and 54 years of age, predominantly women, employed full-time, mostly out-of-state, and married with children (see Table B.2).

| **Table B.2** | Demographic Characteristics of Survey Respondents* |

Row Pct Total	Group 1: Beginning (*n* = 78)	Group 2: Matriculated (*n* = 78)	Group 3: Graduated (*n* = 26)	Group 4: Withdrawn/ Inactive (*n* = 25)	Total
Age					
26–35	45.7	31.4	5.7	17.1	100.0
36–45	41.6	45.5	6.5	6.5	100.0
46–54	35.7	32.9	18.6	12.9	100.0
Over 55	16.7	37.5	25.0	20.8	100.0
Total	77	78	26	25	206
Gender					
Male	33.3	38.7	15.1	12.9	100.0
Female	40.2	37.5	10.7	11.6	100.0
Total	76	78	26	25	205

(Continued)

Table B.2 (Continued)

Row Pct Total	Group 1: Beginning (*n* = 78)	Group 2: Matriculated (*n* = 78)	Group 3: Graduated (*n* = 26)	Group 4: Withdrawn/ Inactive (*n* = 25)	Total
Employment					
Full-time	38.0	37.5	12.0	12.5	100.0
Part-time	35.7	42.9	21.4	0	100.0
Unemployed	0	0	0	100.0	100.0
Total	78	78	26	25	207
NE Residency					
In-state	30.6	37.1	16.1	16.1	100.0
Out-of-state	41.3	37.0	10.9	10.9	100.0
International	28.6	57.1	14.3	0	100.0
Total	78	78	26	25	207
Family status					
Married with kids under 18	39.2	36.7	12.5	11.7	100.0
Married with kids over 18	34.9	44.2	11.6	9.3	100.0
Single with kids under 18	44.4	33.3	0	22.2	100.0
Single, never married	22.2	44.5	11.1	22.2	100.0
Single, divorced or separated	50.0	16.7	25.0	8.3	100.0
Single person, widowed	0	100.0	0	0	100.0
Married without children	14.3	57.1	14.3	14.3	100.0
Total	75	77	25	24	201

*Missing data is excluded.

Scale Items Frequencies Analysis

Most of the participants were satisfied with their academic experiences in the program. The amount of satisfaction was the greatest among the Graduated participants (92.3%), while satisfaction increased from the Beginning group (57.7%) to the Matriculated group (71.8%). Only 20% of the Withdrawn/Inactive group reported the program met their needs, and another 20% expressed negative feelings about the program. The majority of participants in the three matriculated groups positively rated their involvement with the online courses and agreed that online courses were more challenging academically. Across the groups, the participants gave more positive ratings to instructors' accessibility and promptness of the feedback, rather than the quality of the feedback and instructors' willingness to accommodate to distance learners' needs.

Most participants were comfortable learning in the online environment (84.3%). Across the groups, the Graduates expressed the highest comfort level with online learning (96.2%), while the Withdrawn/Inactive group was the least comfortable (47.8%). More participants from the Graduated (100.0%) and the Matriculated (81.3%) groups, than from the Beginning (68.8%) and the Withdrawn/Inactive (39.1%) groups were comfortable with participating in online discussions and the course workload. The same pattern of increased comfort level from the Beginning group to the Graduated group was observed when participants rated their learning in the distributed environment as compared to a face-to-face setting. However, the participants differentially benefited from the virtual community. Only two-thirds of the respondents claimed they could establish long-term social relationship with their fellow-students online. The Withdrawn/Inactive group was the least satisfied, had low comfort level (47.8%), and was more negative in rating the effectiveness of learning in the distributed environment (30.4%).

Participants had different experiences with academic advising. The Graduated group had more positive experiences (76.0%), than any other group. Across all the items, the Matriculated participants rated their experiences with academic advising more positively than the Beginning group, which might be due to the fact that they had more opportunities to experience a variety of relations with their academic advisor than those who had completed less than 30 credit hours in the program. In the Withdrawn/Inactive group, fewer participants rated their academic advisor positively (38.0%).

All the participants, except for the Withdrawn/Inactive group (32.0%), were highly motivated to pursue the doctoral degree in the distributed environment. The Graduates were the most motivated group (100.0%), while the Matriculated group (93.6%) was a little more motivated, than the Beginning group (76.9%). More than 50% of the participants were satisfied with the institutional support services. However, their satisfaction differed depending on the particular service and the level of students' matriculation in the program. The Withdrawn/Inactive group was the least satisfied (48.0%).

More than 70% of the participants agreed they had favorable family conditions to support their efforts to pursue the doctoral degree via distributed means. Across all the groups, the Graduated group received the most support (80.8%) and the Withdrawn/Inactive group the least (65.0%). There was more satisfaction for the Matriculated group (77.6%) than for the Beginning group (77.6%). More Graduates also believed their friends encouraged them in their study efforts (60.0%). About 65.6% of the participants received encouragement from their employers to pursue the doctoral degree. The Graduated participants were the most encouraged (76.9%), while the Matriculated group received the least support (63.0%). 61.1% of the Withdrawn/Inactive participants positively rated their employer.

Discriminant Function Analysis

The analysis yielded three discriminant analysis functions. Based on the Wilks' Lambda test, only the first function was statistically significant (χ^2 = 98.858; df =27; ρ = .000), meaning only this function discriminated for this set of variables (Tabachnick and Fidell, 2000). The standardized coefficients for the first discriminant function indicated all nine predictor variables provided their relative unique contribution to group differences as related to students' persistence in the program (see Table B.3).

Table B.3 Standardized Canonical Discriminant Function Coefficients

	Function		
	1	2	3
Program	1.187	0.458	0.187
Online learning environment	−0.078	0.588	0.065
Faculty	0.187	0.425	−0.608
Self-motivation	0.224	−0.427	0.176
Student support services	−0.341	0.209	0.016
Employment	0.116	0.635	0.151
Virtual community	0.105	0.786	0.163
Academic advisor	−0.180	−0.129	1.076
Family	0.103	−0.080	0.455

The discriminant variate that best discriminated the four groups was represented by the following linear relationship equation:

V =1.187 * program − 0.078 * online learning environment + 0.105 * virtual community + 0.187 * faculty − 0.341 * student support services − 0.180 * academic advisor + 0.224 * self-motivation + 0.103 * family and significant other + 0.116 * employment

The variable "program" (1.187) contributed the most to the participants' being in a particular group as related to their persistence in the ELHE program. No other variable had a similarly high coefficient. The variable "student support services" (−0.341) had the second largest contribution to the group differences. It was followed by "self-motivation" (0.224), "faculty" (0.187), and "academic advisor" (−0.180). Other variables had low coefficients and contributed very little.

Based on the structure coefficients for the three discriminant functions, five variables "program", "online learning environment", "faculty", "self-motivation", and "student support services" had a statistically significant correlation with the discriminant function, and hence, contributed to discriminating the participants as related to their persistence (see Table B.4).

Appendix B

| Table B.4 | Structure Matrix in Discriminant Function Analysis |

	Function		
	1	2	3
Program	0.905*	−0.066	0.030
Online learning environment	0.526*	0.037	−0.160
Faculty	−0.486*	0.245	−0.086
Self-motivation	0.482*	−0.331	0.005
Student support services	0.202*	0.097	−0.046
Employment	−0.111	0.542*	0.255
Virtual community	−0.438	0.521*	0.106
Academic advisor	−0.447	−0.034	0.690*
Family	−0.041	0.190	0.339*

Pooled within-groups correlations between discriminating variables and standardized canonical discriminant functions variables ordered by absolute size of correlation within function. *Largest absolute correlation between each variable and any discriminant function.

"Program" ($r = 0.905$) and "online learning environment" ($r = 0.526$) had the highest correlations and made the most contribution to discriminating the four matriculated groups, followed by "faculty" ($r = −0.486$), "self-motivation" ($r = 0.482$), and "student support services" ($r = 0.202$). Those differences in function and correlation coefficients made it somewhat difficult to interpret the discriminant function, especially since only one function was generated. However, both statistics indicated the top variable was "program". So, we named this function "ELHE program" and concluded that the nature and the context of the program contributed to discriminating the participants as related to their membership in one of the matriculated groups. This discriminant function also indicated that 88.7% of the participants were classified correctly. "Virtual community", "academic advisor", "family and significant other", and "employment" made no significant contribution to the discriminant function.

Functions at group centroids revealed that on the discriminant function the Withdrawn/Inactive group (1.654) differed from the other three participant groups the most. The Graduate group (−.960) differed from both the Beginning and the Matriculated groups, though less from the Matriculated group and the most from the Withdrawn/Inactive group. The Matriculated group (−.410) differed notably from the Beginning group (.200) (see Table B.5).

Membership in the group	Function		
	1	**2**	**3**
Group 1: Beginning	0.200	0.137	−0.177
Group 2: Matriculated	−0.410	−0.224	0.005
Group 3: Graduated	−0.960	0.302	0.284
Group 4: Withdrawn/Inactive	1.654	−0.043	0.242

Table B.5 Functions at Group Centroids in Discriminant Function Analysis

Unstandardized canonical discriminant functions evaluated at group means.

Qualitative Phase

The analysis of each case and across four cases yielded four themes related to the participants' persistence in the ELHE program: quality of academic experiences, online learning environment, support and assistance, and self-motivation. The description of each case follows.

Gwen

Gwen was 40 years old and in her third year in the ELHE program. She was Dean of Students in a small private college in the Midwest. She was single and had a cat Sam, who was her close friend. At the time of the interview, she had successfully completed 30 credit hours, of which 18 were taken online.

Quality of Academic Experiences

Gwen's persistence in the program was positively affected by the tight structure of the program and ability to plan her coursework. The coursework reportedly challenged Gwen's critical thinking and gave her the opportunity to learn from others: "It . . . helped me to think differently, because I have

to put that all in writing and share it with everyone." It was also relevant to her professional life. The quality of the coursework was directly related to an instructor's involvement with the course and the feedback he/she provided.

On the other hand, Gwen did not receive any quality feedback from her academic advisor: "I haven't found my advisor to be fulfilling in that role." On the survey in the first phase of the study, she rated advising negatively. Communication with the advisor was rare and not informative. Analysis of the e-mail communication between Gwen and her advisor revealed that approximately 70% of Gwen's messages were left unanswered. Although low quality advising was frustrating for Gwen, she was determined to continue with her efforts to pursue the degree via DE: "I'm not going to let [the advisor] stop my persistence or stop my progress in the program." At the time of the study, Gwen decided to initiate another attempt to switch the academic advisor. The request was being honored.

Online Learning Environment

Learning via distance was convenient for Gwen and provided a lot of flexibility. An intensive work schedule did not allow her to leave work during the day, so the ability to study at her own pace and time positively affected her matriculation in the program: "You have the opportunity to do things . . . when they work for you." Learning online fit Gwen's learning style. She liked to write and was cognizant enough to participate extensively in written communications with other students. The online format also gave her the opportunity to learn from other students' work. Gwen was comfortable not seeing her classmates and professors and created mental images of them based on their writings: "I'd be getting an idea of a person's looks or image by their work." She believed a virtual community was established among the students, but it depended on the nature of a course and was limited to one course.

Support and Assistance

Support and encouragement from faculty and students was stimulating. Support from peers ranged from encouragement on a particularly challenging assignment to sharing personal stories and school related experiences. Gwen especially benefited from learning about other distance doctoral students and their problems and concerns: "It's been neat to just connect with other students in the program and learn that they're having similar experiences or, they're just as busy in trying to make everything happen." Advice from the faculty was assignment specific, but also related to the content and logistics of the program. Having been left without an active advisor, Gwen was comfortable asking other instructors academic and dissertation related questions: "They've been very open." Institutional support services played an important role in Gwen's persistence and she highly rated those services on the survey. She also received constant support from her new employer and her colleagues, as well as her parents and three sisters. The photos she provided reflected a loving and caring family, attentive to each other's needs. A cat, named Sam, was another source of support.

Gwen admitted both taking care of Sam and his calm attitude kept her "sane and balanced."

Self-Motivation

Gwen was highly motivated to earn a doctoral degree and it positively influenced her persistence in the program. For her securing the terminal degree was both a dream and a personal challenge. She was aware that the process was not smooth and there could be a lot of challenges: "I had just known upfront that it takes a lot of initiative and self thrive to make things happen." Gwen admitted even negative experiences with academic advising would not impact her desire to persist and finish the program. The very idea of moving through the program and being close to completion of her course work was stimulating: "Knowing that . . . almost within the next year I'll be starting a new phase of the program . . . keeps me motivated."

Lorie

Lorie was 43 years old and in her fourth year in the program. She worked as Academic Dean at a private business school on the Eastern Coast. Lorie had been married for 23 years and had a 23-year old son, who was a college senior. She successfully completed 45 credit hours of course work via distributed means. At the time of the study she was working on her dissertation and writing the comprehensive examination.

Quality of Academic Experiences

Lorrie's persistence in the ELHE program was affected by its high quality. On the survey, she indicated program quality, prestige, and offerings as factors contributing to her persistence. Lorie claimed she was learning more online than if she were in a conventional classroom: "I anticipated that maybe I wouldn't learn the depth that I was accustomed to being in the classroom . . . But much to my surprise, I found that it was better." She also benefited from the opportunity to learn from other students and tried to read and respond to everybody in class. Lorie found the course work relevant to what she was doing in her professional life. She benefited most from courses when instructors were acting as facilitators, encouraging students to seek knowledge and find the answers themselves. With few exceptions Lorie received positive and constructive feedback from the instructors and it fulfilled her expectations: "It was exactly what I needed to hear."

The quality of advising evolved along with Lorie's matriculation in the program. When her academic advisor retired, it took nearly a month to get the new advisor to respond to Lorie's e-mail messages. Subsequently, the advisor became more responsive and attentive to her needs. Lorie claimed her advisor had a crucial role in the dissertation stage of her program: "I've never done this before . . . and [advisor] knows the process, and exactly what the committee is looking for, and what works, and what doesn't."

Online Learning Environment

The distributed learning environment offered Lorie convenience and flexibility of learning and positively enhanced her persistence. "I guess that's probably the thing that supported me, that allowed me to stay in the program, because I travel a lot." A high comfort level with technology made it easy for Lorie to learn in this environment. She also enjoyed writing, was comfortable developing essay-type responses to course assignments and participating in online discussions. She purposefully involved herself in discussions with students she had taken classes with, because she knew their "mannerisms, behavior and responsiveness." Examination of selected archival Lotus Notes classes Lorie had taken revealed she typically interacted with the same group of students. Lorie believed a learning community was established among the virtual

students, but it was limited to a particular course and built around some course issues: "It was a community of learners that had a particular interest in a particular subject matter." However, with some students the relationship extended beyond online interactions and later Lorie was able to meet with two students when she traveled to the states they lived in.

Support and Assistance

Lorie's efforts to pursue the degree via DE were supported at different levels. Because she had to travel a lot for her work, the instructors were responsive and willing to accommodate to Lorie's needs. Support from other students in the program was essential, but limited, although she admitted having good relationships with other students and rated peer support high on the survey. Support from the academic advisor came in the form of guidance with "how-to kinds of things." She pointed out student support services played an important role in her persistence in the program, despite not being highly visible. Unfortunately, Lorie did not provide any information related to support from her family and employer.

Self-Motivation

Motivation played an important role in Lorie's persistence in the program. She had always dreamed of having a doctorate, and her intrinsic motivation was supported by a sense of responsibility for the process and by the very nature of the online learning environment, where one's work was exposed to and evaluated by everybody in class. She also knew her classmates depended on her participation in online discussions or her involvement in virtual group projects: "I knew . . . without [my piece of the puzzle] we were all going down." The fact Lorie enjoyed what she was doing in the program added to her intrinsic motivation. She found the process of learning exciting and fascinating: "I enjoyed it. It was like almost my entertainment and my recreation in a twisted way, I guess." A dissertation fellowship added extrinsic motivation to Lorie's persistence in finishing the program.

Larry

Larry was 45 years old when he graduated with the PhD degree from the ELHE program in the Spring of 2001. He successfully completed the program in four years and did most of the coursework online. He was then Dean of Language and Letters in a private religious university in a northwestern state. Larry had been married for more than 25 years and had four children, two graduated from college and one son still in high school.

Quality of Academic Experiences

Larry's persistence in the program was positively affected by its quality. The program was structured and well laid out, "I knew exactly what I needed to do." The course work was relevant and the content covered distinct dimensions of an administrator's work and issues: "The things I was learning . . . were just as current as issues that we were facing on our campus." The emphasis of the program on engaged learning and written communication made it even more appealing to Larry. The idea of learning from colleagues from all over the country and other nations in addition to books and other data sources was beneficial. This idea was also reflected in the professional performance portfolio Larry submitted to his advisor as part of the degree requirement.

Faculty feedback varied in its quality and for Larry sometimes lack of faculty commitment to online students was disappointing. He assigned a big role to his academic advisor in his successful matriculation in the program. The advisor provided high quality professional advice and was an instructor in a third of Larry's courses: "Very good personal encouragement and advice on many dimensions." Larry also received quality feedback from his dissertation committee members and

believed their role was central in the final stages of his program.

Online Learning Environment

The online format of the ELHE program positively affected Larry's persistence. On the survey, Larry chose family, work schedule, convenience and flexibility of the program offerings as factors important for his decision to persist in the program. Absence of time and place constraints gave Larry the convenience of adhering to his work routine and the opportunity to be with his family and his teenaged children even while taking classes: "I was able to work during the day, come home and have dinner with my family, and then sit in my office during the evening at my home and do my course work." This flexibility gave him emotional freedom to pursue the degree.

Larry's comfort level with online learning was very high. Because he was trained as a journalist and liked writing, he never experienced any problems interacting with his classmates in the discussion threads, or communicating with instructors via electronic means. The structure of the program and the delivery method provided a nice fit to his background, talents, and skills, making it easier to be successful in the program: ". . . if I were in another program, I think it would have been very difficult." Larry believed a community of virtual learners had been established, though it was not sustained over the time: "It was really interesting our first semester together, how much time we spent in the cafeteria talking to each other and getting to know each other a little bit better, and how that over time seemed to fade away." The students recognized how demanding it was for everybody to have a full-time position and to pursue a doctoral degree, so the role of the community was not strong.

Support and Assistance

Larry received support and encouragement at different levels. High quality advising and personal friendship with academic advisor created a supporting niche and helped Larry complete the program. Instructors were always ready to waiver the assignment due date understanding the challenges of online learning. Relations with classmates were built on mutual respect and recognition, and the students were sensitive to Larry's religious background and respected his viewpoints. Continuous assistance from different university support services also helped Larry move through the program. Technology help with the course software and platform problems was for the most part "timely", library resources were "invaluable", and the registration and records department staff was always "beyond helpful." Larry also highly rated institutional support services on the survey.

Support also came from sources external to the program, such as family and work. Larry's family had created a supportive environment for him and encouraged his efforts in pursuing the doctorate degree. Larry assigned his mother one of the major roles in his getting the doctorate: ". . . she's probably my number one supporter in terms of 'I'm so proud of you'." The president of the university where Larry was employed also provided constant encouragement and help, including emotional support, release time, and financial assistance.

Self-Motivation

The innovative character of the ELHE program and the notion of pursuing advanced graduate studies via DE constituted specific value for Larry and raised his motivation. The fact of being among the few faculty with a doctoral degree at the institution that did not have a doctoral requirement added to Larry's recognition and self-esteem. Larry assigned a big role to himself and his personal motivation in his efforts to pursue a doctorate via DE. Only once after successfully finishing all the course work and passing his comprehensive examination, did Larry considered quitting the program: "I was getting weary of the grind for the two solid years, year round. . . Just to finish my coursework and my comps. And then you look at that mountain of a dissertation and you're thinking, do I have it in me to even complete that?" It took

Larry some "real internal motivation to get going again" in addition to the encouragement from the academic advisor, his family and university president.

Susan

Susan was 54 years old when she withdrew from the ELHE program. She worked as a registrar at a small private religious college in one of the northern states. She successfully completed two online courses in the program and both were related to her major. At the time of the study she had completed two years of a three year doctoral program at a small private university within 40 miles of her home. She was a single person with no children.

Quality of Academic Experiences

Though Susan took only two courses in the program she believed its quality was high and it was tailored to meet students' needs. She appreciated the broad content of the program and the opportunity to choose the area of concentration later. She was mostly satisfied with the feedback she was getting from the faculty regarding her course work and the promptness of their responses. She also benefited from her interactions with the academic advisor. Though Susan did not get far into the program and did not have an opportunity to discuss the future dissertation, she received good and quick advice from her advisor: "When I wrote a couple of times about different things, [the advisor] was quick to answer and gave me good advice." On the survey, Susan highly rated advising. At the same time, Susan was not satisfied with the quality of other doctoral students' postings and feedback. She believed the students did not possess the appropriate writing skills so important in the program with the focus on written interaction: "It was frustrating to try to respond to those people. . . They really didn't write very well. They didn't express themselves that well." She also did not like the nature of the discussion going online. She thought it was primarily academic and more focused on the exchange of facts, but not the opinion.

Online Learning Environment

Convenience and freedom of time was one of the biggest attractions for Susan in the ELHE program. The focus on writing did not bother her and she was comfortable developing essay-type responses to assignments and responding to other students' postings. However, the asynchronous format of the online courses did not match Susan's learning style. She missed the real time component of face-to-face interactions and could not comply with it: "The whole format of posting my response and then reading other people's responses and responding to them. . . that was very frustrating to me." On the survey, Susan indicated that the online format was the primary factor influencing her decision to withdraw from the program.

Susan was also concerned with not seeing other students and instructors and not being able to observe their body language. In her new campus-based program this component was present and, reportedly, positively affected her persistence. She also believed there was not much community building in the courses she took. On the survey, Susan indicated lack of personal contact with fellow students as the biggest barrier for her in distance learning. Exploration of two Lotus Notes archival courses she had taken showed little social interaction in the course Virtual Cafeteria. Susan herself did not invest a lot of effort into establishing the online community either. Those two components, online learning environment and lack of personal interaction, were the only reasons for Susan not to continue with the program: "The problem was not with [the university] and it wasn't really with the program. It was with the method. And that would be my primary concern and my primary reasons for leaving the program."

Support and Assistance

Although Susan took only two classes in the program, she sensed the supportive atmosphere created by the faculty, students, and institutional support services. The feedback she received from the faculty, especially personal encouraging notes in one class, was helpful to stay focused on the task. Both instructors were also willing to accommodate to her needs. Susan received quick assistance with the technological problems: "When I contacted them, I did get answers pretty quickly." When she was getting set up to take her first course in Lotus Notes, she got all the help she needed and in a timely fashion. That created a positive atmosphere for her to begin the program.

Self-Motivation

In spite of the fact Susan withdrew from the ELHE program, she was highly motivated to earn a doctoral degree. When Susan realized pursuing the degree in the distributed learning environment did not fit her learning style, she began looking for an alternative doctoral program, where she could have real time communication and meet other doctoral students in person. At the time of the study Susan was working on her EdD in Leadership at another university. Every week, she drove 40 miles one way to meet with her cohort. In addition to enjoying the format of her new program, Susan claimed she had a strong personal responsibility for earning the degree. This sense of responsibility and a long-term wish to have a doctorate acted as a driving force for Susan as she commuted weekly to the class and complied with whatever other difficulties she had to face: "It's me, or it ain't going to get done."

Cross Case Analysis

Four similar themes related to the participants' persistence in the ELHE program emerged in the analysis across four cases: quality of academic experiences, online learning environment, support and assistance, and self-motivation. In spite of being common for all participants, those themes differed in the number and similarity of sub-themes and categories comprising them (see Table B.6).

Table B.6 Themes, Sub-Themes, and Categories Across Cases

Themes, Sub-Themes	Gwen	Lorie	Larry	Susan
Quality				
University		Distance education	Research one	
Program	Well-structured	Well-structured	Well-structured	
	Relevant	Relevant	Relevant	
	Scholarly	Scholarly	Scholarly	
	Learning from others	Learning from others	Learning from others	
	Challenging	Challenging		
		Broad content		
	Delivery	Depth	Clarity of expectations	Broad content
	Good fit	Well-known	Engaged learning	Good
	Reputation		Written dialog	Students' needs
	High standards		Laid out	
Faculty	Feedback	Feedback	Feedback	Feedback
	Involvement	Involvement	Involvement	Involvement
	Prompt			Prompt
		Facilitating	Interactions	
		Readiness to teach online	Commitment	
Students	Feedback		Feedback	Feedback
	Professional		Interactions	Writing skills
	Positive		Varied	Fact based discussion
Advising	Negative	Need	Professional	Helpful
	Useless	Varied	Involvement	Prompt

Themes, Sub-Themes	Gwen	Lorie	Larry	Susan
	Lack of guidance	Knowledge of the process	Diligent	
	Communication		Champion dissertation	
	Switching advisor			
Dissertation Committee Members			Second opinion	
Online learning environment				
	Convenience	Convenience	Convenience	Convenience
	Flexibility	Flexibility	Flexibility	Flexibility
	Learning style	Learning style	Learning style	Learning style
	Non-physical presence	Non-physical presence	Non-physical presence	Non-physical presence
	Online community	Online community	Online community	Online community
	Comfort with technology	Comfort with technology	Comfort with technology	
		Work schedule	Work schedule	Work schedule
	Mental images	Class size	Emotional relief	Writing component
	Learning via distance	Familiar students	Staying with family	Non-real time
		Meeting in person		Involvement
Support				
University			Cooperation	
Faculty	Willing to accommodate	Willing to accommodate	Willing to accommodate	Willing to accommodate

Table B.6 (Continued)

Themes, Sub-Themes	Gwen	Lorie	Larry	Susan
	Varied	Receptive	Personal relationship	Personal notes
	Responsive			
	Advice			
	Open			
Students	Encouragement		Encouragement	Encouragement
	Sensitive		Sensitive	
	Polite	Using for references	Respect	
	Personal experiences	Limited to course activities	Recognition	
	Sympathies		Best wishes	
	Congratulations			
Academic Advisor	None	Assistance-guidance	Assistance	No need for assistance
		"How-to"	Friendly	
			Encouragement	
			Personal interest	
			Accommodating	
Student support services	Prompt	Prompt	Prompt	Prompt
	Helpful	Not helpful	Helpful	Helpful
		Smooth		Smooth
	Convenient	Simple	Timely	Straightforward
	Always worked		Easily solved	
	Friendly		Attention	

Themes, Sub-Themes	Gwen	Lorie	Larry	Susan
			Qualified	
Family	Encouragement		Encouragement	
	Pride		Pride	
	Care		Supportive environment	
	Attention			
Employment	Time off		Time off	
	Life learning		Encouragement	
	Sharing experiences		Advice	
			Extra credit	
			Pushing	
Pet	Watching silently			
Self-motivation				
	Responsibility	Responsibility	Responsibility	Responsibility
	Enjoyed	Enjoyed	Enjoyed	Enjoyed
	Exposure	Exposure	Exposure	
	Dream	Dream		Wish
	Balancing	Balancing		
		Dissertation	Dissertation	
	Personal challenge	Dependability	Career advancement	Accreditation
	Credentials	Frustration	Recognition	
	Personal drive	Fellowship	Compensation	
	Extra effort		Experience distance learning	
	Finishing coursework		Doctoral work	
	Staying positive			

Overall, there were more similarities between the participants who were still in the program, although at different stages, than with those who graduated or withdrew from the program. Factors deemed important for these four participants as related to their persistence in the ELHE program were:

Quality of Academic Experiences

This included quality of the program and relevance of the course work, focus on engaged learning, quality of faculty and student feedback and their involvement with online courses, quality of academic advising and an advisor's commitment to students.

Online Learning Environment

The online environment offered students convenience and flexibility of learning, although it differentially affected students' persistence. The students who persisted had a high comfort level with technology, good writing skills and were comfortable interacting with other students online. The virtual community was not very important because it varied with each class and often was limited to a particular course.

Support and Assistance

A supporting and encouraging environment, created by both internal and external entities to the program, positively affected students' persistence. The internal sources of support included: faculty responsiveness and willingness to accommodate to distance learners' needs; peer support and encouragement; academic advisor's assistance and guidance; the institutional student support services infrastructure. Support and encouragement from sources external to the program included families, employment, and pets.

Self-Motivation

This included intrinsic motivation to pursue the doctoral degree in the distributed learning environment, such as personal challenge, responsibility, love for learning, and experiencing the new learning format. Extrinsic factors cited were: career advancement, earning the credentials, recognition, and increase in pay.

● DISCUSSION

The purpose of this mixed methods sequential explanatory study was to identify factors contributing to students' persistence in the ELHE program. In the quantitative phase, five external and internal to the program factors ("program", "online learning environment", "student support services", "faculty", and "self-motivation") were found to be predictors to students' persistence in the program. The qualitative follow up multiple case study analysis revealed that four reasons were pivotal: (1) quality of the program and other related academic experiences; (2) the very nature of the online learning environment; (3) support and assistance from different sources; and (4) student self-motivation. The quality of academic experiences had the most favorable affect on the participants' persistence in the program. Support and assistance they received contributed to their matriculation, while the online format was the cause for quitting the program for one participant. All participants were equally motivated to get the degree.

The way quantitative and qualitative findings highlighted the quality of the program and participants' academic experiences in it, the importance of student support infrastructure, and self-motivation to pursue the doctoral degree in the distributed learning environment were consistent with the basic ideas of Tinto's Student Integration Theory (1975, 1993). At the same time, relative importance of the external factors to doctoral students' persistence did not fully support Bean's Student Attrition Model (1980, 1990), which claimed factors external to an institution equally affected students' matriculation in college. However, Bean's model was specifically tailored to the undergraduate student population. For doctoral students pursuing the degree in the ELHE program, external factors might have played a secondary role to the internal factors related to the program and the online learning environment. The qualitative and the quantitative findings in this study supported the principle components of Kember's (1990, 1995) Model of Dropout from Distance Education Courses. Although Kember's model was limited to mostly undergraduate non-traditional students and individual DE courses, the idea of academic and social integration as embracing all facets of DE course offerings found reflection in this study. The quality of the program and academic experiences learning in the online environment, the importance of student support infrastructure, and student goal commitment were integral components of students' persistence in the ELHE program.

Program-Related Factors

Program

Quantitatively, most of the participants were satisfied with their academic experiences, the relevance and usefulness of the program, and how the program met their needs. The amount of satisfaction, however, was the greatest among the graduated participants and the lowest among the Withdrawn/Inactive group. A multiple case study analysis revealed all participants had high quality experiences in the program. This quality was reflected in the scholarly character of the program, its high standards, clarity of expectations, relevance, good structure and the opportunity to learn from others. The challenging character of the program, its broad content, and focus on

engaged learning also were recognized. Quality of interactions with students and their feedback differentially affected the participants' persistence. Those who successfully matriculated in the program received more meaningful and constructive peer feedback.

These findings were consistent with the limited research on the structure and content of a doctoral program and its impact on students' persistence. Usually students' academic experiences in the program were combined with other academic or institutional related factors, such as departmental orientation, relationship between course work and research skills, attitudes towards students, and student participation (Ferrer de Valero, 2001; Golde, 1998). Distance students usually are at a loss for recognizing and copying with such ambiguity, and must rely upon guidance from a concerned academic advisor or other students. In a fewer studies devoted to the quality of doctoral student experiences in DE programs (Huston, 1997; Sigafus, 1996; Wilkinson, 2002) the program structure was reported to be one of the contributing factors that positively affected students' experiences. Being able to anticipate or know the "roadmap" provided students with a sense of control. In a qualitative study of one course offered in the ELHE program (Ivankova and Stick, 2005), the focus of the program on engaged learning was cited as one of its quality indices. The participants believed they benefited more due to meaningful interactions between and among the students and instructors.

Online Learning Environment

The quantitative results indicated a majority of the participants were comfortable learning in the online environment, were satisfied with their online learning experiences, and believed learning was at least as effective as in a face-to-face classroom. The more matriculated in the program the participants were, the more positively they rated their online learning experiences. The qualitative findings revealed the participants were attracted by such characteristics of the online environment as its being location and time free, which allowed keeping both work and family schedules intact while taking classes. A second important characteristic was relative flexibility of learning at one's pace and time within the prescribed parameters of the course. However, the online format differentially affected the participants' persistence. For those who successfully matriculated in the program, the asynchronous format positively affected their progress, because, reportedly, it matched their learning style preferences. Factors impeding persistence included the non-real time format of the course related interactions and the focus on written versus oral communication.

These findings are supported by other studies that explored advantages and disadvantages of online learning, although not directly related to the issue of persistence. Flexibility to pursue education at personally convenient times was reported as a great advantage of learning at a distance (Quintana, 1996; Simonson, Smaldino, Albright, and Zvacek, 2000), while the learner-centered focus of online format was argued to lead to increased interaction and more active involvement (Chute, Thompson, and Hancock, 1999; Moore and Kearsley, 2005). The capacity to support interaction in an asynchronous format provided an opportunity for reflection and deliberation not found in any synchronous learning environment, including face-to-face classrooms (Anderson and Garrison, 1998; Berge and Collins, 1995; Hart and Mason, 1999). In addition, text-based communication contributed to a social "equalizing" effect with less stereotyping and more equitable participation (Harasim, 1990).

Virtual Community

Statistically, "virtual community" did not contribute to the function discriminating among the participant groups. Overall, half of the participants were satisfied with the online community, and two-thirds of the participants believed they were able to establish long-term social relationship with their fellow-students online. Those who had withdrawn or were inactive in the program, more negatively rated their community experiences. The qualitative analysis revealed that although the participants found the virtual community helpful, it was not a very important part of their academic experiences. No participant indicated a strong relationship between the community and his/her persistence in the program, because the community varied with each course, was limited to the course activities, and depended on one's willingness to participate in it. However, within some courses students managed to create a supportive and encouraging environment, both at the academic and personal level. Thus, social integration for those students was bounded by a particular course and particular activities.

These findings, to some extent, contradicted extensive research on the topic of community building in the online learning environment. Hiltz (1998) argued it was possible for people with shared interests to form and sustain relationships and communities through the use of computer-mediated communication. Community building in such an environment was based on collaborative learning and cooperation between and among the participants (Curtis and Lawson, 2001; Harasim, Hiltz, Teles, and Turoff, 1995; Palloff and Pratt, 2003). However, these and other studies mostly explored community building in single distance courses. Although an established virtual community reportedly helped keep students in a course (Brown, 2001; Eastmond, 1995; Garrison, 1997; Hiltz, 1998; Ivankova and Stick, 2005; Palloff and Pratt, 2003), community development was not studied from the angle of students' persistence in the entire program, and specifically a doctoral program. The results from the current study were interpreted as meaning community was a transitory phenomenon and was viewed as one of many "communities" the participants functioned in.

Academic Advisor- and Faculty-Related Factors

Academic Advisor

Although statistically an academic advisor did not have any significant effect on the participants' persistence in the program, about two-thirds of the participants were satisfied with the relationships they had with an academic advisor. More matriculated students had more positive experiences than the Beginning or Withdrawn/Inactive participants. Case study analysis showed that the quality of advising differed across the four participants. In case of the graduated participant the academic advisor's involvement was very high and was reflected in good professional advice, diligent feedback, and guidance with the dissertation. For another participant, who was approaching the dissertation stage in the program, advising was limited to providing knowledge of the process. The one, who had withdrawn from the program, had little exposure to advising, but what had been provided was deemed helpful and prompt. For the fourth participant, who was in the first half of the program, the academic advising experience was negative. Reportedly, there was lack of guidance, communication, and whatever little feedback was provided turned out to be of questionable value. Efficient academic advising also was associated with support and assistance in academic and personal problems, and encouragement toward earning the degree.

The fact that an academic advisor did not significantly affect students' persistence in this study was not consistent with other research on doctoral students' persistence. Ferrer de Valero (2001), Girves and Wemmerus (1988), Golde and Dore (2001), and Lovitts (2001) found that positive relations between a student and academic advisor were important for doctoral students' persistence in traditional campus-based programs. Doctoral students' withdrawal from a program was also reported to be due, in part, to inadequate or inaccurate advising, lack of interest or attention on the part of an advisor, and unavailability of an advisor (Bowen and Rudenstine, 1992; Golde, 2000). The inconsistencies of these findings might be explained by different doctoral student populations studied. Presumably, DE students were more self-sufficient and more focused on earning their degree. Being educational administrators in their professional lives, they might have been more organized and disciplined to persist in their efforts, and for many earning a doctoral degree was a necessary credential for keeping a job or getting promoted. In addition, there were other members of the program faculty always ready to provide the necessary guidance and assistance when an assigned academic advisor was not available.

Faculty

In the quantitative analysis, "faculty" was found to significantly contribute to the function discriminating among the four groups as related to their persistence. The degree of satisfaction with different aspects of instructors' teaching in the distributed environment varied. The participants were more satisfied with instructors' accessibility and promptness of feedback, than the quality of their feedback and their willingness to accommodate to distance learners' needs. The qualitative findings revealed that the quality of feedback depended on the readiness of faculty to teach online, their involvement with a course, and commitment to students. Students' persistence was positively affected by support and encouragement they received from the faculty and their ability to provide personal assistance. Such responsiveness was especially important in the absence of any assistance or guidance from an academic advisor.

These findings were supported by other studies of doctoral students' persistence. Lack of persistence in traditional doctoral programs often was attributed to lack of support and encouragement from a department and departmental faculty (Ferrer de Valero, 2001; Golde, 2000; Hales, 1998; Lovitts, 2001; Nerad and Cerny, 1993). Students who perceived support from their faculty were more likely to complete their degrees. However, little research has been conducted on the role of faculty in DE doctoral students' persistence. For example, in Sigafus's (1996) study faculty was cited as the most helpful source of support for those students.

Institution-Related Factors

Statistically "student support services" significantly affected the participants' matriculation in the program. Although more than half of the participants were satisfied with the institutional support services, their satisfaction differed depending on the particular service. The degree of satisfaction was not always consistent across the three matriculated groups, with the exception of the Withdrawn/Inactive participants who were the least satisfied. The case study analysis revealed that although the participants differed in the type and number of services they used and this need depended on the student's status in the program, the support infrastructure was friendly, convenient, and timely, and the procedures were convenient, smooth, and simple.

The importance of having a good support infrastructure for DE students was well established in the literature (King, Seward, and Gough, 1980; Moore and Kearsley, 2005; Rumble, 1992; Simpson, 2000). Availability and access to student support services were found to be a critical factor in distance students' academic success (Biner, Dean, and Mellinger, 1994; Tinto, 1993; Voorhees, 1987). However, no studies were located that explored the role of institutional support infrastructure in doctoral students' persistence in the distributed learning environment or programs like ELHE.

Student-Related Factors

Quantitatively, "self-motivation" had a significant affect on students' persistence in the program. All participants, except for the Withdrawn/Inactive group, were highly motivated to pursue the doctoral degree via distributed means. Not surprisingly, the Graduates were the most motivated group, while the Matriculated group was more motivated than the Beginning group. The case study analysis revealed that motivation was a strong factor for successful matriculation in the distributed environment. Intrinsic motivation included love for learning, personal challenge, a life long dream, and experiencing the new learning format. Responsibility was sustained by the fact everybody's work was being judged and evaluated by everybody in a class. Balancing work and studies was a challenge to motivation, but the unstructured process of dissertation work, perhaps, was the most daunting. Extrinsic factors also were important for staying on task; however, they were more important for male than female participants.

These findings were supported by other studies of doctoral students' persistence with regards to their motivation to complete the degree. Ferrer de Valero, (2001), Lovitts (2001), and Reynolds (1998) demonstrated that self-motivation was an important factor in obtaining the doctorate in campus-based programs. Students who had a "never give up" attitude, or had positive views of themselves, were more likely to complete the doctorate, especially during the tenuous time between course completion and dissertation work. Motivation and assumption of the responsibility for the learning process were especially important for distance doctoral students. Intrinsic motivation was reported as a significant predictor of success for such students

(Huston, 1997), while personal responsibility was found to be a contextual factor helping students matriculate successfully in the online environment (Scott-Fredericks, 1997).

External Factors

Based on the quantitative analysis, external factors, such as "family and significant other", and "employment" did not significantly affect students' persistence in the ELHE program, although two-thirds of the participants reported being supported by family, significant others, friends, and employers in their efforts to study in the distributed environment. The graduated participants received the most support among the four groups; however, they also claimed to be the most challenged by pressing job responsibilities and work schedules. The qualitative findings revealed different participants had different sources of external support: for some it was family and employment, for others family and pets, and for some there was no apparent support from external sources.

These findings were partially consistent with previous research. Frasier (1993), Girves and Wemmerus (1988), and Siegfried and Stock (2001) also indicated marital status did not affect doctoral students' persistence in campus-based programs. In the AHA Survey of Doctoral Programs in History (The American Historical Association, 2002), only 4% of the history major students indicated family reasons were among the most important factors causing them to drop out from doctoral programs. On the other hand, Golde (1998) found family commitments were crucial barriers leading some participants to quit the program. For traditional campus based doctoral students keeping priorities straight and balancing work and family is more difficult and might result in procrastination or withdrawal from the program. This study focused on doctoral students pursuing degrees in the distributed

environment, which offered convenience, flexibility, and the opportunity to keep regular work and family schedules. Free from the constraints of the traditional classroom, DE students could establish priorities, chose suitable time for studies, and enjoy full-time employment. Limited research on the affect of external factors on doctoral students' persistence in the distributed environment also suggested families, friends, and employers among the most helpful sources of support (Huston, 1997; Riedling, 1996; Sigafus, 1996).

Implications and Recommendations

Recognizing that many institutions of post-secondary and higher education offer graduate and professional degrees via distributed means, the results of this study are aimed at numerous stakeholders: policy makers and educational administrators, graduate program developers and instructional designers, institutional faculty and staff, and students, who currently pursue their doctoral degrees in the distributed environment or consider doing so. Knowing the predictive power of external and internal factors to students' persistence in the distributed learning environment may assist programs in developing strategies to enhance doctoral persistence and eventually degree completion. Specifically, the implications of this study include:

1. The scholarly and challenging character of the program, its relevance and applicability to students' professional activities, high standards and focus on an individual may lead to a more successful matriculation in the program. A distributed program meeting such requirements may have a greater potential for attracting promising applicants, nurturing their scholastic development, and ultimately improving their persistence and graduation rates.

2. To benefit from learning in the distributed environment, students need to be comfortable with technology and have good writing skills. Text-based learning should match their learning style preferences and they should be comfortable interacting with other students and instructors online. Students considering or applying to a distributed program should be informed upfront of the program format and what the expectations are in terms of performance.

3. Students benefit from online courses when an instructor acts as a facilitator of learning, is actively involved with the course, and provides the necessary encouragement and assistance. To fulfill this role, faculty should be prepared to teach online, be ready to provide constant and timely quality feedback, and be flexible to accommodate to distance learners' needs.

4. Institutional student support infrastructure should be in place to assist distance learners with all their needs, problems and concerns. Such infrastructure should include all possible services distance learners might encounter during their matriculation process. Of particular importance is prompt and qualified assistance with possible technology problems, obtaining the course materials, and gaining access to the library reserves and other resources.

5. Students who want to succeed in a distributed learning environment need to be highly motivated, disciplined and organized to successfully balance studies, work, and families. Students' intrinsic motivation should be supported and encouraged by the program quality, user-friendly online format, favorable learning environment, as well as external to the program factors. Extrinsic motivation also is important, but could be different in each particular case.

6. The quality and responsiveness of academic advising in distributed doctoral programs need to be at a high level. Students should receive professional advising and guidance from their academic advisor throughout the entire program. Reasonably consistent contact between a student and an advisor helps ensure a continued progress in a program. Assistance with academic problems and personal encouragement should be part of a distance advisor–advisee relationship.

7. Online community may enhance students' progress, if it is established and supported throughout the entire program. Faculty may take a lead in launching and facilitating informal interactions with the class alongside with other academic activities. Schools and departments also should reflect upon more strategies to virtually bring distance learners together, such as summer residencies, listservs, and virtual student organizations.

This study provided only one perspective on persistence in the distributed doctoral program— that of the students themselves, excluding other internal and external constituents. Also, the marginal reliability estimates of the two subscales measuring "family and significant other" and "employment" are recognized as the limitation to the related findings. Being the only research on students' persistence in a distributed doctoral program, this study leaves some unanswered questions and opens a door for future research on students' persistence in such environments. In-depth exploration of distance students' persistence might help their journey be less stressful and more efficient. The results would be productive for students, institutions, and society.

NOTES

1. The study design was reported elsewhere (Ivankova, 2004; Ivankova, Creswell, and Stick, 2006).

2. A detailed explanation of the case selection procedure for the qualitative phase of this study was reported elsewhere (Ivankova et al., 2006).

REFERENCES

Anderson, D., and Garrison, D. R. (1998). Learning in a networked world: New roles and responsibilities. In: Gibson, C. C. (ed.), *Distance learners in higher education: Institutional response for quality outcomes.,* Atwood Publishing, Madison, WI, pp. 97–112.

Austin, A. (2002). Preparing the next generation of faculty: Graduate school as socialization to the academic career. *The Journal of Higher Education* 73(1): 94–122.

Bair, C. R., and Haworth, J. G. (1999). Doctoral student attrition and persistence: A meta-synthesis of research. Paper presented at the annual meeting of the Association for the Study of Higher Education, San Antonio, TX. (ERIC Document Reproduction Service No. ED 437 008).

Bauer, W. C. (1997). Pursuing the Ph.D.: Importance of structure, goal-setting and advising practices in the completion of the doctoral dissertation (Doctoral dissertation, University of California, Los Angeles, 1997). Dissertation Abstracts International, 58:2096.

Bean, J. P. (1980). Dropouts and turnover: The synthesis and test of a causal model of student attrition. *Research in Higher Education* 12: 155–187.

Bean, J. P. (1990). Why students leave: Insights from research. In: Hossler, D. (ed.), *The strategic management of college enrollments,* Jossey-Bass, San Francisco CA, pp. 147–169.

Berge, Z. L. and Collins, M. P., (Eds.) (1995). *Computer mediated communication and the online classroom: Overview and perspectives, 1,* Hampton Press, Cresskill, NJ.

Biner, P. M., Dean, R. S., and Mellinger, A. E. (1994). Factors underlying distance learner satisfaction with televised college-level courses. *The American Journal of Distance Education* 9(1): 60–71.

Bowen, W. G., and Rudenstine, N. L. (1992). *In pursuit of the PhD,* Princeton University Press, Princeton, NJ.

Brien, S. J. (1992). The adult professional as graduate student: A case study in recruitment, persistence, and perceived quality (Doctoral dissertation, Northern Illinois University, 1992). *Dissertation Abstracts International* 53: 2203.

Brown, R. E. (2001). The process of community-building in distance learning classes. *Journal of Asynchronous Learning Networks* 5(2): 18–35.

Carr, S. (2000). As distance education comes of age, the challenge is keeping the students. The Chronicle of Higher Education, 23, A1. Retrieved May 25:2002, from http://www.chronicle.com/free/v46/i23/23a00101.htm.

Cesari, J. P. (1990). Thesis and dissertation support groups: A unique service for graduate students. *Journal of College Student Development* 31: 375–376.

Chute, A. G., Thompson, M. M., and Hancock, B. W. (1999). *The McGraw-Hill handbook of distance learning,* The McGraw-Hill Companies, Inc, New York, NY.

Creswell, J. W. (1998). *Qualitative inquiry and research design: Choosing among five traditions,* Sage Publications, Thousand Oaks, CA.

Creswell, J. W. (2005). *Educational research: Planning, conducting, and evaluating quantitative and qualitative approaches to research,* (2nd Ed.), Merrill/Pearson Education, Upper Saddle River, NJ.

Creswell, J. W., and Miller, D. (2002). Determining validity in qualitative inquiry. *Theory into Practice* 39(3): 124–130.

Creswell, J. W., Plano Clark, V. L., Gutmann, M., and Hanson, W. (2003). Advanced mixed methods research designs. In: Tashakkori, A., and Teddlie, C. (eds.), *Handbook on mixed methods in the behavioral and social sciences.,* Sage Publications, Thousand Oaks, CA, pp. 209–240.

Curtis, D. D., and Lawson, M. L. (2001). Exploring collaborative online learning. *Journal of Asynchronous Learning Networks* 5(1): 12–34.

Diaz, D. P. (2000). Comparison of student characteristics, and evaluation of student success, in an online health education course. Unpublished doctoral dissertation, Nova Southeastern University, Fort Lauderdale, Florida. Retrieved May 22:2002, from http://www.LTSeries.com/LTS/pdf_docs/dissertn.pdf.

Eastmond, D. V. (1995). *Alone but together: Adult distance study through computer conferencing,* Hampton Press, Cresskill, NJ.

Ferrerde Valero, Y. (2001). Departmental factors affecting time-to-degree and completion rates of doctoral students at one land-grant research institution. *The Journal of Higher Education* 72(3): 341–367.

Finke, W. F. (2000). *Lifelong learning in the information age: Organizing net-based learning and teaching systems,* Fachbibliothek-Verlag, Bueren, Germany.

Frasier, E. R. M. (1993). Persistence of doctoral candidates in the college of education, University of Missouri-Columbia (Missouri) (Doctoral dissertation, University of Missouri, Columbia, 1993). *Dissertation Abstracts International* 54: 4001.

Garrison, D. R. (1997). Computer conferencing: The post-industrial age of distance education. *Open Learning* 12(2): 3–11.

Geiger, R. (1997). Doctoral education: The short-term crisis vs. long-term challenge. *The Review of Higher Education* 20(3): 239–251.

Girves, J. E., and Wemmerus, V. (1988). Developing models of graduate student degree progress. *Journal of Higher Education* 59(2): 163–189.

Golde, C. M. (1998). Beginning graduate school: Explaining first-year doctoral attrition. In: Anderson, M. S. (ed.), *The experience of being in graduate school: An exploration. New Directions for Higher Education, 101,* Jossey-Bass Publishers, San Francisco, CA, pp. 55–64.

Golde, C. M. (2000). Should I stay or should I go? Student descriptions of the doctoral attrition process. *The Review of Higher Education* 23(2): 199–227.

Golde, C. M. (2001). *Overview of doctoral education studies and reports: 1990-present.* Retrieved November 15, 2002, from http://www.carnegie foundation.org.

Golde, C. M., & Dore, T. M. (2001). *At cross purposes: What the experiences of doctoral students reveal about doctoral education.* A report for The Pew Charitable Trusts. Philadelphia, PA. Retrieved November 15, 2002, from http://www.phd-survey .org.

Green, J. C., Caracelli, V. J., and Graham, W. F. (1989). Toward a conceptual framework for mixed-method evaluation designs. *Educational Evaluation and Policy Analysis* 11(3): 255–274.

Hales, K. S. (1998). The relationship between personality type, life events, and completion of the doctorate degree (Doctoral dissertation, Texas A&M University-Commerce, 1998). *Dissertation Abstracts International* 59: 1077.

Hanson, W. E., Creswell, J. W., Plano Clark, V. L., Petska, K. P., and Creswell, J. D. (2005). Mixed methods research designs in counseling psychology. *Journal of Counseling Psychology* 52(2): 224–235.

Harasim, L. (1990). Online education: An environment for collaboration and intellectual amplification. In:

Harasim, L. (ed.), *Online education: Perspectives on a new environment.,* Praeger, New York, pp. 39–64.

Harasim, L., Hiltz, S., Teles, L., and Turoff, M. (1995). *Learning networks,* MIT Press, Cambridge, MA.

Hart, G., and Mason, J. (1999). Computer-facilitated communications in transition. In: Feyten, C. M., and Nutta, J. W. (eds.), *Virtual instruction. Issues and insights from an international perspective,* Libraries Unlimited, Inc, Englewood, CO, pp. 147–171.

Haworth, J. G. (1996). Assessment in graduate and professional education: Present realities, future prospects. In: Haworth, J. G. (ed.), *Assessing graduate and professional education: Current realities, future prospects. New Directions for Institutional Research, 92,* Jossey-Bass Publishers, San Francisco, CA, pp. 89–97.

Hiltz, S. R. (1998). Collaborative learning in asynchronous learning networks: Building learning communities. (ERIC Document Reproduction Service No. ED 427 705).

Holmberg, B. (1995). *Theory and practice of distance education,* Routledge, New York, NY.

Huston, J. L. (1997). Factors of success for adult learners in an interactive compressed video distance learning environment (Doctoral dissertation, University of Kentucky, 1997). *Dissertation Abstracts International* 58: 1199.

Ivankova, N. V. (2004). *Students' persistence in the University of Nebraska-Lincoln distributed doctoral program in Educational Leadership in Higher Education: A mixed methods study.* Unpublished doctoral dissertation, University of Nebraska, Lincoln.

Ivankova, N. V., Creswell, J. W., and Stick, S. L. (2006). Using mixed methods sequential explanatory design: From theory to practice. *Field Methods, 18*(1): 3–20.

Ivankova, N. V., and Stick, S. L. (2002). *Students' persistence in the Distributed Doctoral Program in Educational Administration: A mixed methods study.* Paper presented at the 13th International Conference on College Teaching and Learning, Jacksonville, FL.

Ivankova, N. V., and Stick, S. L. (2003). Distance education doctoral students: Delineating persistence variables through a comprehensive literature review. *The Journal of College Orientation and Transition* 10(2): 5–21.

Ivankova, N. V., & Stick, S. L. (2005, Fall). Collegiality and community-building as a means for sustaining student persistence in the computer-mediated asynchronous learning environment. *Online Journal of*

Distance Learning Administration, 8(3), http://www.westga.edu/~distance/jmain11.html.

Johnson, B., and Turner, L. A. (2003). Data collection strategies in mixed methods research. In: Tashakkori, A., and Teddlie, C. (eds.), *Handbook on mixed methods in the behavioral and social sciences.*, Sage Publications, Thousand Oaks, CA, pp. 297–320.

Johnson, E. M., Green, K. E., and Kluever, R. C. (2000). Psychometric characteristics of the revised procrastination inventory. *Research in Higher Education* 41(2): 269–279.

Kember, D. (1990). The use of a model to derive interventions which might reduce drop-out from distance education courses. *Higher Education* 20: 11–24.

Kember, D. (1995). *Open learning courses for adults: A model of student progress,* Educational Technology Publications, Englewood Cliffs, NJ.

King, B., Sewart, D., and Gough, J. E. (1980). Support systems in distance education. *Open Campus* 3: 13–38.

Kline, R. B. (1998). *Principles and practice of structural equation modeling,* Guilford, NY.

Kowalik, T. F. (1989). What do we know about doctoral student persistence. *Innovative Higher Education* 13(2): 163–171.

Lincoln, Y. S., and Guba, E. G. (1985). *Naturalistic inquiry,* Sage, Beverly Hills, CA.

Lovitts, B. E. (2001). *Leaving the ivory tower: The causes and consequences of departure from doctoral study,* Rowman and Littlefield Publishers, Inc, New York, NY.

Lovitts, B. E., and Nelson, C. (2000). The hidden crisis in graduate education: Attrition from PhD programs. *Academe* 86(6): 44–50.

McCabe-Martinez, M. C. (1996). A study of perceptions of factors that enhanced and impeded progress toward the completion of the doctoral degree in education for Hispanic students employed in the public school systems (Doctoral dissertation, Boston College, 1993). *Dissertation Abstracts International* 57: 2900.

McMillan, J. H. (2000). *Educational research: Fundamentals for the consumer,* (3rd Ed.), Addison Wesley Longman, New York, NY.

Miles, M. B., and Huberman, A. M. (1994). *Qualitative data analysis: A sourcebook,* (2nd Ed.), Sage Publications, Thousand Oaks, CA.

Moore, M. G., and Kearsley, G. (2005). *Distance education: A systems view,* (2nd Ed.), Wadsworth Publishing Company, Belmont, CA.

NCES (National Center for Education Statistics). (2002). *Digest of Education Statistics,* 2002 Washington,

DC: Institute of Education Sciences, US Department of Education. Retrieved March 15, 2004, from http://www.nces.ed.gov/programs/digest/d02/ch_3.asp#1.

Nerad, M., and Cerny, J. (1993). From facts to action: Expanding the graduate division's educational role. In: Baird, L. (ed.), *Increasing graduate student retention and degree attainment. New Directions for Institutional Research, 80,* Jossey-Bass Publishers, San Francisco, CA, pp. 27–39.

Nerad, M., and Miller, D. S. (1996). Increasing student retention in graduate and professional programs. In: Haworth, J. G. (ed.), *Assessing graduate and professional education: Current realities, future prospects. New Directions for Institutional Research, 92,* Jossey-Bass Publishers, San Francisco, CA, pp. 61–76.

Nolan, R. E. (1999). Helping the doctoral student navigate the maze from beginning to end. *Journal of Continuing Higher Education* 48(3): 27–32.

NSF (National Science Foundation, Division of Science Resources Studies). (1998). *Summary of workshop on graduate student attrition,* NSF 99–314. Arlington, VA.

Palloff, R. M., and Pratt, K. (2003). *The virtual student: A profile and guide to working with online learners,* Jossey-Bass Publishers, San Francisco, CA.

Parker, A. (1999). A study of variables that predict dropout from distance education. *International Journal of Educational Technology,* 1(2). Retrieved June 23, 2002, from http://www.outreach.uiuc.edu/ijet/v1n2/parker/index.html.

Quintana, Y. (1996). *Evaluating the value and effectiveness of Internet-based learning.* Retrieved July 25, 2001 from: http://www.isoc.org/inet96/proceedings/c1/c1_4.htm.

Reynolds, K. A. (1998). Factors related to graduation of doctoral students in the higher education program at Southern Illinois University—Carbondale (Southern Illinois University at Carbondale) (Doctoral dissertation, Southern Illinois University at Carbondale, 1998). *Dissertation Abstracts International* 60: 0673.

Riedling, A. M. (1996). An exploratory study: Distance education doctoral students in the field of educational policy studies and evaluation at the University of Kentucky (Doctoral dissertation, University of Louisville, 1996). *Dissertation Abstracts International* 57: 4337.

Rumble, G. (1992). *The management of distance learning systems,* International Institute for Educational Planning, Paris: UNESCO.

Scott-Fredericks, G. L. (1997). The graduate student experience in computer-mediated classes: A grounded theory study (Lotus Notes, distance education) (Doctoral dissertation, University of Nebraska, Lincoln, 1997). *Dissertation Abstracts International* 58: 4625.

Siegfried, J. J., & Stock, W. A. (2001, Spr). So you want to earn a Ph.D. in economics? How long do you think it will take? *Journal of Human Resources,* 36(2), 364–378.

Sigafus, B. M. (1996). The complexities of professional life: Experiences of adult students pursuing a distance learning doctoral program in educational administration (Doctoral dissertation, University of Kentucky, 1996). *Dissertation Abstracts International* 57: 2310.

Simonson, M., Smaldino, S., Albright, M., and Zvacek, S. (2000). *Teaching and learning at a distance: Foundations of distance education,* Prentice-Hall, Inc, Upper Saddle River, NJ.

Simpson, O. (2000). *Supporting students in open and distance learning,* Kogan Page Limited, London, UK.

Stake, R. E. (1995). *The art of case study research,* Sage, Thousand Oaks, CA.

Stick, S., and Ivankova, N. (2004, December). Virtual learning: The success of a world-wide asynchronous program of distributed doctoral studies. *Online Journal of Distance Learning Administration,* 7 (4), http://www.westga.edu/~distance/jmain11.html.

Tabachnick, B. G., and Fidell, L. S. (2000). *Using multivariate statistics,* Allyn & Bacon, New York, NY.

Tashakkori, A., and Teddlie, C. (1998). *Mixed methodology: Combining qualitative and quantitative approaches. Applied Social Research Methods Series, 46,* Sage Publications, Thousand Oaks, CA.

Tashakkori, A. and Teddlie, C., (Eds.) (2003). *Handbook on mixed methods in the behavioral and social sciences,* Sage Publications, Thousand Oaks, CA.

The American Historical Association. (2002). *Preliminary results of the ANA survey of doctoral programs in history.* Retrieved on August 15, 2003, from http://www.theaha.org/projects/grad-survey/Preliminary4.htm.

Thompson, M. M. (1998). Distance learners in higher education. In: Gibson, C. C. (ed.), *Distance learners in higher education: Institutional responses for quality outcomes,* Atwood Publishing, Madison, WI, pp. 11–23.

Tinto, V. (1975). Dropout from higher education: A theoretical synthesis of recent research. *Review of Educational Research* 45: 89–125.

Tinto, V. (1993). *Leaving college: Rethinking the causes and cures of student attrition* (2nd Ed.), The University of Chicago Press, Chicago, IL.

Tinto, V. (1998). Colleges as communities: Taking research on student persistence seriously. *The Review of Higher Education* 21(2): 167–177.

Verduin, J. R. Jr., and Clark, T. A. (1991). *Distance education: The foundations of effective practice,* Jossey-Bass Publishers, San Francisco, CA.

Voorhees, R. A. (1987). Toward building models of community college persistence: A logic analysis. *Research in Higher Education* 26(2): 115–129.

Wilkinson, C. E. (2002). A study to determine the predictors of success in a distance education doctoral program (Doctoral dissertation, Nova Southeastern University, 2002). *Dissertation Abstracts International* 63: 2165.

Yin, R. (2003). *Case study research: Design and methods,* (3rd Ed.), Sage Publication, Thousand Oaks, CA.

Received May 3, 2005.

Appendix C—An Example of the Exploratory Sequential Design

Exploring the Dimensions of Organizational Assimilation

Creating and Validating a Measure

Karen Kroman Myers and John G. Oetzel

The purpose of this study was to create and validate a measure of organizational assimilation index. Organizational assimilation describes the interactive mutual acceptance of newcomers into organizational settings. Members from the advertising, banking, hospitality, university, nonprofit, and publishing industries participated in two phases of research. In the first phase, 13 interviewees suggested six dimensions of organizational assimilation: familiarity with others, organizational acculturation, recognition, involvement, job competency, and adaptation/role negotiation. The second phase involved analysis of a survey of 342 participants that appeared to validate the six dimensions. The OAI's construct validity was tested and supported through the use of three other scales. Job satisfaction and organizational identification related positively to assimilation, while propensity to leave related negatively.

KEY CONCEPTS: organizational assimilation, organizational assimilation index, organizational socialization, newcomer integration

Karen K. Myers (M.A., University of New Mexico, 2001) is a doctoral student in the Hugh Downs School of Human Communication at Arizona State University. John Oetzel (Ph.D., University of Iowa, 1995) is an associate professor in the Department of Communication and Journalism at the University of New Mexico. This manuscript was derived from the first author's thesis under the guidance of the second author. The authors wish to thank several individuals for their insightful comments: Dirk Gibson, Jacqueline Hood, Robert McPhee, Pam Lutgen-Sandvik, David Seibold who responded to an earlier version of this paper when it was presented at the 2002 annual meeting of the International Communication Association in Seoul, South Korea, and two anonymous reviewers. Address correspondence to the first author at The Hugh Downs School of Human Communication, Arizona State University, PO Box 871205 Tempe, AZ 85287-1205.

SOURCE: Myers, K. K., & Oetzel, J. G. (2003). Exploring the dimensions of organizational assimilation: Creating and validating a measure. *Communication Quarterly, 51*(4), 438–457. Reprinted with permission of Taylor & Francis Group, LLC.

Fisher (1986) highlighted about the lack of research devoted to the processes involved in organizational entry. In recent years, however, our understanding of the processes related to organization entry has benefited from several lines of research, including processes organizations use to orient and mold recruits into productive organizational members (Chao, O'Leary-Kelly, Wolf, Klein, & Gardener, 1994; Schein, 1968; Van Maanen & Schein, 1979), socialization turning points (Bullis & Bach, 1989), methods of organizational orientation and training (Holton, 1996; Jones, 1986), efforts exerted by newcomers themselves through adaptation (Ashford & Taylor, 1990), behavior self-management (Saks & Ashforth, 1996), coping strategies (Teboul, 1997; Waung, 1995), early involvement (Bauer & Green, 1994), information seeking (Miller & Jablin, 1991; Morrison, 1993; Ostroff & Kozlowski, 1992), and role negotiation (Kramer & Miller, 1999; Miller, Jablin, Casey, Lamphear-Van Horn, & Ethington, 1996).

Organizational assimilation refers to "the processes by which individuals become integrated into the culture of an organization" (Jablin, 2001, p. 755). Although some scholars note that "assimilation" emphasizes individuals' giving up their individualities to fit in with their new collectives (see Bullis, 1999; Clair, 1999; Turner, 1999), we feel it is a useful aspect of newcomer entry because becoming an effective member of an organization involves not just organizational efforts to socialize new members nor, in turn, efforts of recruits to become accepted by organizational incumbents. Successful assimilation involves both organizations and newcomers.

Despite the prevalence of research on these processes, there is no measure of organizational assimilation. The stage model of organizational assimilation (Jablin, 1987, 2001) so prevalent in scholarly literature, and most notably in organizational

communication textbooks, suggests that employees are assimilated in a kind of linear progression, wherein it is implicitly assumed that they continue to feel increasingly a part of the organization until they exit. To the contrary, it makes sense to presume that the extent to which one feels him or herself to be a valuable part of an organization is likely to vary over time in accordance with unmet expectations, environmental shifts, changes in responsibility, promotions, burnout, and a wide variety of experiences that constitute organizational life.

The absence of an instrument to measure the rise and fall of assimilation is perhaps one reason these linear assumptions endure. Moreover, it appears obvious that some newcomers and their organizations vary not only with the efficiency of their assimilation processes generally, but also with regard to specific dimensions of the phenomenon. For example, individuals may feel assimilated in some aspects of the organization because they are involved in their work, but may have failed to develop productive working relationships in the environment. What is needed then is an instrument to assess members' level of organizational assimilation.

Such a measure could provide information to management indicating not only whether assimilation deficiencies exist but also what specific dimensions of assimilation are most lacking. Further, the measure could enable scholars to focus on the various dimensions of assimilation, as well as the impact of antecedent phenomena. The study reported here proceeded in two stages. In the first stage, we explored dimensions of organizational assimilation to define the processes involved in transitioning from newcomer to organizational member. Stage one concluded with the development of an instrument that permits operationalization of the dimensions of organizational assimilation. The second stage consisted of efforts to validate the measure, which we call the Organizational Assimilation Index (OAI).

Dimensions of
Organizational Assimilation

As previously mentioned, several studies have examined processes related to assimilation. These studies provide insight into newcomer integration. We review three exemplars that provide a framework for the creation of a measure of organizational assimilation. We then discuss related concepts with validated measures, which we used to establish the construct validity of the new measure.

Stage models depict at least three stages of organizational assimilation (Jablin, 2001). Anticipatory socialization refers to all socialization efforts prior to organizational entry. The encounter stage begins as the newcomer enters the organization and begins to become socialized through training and orientation. Finally, the metamorphosis stage is viewed as long-term settling in, during which newcomers make the transition to become full members of the organization. However, the question of when a newcomer passes through the stages of assimilation is problematic. Jablin noted that many organizations arbitrarily designate the newcomer-member transition to end after the member has been with the organization between three and six months. This designation appears to ignore the fact that some newcomers assimilate more quickly than others and that employees might assimilate in one aspect of organizational life quicker than in other aspects (Ostroff & Kozlowski, 1992). Therefore, we looked for studies that focused on the content of assimilation.

In a multi-phase study, Chao et al. (1994) attempted to identify "what is learned during socialization" (p. 730). They found six content areas of newcomer socialization (organizational history, language, politics, people, goals and values, and performance proficiency) and examined the relationship of those factors to career outcomes. Next, they compared levels of these six dimensions for individuals who had not switched jobs, others who had

not switched jobs within the same organization, and individuals who had switched organizations. The results revealed that workers who had not switched jobs and remained in their original organization were highest on five of the dimensions. The next highest group consisted of individuals who had only switched jobs, not organizations.

Bullis and Bach (1989) interviewed 28 new graduate students and asked them to describe turning points, specific messages that had long-term impact on their relationship with their new academic departments. The participants described 15 different turning point events, such as moving in, getting away, receiving formal recognition, and doubting one's self. At two weeks and eight months following their entry, the students identified turning points they had experienced and completed Cheney's (1983) Organizational Identification Questionnaire. The authors reported that instances of socializing positively affected identification, whereas instances of disappointment had a negative impact. Because many of the turning points involved communication indicating the newcomer's acceptance within the organization, Bullis and Bach's (1989) study takes us one step closer to understanding processes associated with newcomer assimilation.

Although these exemplars illustrate some of the content and processes of organizational assimilation, the researchers have not explicitly described all of the dimensions of organizational assimilation. From these studies, it appears that getting to know others in the organization and receiving recognition are relevant factors, but other dimensions and tactics may also contribute to an individual's assimilation. In the first stage of this study, we explored and identified an exhaustive list of dimensions of assimilation. This stage helped to establish content validity of the instrument. We addressed the following research question:

RQ: What are the dimensions of organizational assimilation?

In the second stage of this study, we sought to validate an instrument (the Organizational Assimilation Instrument [OAI]) for measuring organizational assimilation. The answer to the first research question provided specific items for use in operationalizing organizational assimilation. We included three scales to test for construct validity of the OAI: job satisfaction, organizational identification, and propensity to leave. *Job satisfaction* is the level of affinity one feels for a job and company (Brayfield & Rothe, 1951). To capture this attribute, Brayfield and Rothe created the Job Satisfaction Index. *Organizational identification* refers to a member's perception that the organization's values and interests are of primary concern with evaluating decision alternatives (Tompkins & Cheney, 1983). Cheney (1983) speculated that individuals with organizational identification make decisions that affect the organization on the basis of that identification and created the organizational identification questionnaire (OIQ) to assess the level of employee identification with an organization. *Propensity to leave* refers to the likelihood that an individual will sever ties with an organization (Lyons, 1971). Lyons developed an index that has questions related to one's desire to remain in an organization's employ, likelihood of his or her rejoining the organization if he or she were forced to spend time away from the organization (for example due to illness), and how long the person wishes to remain in the organization.

We predicted that job satisfaction and organizational identification would correlate positively with organizational assimilation and that propensity to leave would correlate negatively with organizational assimilation. Further, we expected the dimensions of organizational assimilation to be better predictors of the job satisfaction, organizational identification, and propensity to leave than tenure. Since tenure is often used as an indicator of assimilation, it was necessary to show that the OAI provides information over and above tenure. These expectations led to four hypotheses:

HI: The dimensions of organizational assimilation correlate positively with job satisfaction.

H2: The dimensions of organizational assimilation correlate positively with organizational identification.

H3: The dimensions of organizational assimilation correlate negatively with propensity to leave.

H4: The dimensions of organizational assimilation account for more variance in job satisfaction, organizational integration, and propensity to leave than tenure.

● DETERMINING FACE AND CONTENT VALIDITY

The study unfolded in two phases. The purpose of Phase One was to answer the research question defining the dimensions of organizational assimilation. This was accomplished by asking organizational members to describe their assimilation experiences. Phase Two involved constructing and validating a questionnaire that could be used to measure organizational assimilation. The next section describes the first phase and results.

Participants

To ensure that the measure would be appropriate for assessing assimilation, regardless of industry or organizational level, members of several different types of organizations at various levels within organizations took part in the study. Seven women and six men who represented a wide range of positions within their organizations, tenures, ages, and industries participated in interviews. The participants' positions ranged from entry-level hourly employees to higher-level executives, with tenure from half a month to 109.5 months (M=28.57, SD=34.72). The participants' ages varied from 18 to 61 years (M=37.08, SD=11.54). They included individuals from the hospitality, university, high-technology, and advertising industries. The organizations were located in two cities in the southwestern United States.

Data Collection

Interview protocol. We designed the questions in such a way as to help the interviewees think about how their status as organizational members had changed since their first day in the organization and what processes led to those changes (See Appendix A for the interview protocol.). The first two questions required participants to indicate whether they felt more a part of the organization now (at the time of the interview) than they did on their first day, and if so, to describe the changes. The respondents were then to think of a situation that may have caused them to feel that they were becoming accepted as members of the organization and communication they received during the situation (Questions 3–6). The next two questions (7–8) entailed describing people who were or were not assimilated and the communicative behavior that negatively relates to assimilation. For question #9, participants described communication from co-workers relating to assimilation. Questions #10 and #11 concerned knowledge about "fitting in." The final question related to strategies individuals reportedly used to assimilate into the organization. These questions helped to identify specific types of communicative behavior relating to the participants' assimilation into the organization for different periods during their tenure, as well as that were indicative of the success level.

Interview procedures. The participants took part individually at their respective organizations, except for one person who was interviewed over the telephone. Participation was voluntary, and the participants understood that they could refuse to answer any questions and terminate the interview at any time. Interviewees were informed that the purpose of the research was to explore how a new employee becomes a part of the organization. Participants had assurance that the interviews would not be shared with management, and, if quoted in the research results, pseudonyms would replace their actual names. We asked participants to think about how they attempted to assume membership in their organizations. Interviewees were to reflect on their experiences and tell stories that they believed illustrated their characterizations. Questions were often followed by additional questions to probe for detailed explanations. The interviewer took detailed field notes. All participants gave the interviewer permission to tape-record the sessions. Interviews continued until the point of theoretical saturation was reached (Lindlof, 1995), which indicated diminishing original insights. The interviews ranged in length from 20 to 50 minutes, and each was later transcribed for use in analysis of participants' responses.

Data Analysis

The analysis for Phase One served to answer the research question concerning the dimensions of organizational assimilation. Using Glaser and Strauss's (1967) method of constant comparison, and Miles and Huberman's (1994) suggestions for coding qualitative data, we identified and categorized all processes that the participants described or referred to in the interviews that pertained to their attempts to learn about and fit into the organization. We completed this process in several iterations. First, we read the transcriptions to obtain an overall flavor of the interviewees' responses. Next to each line or paragraph, we generated labels to reflect our initial coding. From these labels, we developed a general category scheme of the participant responses.

Second, we began to identify themes by sorting the initial scheme into concrete categories and subcategories. The categorization reflected similarity of responses (in regard to the assimilation process) and frequency of responses. At least half the participants had to identify an initial theme for it to be included. Next, we reread the transcripts and field notes and looked for frequently occurring expressions and unexpected counterintuitive material that provided atypical evidence of participant experiences. We categorized the responses according to several initial themes, such as getting to know coworkers' names, learning the organizations' standards, developing job skills, being appreciated by supervisors and coworkers, and understanding how they fit into their organizations.

Third, we reviewed these themes to determine how they fit into existing assimilation theory or how they might contribute to an understanding of the assimilation process. During this step, we used two criteria (Patton, 1990): Does the information confirm current organizational assimilation theory, and does it offer new insights into and interpretations of member assimilation? Continuing with this third step, we also thought about members' underlying purposes of the processes described and how those actions may have fulfilled an objective of assuming a role or becoming a part of the organization. As a result, we combined and renamed the initial themes into six dimensions of assimilation. Finally, we reread the responses and categorized them into one of the six themes to ensure goodness of fit. After this step, we determined that the resulting six dimensions adequately reflected the responses provided by participants.

Results

Familiarity with others included getting to know coworkers, making friends with coworkers, feeling comfortable with coworkers, feeling and expressing a general friendliness, learning how to interact with coworkers, speaking up at meetings, demonstrating a willingness to interact with coworkers, deriving emotional support from organizational members, and generally feeling a sense of community. Rodney,

head waiter at a hotel restaurant, thought that his bond with the organization was stronger because of relationships he had formed with coworkers. "I've gotten to know the people who have been around for a while. If we happen to go out and grab a cocktail after work or something, you build more of a social relationship. It is kind of a commitment to the [organization] just because the people become more enduring to you."

Acculturation, or learning and accepting the culture, was the second dimension of organizational assimilation. Interviewees described aspects of learning the norms of the organization and "how things get done" within their respective organizations. Kelly, who works in an organizational development position with a technology company, talked about steps her organization takes to introduce newcomers into the organization. She said that her organization's culture is not conducive to an overly friendly assimilation process. Others orient newcomers by teaching them "what they need to watch out for, like what they need to not step into, what are the kinds of things that will get them in trouble." She also noted that newcomers who have not assimilated are more likely to break organizational norms, such as not keeping coworkers informed about projects they are working on. Violating norms would likely cause established members to be less accepting of the newcomer resulting in less acceptance and added stress. Therefore, acculturating was critical to organizational assimilation.

We labeled the third dimension *recognition.* According to the participants, being recognized as valuable, either by superiors or coworkers, and feeling that their work was important to the organization was a significant part of feeling accepted into the organization. Jessica, a hotel manager, identified a defining moment in her assimilation when she was able to use her ability to speak Spanish to rescue her general manager. She proudly remembered, "He came to me and said, 'I need your help. I have all these ladies. They don't speak English.' I translated the meeting for him. I was excited to be able to do this for my general manager."

Some participants suggested that they can tell when someone has not assimilated into the organization because of the employee's level of *involvement* with the organization. When members are involved with the organization, they seek ways to contribute to the organization, often by volunteering to perform extra work or take on added responsibility for the sake of the organization and its members. Margaret, a university instructor, compared two students, one who was thriving in her new environment and one who was not. The thriving student had become involved in many aspects of university life. She had made friends with many others within the department and was deriving social support from her fellow students. The other student was not really sure of her long-term goals and was not involved in any way beyond attending classes. According to Margaret, "She doesn't feel any connection with the people that are in their class." Margaret concluded by speculating that the non-involved student still derives her emotional support from long-time friends in a neighboring community.

Job competency was reportedly another important aspect of becoming accepted into organizations. Assimilated employees apparently know how to do their jobs, and they do them well. As a newcomer, Sarah, an advertising sales representative, said making her first sale was a defining moment. "Oh yeah, the first sale helped a lot!" she laughed. She went on to explain how making the first sale may have helped her feel as though she was capable of performing.

The sixth dimension emerging from the interviews was *adaptation and role negotiation*. Adapting to their new organization and/or negotiating roles within it signified that the newcomer was settling into the organization. Role negotiation involves newcomers' compromising between their expectations and expectations of the company. Adaptation suggests more compromise on the part of the newcomer. When newcomers adapt, they adjust to the organization's standards and environment. As an experienced hotel manager beginning work for a new company, Curtis described the situation that caused him to negotiate his role. "Just before I joined the company, they had undergone a massive refinancing. There was a strong drive to control costs throughout the company. But, I decided I would rather spend a little more and be told I was spending too much than save the company a nickel and have the place get run down." While the company saw his position as "cost cutter," he chose instead to be "guardian of the property." His promotion later in the year may have been an indication of his successful role negotiation.

The interviewees described processes they have used or witnessed to become assimilated into organizations. Analyzing the qualitative data provided six dominant themes these individuals associated with accepting and becoming accepted into organizations. The six themes—*familiarity with others, acculturation, recognition, involvement, job competency, adaptation and role negotiation*—are dimensions or processes associated with becoming full members in organizations. The six dimensions provided a foundation for development of a measure of organizational assimilation. The next section describes the process of creating and validating the Organizational Assimilation Index.

● **CONSTRUCTING A MEASURE OF ORGANIZATIONAL ASSIMILATION AND ESTABLISHING ITS CONSTRUCT VALIDITY**

In Phase Two, we developed 61 items to represent the six dimensions of organizational assimilation. We then asked a sample of employees at several organizations to use them to assess their assimilation experiences.

Participants

To create a generalizable measure of assimilation for members of varied industries, we involved 342 employees in distinctly different industries: lodging, banking, advertising, publishing, hospitality, and a nonprofit service agency. There were four hotels from one company located in Arizona, California, and Washington. The bank, advertising agency, and nonprofit agency were located in a large city in the southwestern United States. The bank included two participating branches.

Since assimilation is a continual process, we encouraged all employees, not just newcomers, from the organizations to take part in the survey. The sample included 114 men and 219 women from at least six ethnic backgrounds: 153 Caucasian, 148 Hispanic, 12 Asian/Pacific Islander, 7 Native American, 4 African American, 3 other. Fifteen did not indicate their ethnicity. Ages of the participants ranged from 17 to 77 (M=32.69, SD=12.56). In respect to education, 32 had some high school, 68 were high school graduates, 159 had some college, 56 were college graduates, 10 were post-graduate, 2 classified their education as "other," and 15 declined to state their level of education. As to level within the organizations, 18 were executives, 75 were supervisors/managers, 224 were hourly employees, and 25 did not disclose. Tenure ranged from two weeks to 40 years (M=2.68 years, SD=3.81). Seventy-four percent of the participants completed the questionnaire in English, and the other 26% did so in Spanish.

Instrument

The 61-item questionnaire had nine to eleven items for each of the six dimensions noted above. Items reflected the specific content of the six themes (see Appendix B). The instrument had three additional scales: Brayfield and Rothe's (1951) Job Satisfaction Scale; Lyons's Propensity to Leave Scale (1971); and six randomly selected items from Cheney's (1983) Organizational

Identification Questionnaire. Brayfield and Rothe generated evidence of content validity by having respondents rate the items along the continuum of satisfied to dissatisfied. The reliability of the index was demonstrated by an even-odd product moment coefficient of .77. The Lyons (1971) scale correlated negatively with measures of work satisfaction and correlated positively with voluntary turnover and job tension. Koberg and Hood (1991) reported a Cronbach's alpha of .83 with Lyons's Propensity to Leave Scale. Cheney provided evidence of the validity of his instrument by interviewing 178 corporate workers for purposes of exploring the process of organizational identification to create the instrument. Further, he found that 52% of the respondents who were looking for jobs elsewhere were low in organizational identification.[1] The data yielded a Cronbach's alpha of .94. In this study, Cronbach's alpha for the three scales was .61, .85, and .68, respectively.[2]

We interspersed items from the scales randomly throughout the instrument. Each was accompanied by a five-point scale in a Likert format (1 = strongly agree, 5 = strongly disagree) for all items. Finally, the questionnaire contained six demographic items related to the respondent's level within the organization: tenure, age, sex, education, and ethnicity.

Procedures

We pretested the questionnaire with two graduate students and two individuals with lower levels of education from the participating organizations. One was a high school graduate, and the other had only completed elementary school. We revised the questionnaire on the basis of their recommendations. Finally, because many of the participants were monolingual, Spanish-speaking, we had the questionnaire and consent form translated into Spanish by a professional translating service. A bilingual manager at one of the participating hotels validated the Spanish translation by back-translating it into English. Because many of the participants were apt to have minimal education and hold non-managerial-type positions, he recommended some changes in wording. For example, such words as "colleague" and "socialize" we replaced with "coworker" and "get together away from work."

The questionnaire was available in English or Spanish to the participants. They secured copies during company meetings. The participants understood that the purpose of the research was to examine the processes involved in people becoming a part of organizations. The employees knew that their participation in the survey was voluntary and that they had assurance that management would not see the responses from any individual participants (only summarized results for their organization). When they finished, they placed their questionnaires in large collection envelopes as a means of further ensuring their anonymity. Most participants completed the questionnaire in about 15 minutes. Six months later, the participants again completed the questionnaire under similar conditions. By asking that participants fill in blanks asking for their middle names and mothers' maiden names, we were able to ensure some level of anonymity, but still to identify 91 participants at follow-up. Those 91 surveys enabled us to estimate the test-retest reliability of the OAI.

Results

We subjected the data from Phase Two to confirmatory factor analysis to establish content validity of the *a priori* dimensions. Second, we used correlation analysis to test the four hypotheses concerning the relationships of the factors to job satisfaction, propensity to leave, organizational identification, and tenure. Support for these hypotheses provided evidence of validity for the instrument in the sense that the dimensions identified showed essentially the same relationship to the dependent variables of interest. Construct validity exists, according to Bailey (1982) when different indices (in this case dimensions) show the same relationship to other

measures as one would expect on the basis of the theory in which they appear.

Factor Analysis. The AMOS version 3.61 structural equation modeling package (Arbuckle, 1997) with maximum likelihood estimation of the covariances of the items enabled us to test the empirical validity of the hypothesized model. We utilized several criteria to determine the inclusion of the items and model fit. First, items had to have a primary factor loading of .40. Second, items had to be unidimensional as demonstrated by the tests of internal consistency and parallelism (Hunter & Gerbing, 1982). Internal consistency requires that the items comprising a scale have a similar statistical relationship to the primary factor. Parallelism requires that the items of a scale have a similar statistical relationship to the other factors. Since AMOS does not directly test for internal consistency or parallelism, we removed items from the model that the modification option of AMOS suggested had a path to another factor. Essentially, this procedure assured that an item only loaded on one factor. For the final model, internal consistency and parallelism were tested using the product rules of internal consistency and parallelism (Hunter & Gerbing, 1982). Third, the items had to have homogeneous content. Fourth, the items needed to show an acceptable level of reliability (Cronbach's alpha).

After removing items from the model in line with the first two criteria, the empirically-derived model corresponded to the six dimensions on the conceptual model, χ^2 (155, $N = 342$) = 365.92, $p < .001$, IFI = .92, CFI = .92, GFI = .90. Because the chi-square test statistic and p-value is biased by sample size and model size (see Kline, 1996; Marsh & Hocevar, 1985), the chi-square to degrees of freedom ratio is a more meaningful summary than chi-square alone. The expected ratio of chi-square to degrees of freedom is 1, and the smaller the ratio, the better the fit. Researchers suggest that a ratio as high as 5 to 1 indicates good fit (Marsh & Hocevar, 1985), but that 3 to 1 is better (Kline, 1996). The ratio in the current instance was 2.36, which suggests an adequate fit. Further, the model fit indices were at or above the recommended .90 (Hoyle & Panter, 1995). Additionally, there were no deviations in internal consistency or parallelism for the items. The dimensions also demonstrated homogeneous item content, and estimated reliability ranged from adequate to good (recognition α = .86, familiarity α = .73, acculturation α =. 73, involvement α = .72, role negotiation α = .64, job knowledge α = .62). Estimates of re-test reliability were similar (recognition α = .85, familiarity α = .77, acculturation α =. 71, involvement α = .70, role negotiation α = .57, job knowledge α = .66). Overall, these results suggest a good set of measures for assessing people's perceptions of their organizational assimilation experiences. Table C.1 displays the items and factor loadings. As an additional check for generalizability, we examined each of the factors' reliability for consistency across the sample. Reliability coefficients did not vary significantly across organizations or organizational level, nor did they vary as a function of the language of the questionnaire or sex of the respondents.

Table C.1 Organizational Assimilation Index

Dimension	Factor Loading
Supervisor Familiarity	
I feel like I know my supervisor pretty well.	.54
My supervisor sometimes discusses problems with me.	.69
My supervisor and I talk together often.	.85
Acculturation	
I understand the standards of the company.	.68
I think I have a good idea about how this organization operates.	.64
I know the values of my organization.	.78
Recognition	
My supervisor recognizes when I do a good job.	.68
My boss listens to my ideas.	.74
I think my supervisor values my opinions.	.86
I think my superior recognizes my value to the organization.	.82
Involvement	
I talk to my coworkers about how much I like it here.	.70
I volunteer for duties that benefit the organization.	.52
I talk about how much I enjoy my work.	.70
I feel involved in the organization.	.66
Job Competency	
I often show others how to perform our work.	.61
I think I'm an expert at what I do.	.44

Dimension	Factor Loading
I have figured out efficient ways to do my work.	.54
I can do others' jobs, if I am needed.	.58
Role Negotiation	
I have offered suggestions for how to improve productivity.	.70
I have helped to change the duties of my position.	.66

To provide evidence for the discriminant validity of the six dimensions of organizational integration, we specified a single factor solution that alternatively assumed that the items represent a single construct. This model provided a poor fit to the data, χ^2 (170, $N = 342$) = 931.28, $p < .001$, IFI = .69, CFI = .69, GFI = .73. The ratio of chi-square to degrees of freedom was 5.47. A comparison of the fit between the two models indicated that the six-factor model exhibited a significantly better fit to the data, χ^2 (15, $N = 342$) = 565.36, $p < .001$. Thus, we were able to reject the assumption that a single factor underlies these measures.

Construct validity of the OAI. Hypotheses One-Three proposed that the dimensions of organizational assimilation would correlate positively with job satisfaction and organizational identification and negatively with propensity to leave (see Table C.2 for correlation matrix). A significant, positive relationship existed between job satisfaction and all six of the dimensions of the OAI. Similarly, a significant, positive relationship between organizational identification and all six of the dimensions of the OAI emerged. Finally, a significant, negative relationship between propensity to leave and all six dimensions of the OAI surfaced. Thus, Hypotheses One, Two, and Three, in receiving support, provide further evidence of the construct validity of the OAI.

The fourth hypothesis proposed that the dimensions of organizational assimilation account for more variance in job satisfaction, organizational identification, and propensity to leave than did tenure. Tenure did not correlate significantly with job satisfaction, organizational identification, or propensity to leave. Thus, the fourth hypothesis was supported.

● DISCUSSION

The purpose of this study was to develop and validate a measure of organizational assimilation. This section discusses the findings of both phases of the study and how the research fulfills the predetermined purpose. We review the factors of organizational assimilation that emerged in the study and note implications, limitations, and future directions.

Dimensions of Organizational Assimilation

In the first phase, we identified six dimensions of organizational assimilation. Confirmatory factor analysis of the Organizational Assimilation Index provided empirical support for these six dimensions. This constituted evidence of content validity. Further, the OAI's construct validity was apparent by the positive association of each of the dimensions with job satisfaction and organizational identification and the negative association of each of the dimensions with propensity to leave. This section examines each of these dimensions.

The first factor involved familiarity with supervisors. Participants in this study depicted the process of getting to know supervisors as the first step of fitting into organizations and said that their feelings toward their organizations changed as a result of becoming acquainted with superiors. This corresponds with Jablin's (1982) description of assimilation as an interactive communicative process and coincides with prior scholarship indicating that

Table C.2 Correlation of Organizational Assimilation Dimensions and Job Satisfaction, Propensity to Leave, and Organizational Identification

Variable	1	2	3	4	5	6	JS	PL	ID	Tenure
1	1.00	.53**	.46**	.30**	.37**	.48**	.33**	-.28**	.43**	.01
2		1.00	.51**	.52**	.53**	.44**	.45**	-.41**	.60**	.08
3			1.00	.41**	.25**	.30**	.24**	-.34**	.46**	.15**
4				1.00	.42**	.35**	.53**	-.63**	.61**	.01
5					1.00	.46**	.37**	-.31**	.50**	-.04
6						1.00	.34**	-.19**	.35**	.01
JS							1.00	-.68**	.62**	.04
PL								1.00	-.74**	-.03
ID									1.00	-.03
Tenure										1.00
M	2.19	1.89	1.94	2.14	1.98	3.50	1.90	1.99	2.20	2.68
SD	.78	.65	.72	.65	.57	1.22	.58	.89	.68	3.81

**p < .01, two-tailed.

(1) = Supervisor Familiarity, (2) = Acculturation, (3) = Recognition, (4) = Involvement, (5) = Job Competency, (6) = Role Negotiation, (JS) = Job Satisfaction, (PL) = Propensity to Leave, (ID) = Organizational Identification.

socialization involves establishing relationships with members within the organization and Chao et al.'s (1994) findings that suggested getting to know people is an important socialization outcome.

The second factor was organizational acculturation. When members are acculturated, they have accepted the organization's culture and are willing to make personal changes in order to integrate into it (Wilkens, 1983). The development of a shared understanding by organizational members is the important difference between those who are genuinely a part of the organization and those who are not (Deal & Kennedy, 1982; Jaques, 1951; Kanter, 1984; Peters & Waterman, 1982). Chao et al. (1994) reached a similar conclusion in noting that becoming familiar with organizational goals and values is such an important socialization outcome that those who perceive they are mismatched between personal values and goals and those of the organization are likely to leave.

The third factor, recognition, involves perceiving one's value to the organization and feeling recognized by superiors. In previous research, recognition has been linked to such positive outcomes as job satisfaction and commitment (Baird & Deibolt, 1976; Garland, Oyabu, & Gipson, 1989; Pincus, 1986). Receiving recognition was also strongly influential in increasing organizational identification in Bullis and Bach's (1989) examination of socialization turning points.

Involvement, the fourth factor, encompassed many aspects of being a part of organizations. Interviewees described various norms for involvement, such as volunteering for extra organizational duties, figuring out ways to accomplish work more efficiently, and feeling involved in the organization. This parallels the findings of Bauer and Green (1994) who determined that early newcomer involvement caused participants to feel more acceptance and less role conflict. Early involvement also translated to higher levels of productivity.

The fifth factor, job competency, related to members' beliefs that they were able to adequately perform their designated duties. As Feldman (1981) noted, "No matter how motivated the employee, without enough job skills, there is little chance for success" (p. 313).

Finally, the results for Phase One suggested that the processes of adaptation and role negotiation would cluster to form the sixth factor. However, confirmatory factor analysis produced evidence for role negotiation only. To participants, role negotiation represented ways newcomers interact with others in the organization in an attempt to compromise on ways a role should be enacted (Miller et al., 1996). Ashforth and Saks (1996) alleged that the employee selection process does not result in a perfect fit between the organization and the recruit. Successful role negotiation requires willingness to compromise on behalf of both the organization and the newcomer (Jablin, 1987).

In sum, the OAI includes measures for six distinct and empirically verified aspects of organizational assimilation: familiarity with supervisors, organizational acculturation, recognition, involvement, job competency, and role negotiation. With the two-phase process, we demonstrated face, content, and construct validity for the OAI. Although each dimension in itself provides valuable information, the full index can more efficiently offer data to organizations about success or inadequacies of new member development programs. Over the long-term, relationships among the six dimensions might become apparent both for the benefit of industry and also contributing to future theories related to assimilation.

Implications

This study contributes to our understanding of the processes associated with organizational assimilation and creates an instrument for its measurement. The need for the index is evidenced by an assumption commonly made about assimilation: the longer members are with an organization, the more assimilated they become. Comments by some participants suggest that levels of assimilation do not always rise. Our results parallel those of Bullis and Bach (1989), who found that organizational identification levels vary throughout members' tenures. Similarly, assimilation levels may rise and fall throughout a member's tenure as a result of a variety of factors, including leadership, management policy, and relationships with coworkers. We are of the view that assessments permitted by using the OAI are a more accurate index of assimilation than tenure, and our results provide strong evidence in its support.

The OAI appears to be useful for diverse organizational types, not specific to any particular industry. Further, it is applicable to members at all levels of organizations. This sort of generalizability should enhance its value in organizational settings.

Organizational leaders may be particularly interested in three implications of this study. First, we believe the findings contribute to understanding communicative processes and outcomes that foster premature turnover. Turnover rates for new workers are at least three times as high as those for workers who have been with the organization for more than four weeks (Wanous, 1992). Organizations incur costs for publicizing job openings, interviewing, and training, with expectations that recruits will be less productive than experienced coworkers and make many mistakes before they are considered proficient in their duties. Appreciating communicative processes associated with assimilation may be beneficial toward encouraging newcomer integration. When organizations become aware of types of communication involved in assimilation, they can take action to encourage those processes. For example, front-line supervisors should be educated on the importance of early and frequent recognition and its potential impact on member assimilation and commitment. This knowledge and practices may lead to longer tenures and reduced turnover costs.

Second, the OAI may assist organizations in improving assimilation efforts. In using it to assess levels of assimilation following implementation of differing socialization tactics, they may be better able to determine which of their socialization methods are most effective in a given organization, department, or job. The result could be better quality assimilation at lower costs.

Finally, in fostering processes likely to cause higher levels of organizational assimilation, employees may demonstrate higher levels of organizational identification and job satisfaction that may encourage a more favorable culture. This, of course, can further the interests of organizations and potentially enhance their performance.

As a final word of caution, the OAI should not be used by organizational leaders to determine "who does not fit in" with the objective of terminating or "weeding out" those who do not fit. Individuals who have not assimilated to a standardized goal may still be productive and valuable assets to the organization. In fact, although our findings suggest becoming competent at performing duties is a dimension of assimilation, we cannot explicitly tie levels of assimilation to increased productivity. However, the instrument may direct organizations toward evaluating methods for inducting recruits and maintaining an atmosphere that encourages acceptance and assimilation (especially of individuals with diverse perspectives and backgrounds).

Limitations and Future Directions

We also recognize certain limitations to this study and suggest that some limitations may provide opportunities for future research. The first limitation relates to the sixth factor. Specifically, instead of adaptation and role negotiation emerging as one factor as predicted, adaptation items appeared to be related to acculturation items. In retrospect, considering the similarity in the processes of acculturation and adaptation, this may be a reasonable outcome. However, the factor that emerged defined as role negotiation contained only two items following factor analysis.

A second limitation is that while organizational assimilation is a dynamic process, this study utilized only participants who were current members of their organization. However, assimilation may begin prior to entry, and some of the effects may linger after organizational exit. Patterns in members' levels of assimilation might be evident in studies that test participants at various intervals. Repeated longitudinal testing of the OAI, and preferably including individuals who had not yet entered the organization and those who had recently exited, would provide researchers with a better understanding of the ongoing nature of the assimilation process.

Finally, we suggest that future research should utilize the OAI to evaluate members' organizational assimilation levels for comparison to other data such as turnover rates, absenteeism, and productivity. This information would be useful in determining how organizational assimilation relates to other organizational outcomes. This would provide valuable information for theoretical, as well as practical use.

● NOTES

1. Our purpose in using the OIQ was to include items which should positively correlate with items to measure organizational integration. We acknowledge Sass and Canary's (1991) and Miller, Allen, Casey, and Johnson's (2000) concern that the OIQ may in fact measure organizational commitment or some other form of organizational unity. Although we respect this position, we do not view this as problematic for use to validate the OAI as commitment or identification should both positively correlate with the assimilation dimensions.

2. We included four items from the Communicator Style Measure (Norton, 1978) as a test of "no correlation." We used the items from the animated sub-scale of the measure. In this study, Cronbach's alpha was .45. Due to the low reliability, this scale was not used to validate the OAI.

● REFERENCES

Arbuckle, J. L. (1997). *Amos users' guide version 3.6.* Chicago: Small Waters Corporation.

Ashford, S. J., & Taylor, M. S. (1990). Adaptation to work transitions: An integrative approach. In G. Ferris & K. Rowland (Eds.), *Research in personnel and human resource management* (Vol. 8, pp. 1–39). Greenwich, CT: JAI Press.

Ashforth, B. E., & Saks, A. M. (1996). Socialization tactics: Longitudinal effects on newcomer adjustment. *Academy of Management Journal, 39,* 149–178.

Bailey, K. D. (1982). *Methods of social research (2nd ed.).* New York: The Free Press.

Baird, J. E., & Deibolt, J. C. (1976). Role congruence, communication, superior-subordinate relations and employee satisfaction in organizational hierarchies. *Western Journal of Speech Communication,* 40, 260–267.

Bauer, T. N., & Green, S. G. (1994). Effect of newcomer involvement in work-related activities: A longitudinal study of socialization. *Journal of Applied Psychology, 79,* 211–223.

Brayfield, A. H., & Rothe, H. F. (1951). An index of job satisfaction. *Journal of Applied Psychology, 35,* 307–311.

Bullis, C. (1999). Mad or bad: A response to Kramer and Miller. *Communication Monographs, 66,* 368–373.

Bullis, C., & Bach, B. W. (1989). Socialization turning points: An examination of change in organizational identification. *Western Journal of Speech Communication, 53,* 273–293.

Chao, G., O'Leary-Kelly, A., Wolf, S., Klein, H., & Gardner, P. (1994). Organizational socialization: Its content and consequences. *Journal of Applied Psychology, 79,* 730–743.

Cheney, G. (1983). On the various and changing meanings of organizational membership: Field study of organizational identification. *Communication Monographs, 50,* 342–362.

Clair, R. P. (1999). Ways of seeing: A review of Kramer and Miller's manuscript. *Communication Monographs, 66,* 374–381.

Deal, T., & Kennedy, A. (1982). *Corporate cultures.* Reading, MA: Addison-Wesley.

Feldman, D. C. (1981). The multiple socialization of organization members. *Academy of Management Review, 6,* 309–319.

Fisher, C. D. (1986). Organizational socialization: An integrative review. *Research in Personnel and Human Resource Management, 4,* 101–145.

Garland, T. N., Oyabu, N., & Gipson, G. A. (1989). Job satisfaction among nurse assistants employed in nursing homes: An analysis of selected job characteristics. *Journal of Aging Studies, 3,* 369–383.

Glaser, B. G., & Strauss, A. L. (1967). *The discovery of grounded theory.* Chicago: Aldine-Athestor.

Holton, E. F. (1996). New employee development: A review and reconceptualization. *Human Resource Development Quarterly, 7,* 233–252.

Hoyle, R. H., & Panter, A. T. (1995). Writing about structural equation models. In R. H. Hoyle (Ed.),

Structural equation modeling: Concepts, issues, and applications (pp. 158–176). Thousand Oaks, CA: Sage.

Hunter, J. E., & Gerbing, D. W. (1982). Unidimensional measurement, second order factor analysis, and causal models. *Research in Organizational Behavior, 4,* 267–320.

Jablin, F. M. (1982). Organizational communication: An assimilation approach. In M. E. Roloff & C. R. Berger (Eds.), *Social cognition and communication* (pp. 255–286). Newbury Park, CA: Sage.

Jablin, F. M. (1987). Organizational entry, assimilation, and exit. In F. M. Jablin, L. L. Putnam, K. H. Roberts, & L. W. Porter (Eds.), *Handbook of organizational communication* (pp. 679–740). Newbury Park, CA: Sage.

Jablin, F. M. (2001). Organizational entry, assimilation, and exit. In F. M. Jablin & L. L. Putnam (Eds.), *The new handbook of organizational communication* (pp. 732–818). Thousand Oaks, CA: Sage.

Jaques, E. (1951). *The changing culture of a factory.* New York: Dryden Press.

Jones, G. R. (1986). Socialization tactics, self-efficacy, and newcomers' adjustments to organizations. *Academy of Management Journal, 29,* 262–279.

Kanter, R. M. (1984). *The changemasters.* New York: Simon & Schuster.

Kline, R. B. (1996). *Principles and practice of structural equation modeling.* New York: Guilford Press.

Koberg, C. S., & Hood, J. N. (1991). Cultures and creativity within hierarchical organizations. *Journal of Business and Psychology, 6,* 265–271.

Kramer, M. W., & Miller, V. D. (1999). A response to criticisms of organizational socialization research: In support of contemporary conceptualization of organizational assimilation. *Communication Monographs, 66,* 358–367.

Lindlof, T. (1995). *Qualitative communication research methods.* Thousand Oaks, CA: Sage.

Lyons, T. F. (1971). Role clarity, need for clarity, satisfaction, tension, and withdrawal. *Organizational Behavior and Human Performance, 6,* 99–110.

Marsh, H. W., & Hocevar, D. (1985). The application of confirmatory factor analysis to the study of self-concept: First and higher order factor structures and their invariance across age group. *Psychological Bulletin, 97,* 562–582.

Miles, M., & Huberman, A. (1994). *Qualitative data analysis.* Thousand Oaks, CA: Sage.

Miller, V. D., Allen, M., Casey, M. K., & Johnson, J. R. (2000). Reconsidering the organizational identification questionnaire. *Management Communication Quarterly, 13,* 625–658.

Miller, V. D., & Jablin, F. M. (1991). Information seeking during organizational entry: Influences, tactics, and

a model of the process. *Academy of Management Review, 16,* 92–120.

Miller, V. D., Jablin, F. M., Casey, M., Lamphear-Van Horn, M., & Ethington, C. (1996). The maternity leave as a role negotiation process. *Journal of Managerial Issues, 8,* 286–309.

Morrison, E. W. (1993). Longitudinal study of the effects of information seeking on newcomer socialization. *Journal of Applied Psychology, 77,* 173–183.

Norton, R. W. (1978). Foundation of a communicator style construct. *Human Communication Research, 4,* 99–112.

Ostroff, C., & Kozlowski, S.W.J. (1992). Organizational socialization as a learning process: The role of information acquisition. *Personnel Psychology, 45,* 849–874.

Patton, M. (1990). *Qualitative evaluation and research methods.* Newbury Park, CA: Sage.

Peters, T. J., & Waterman, R. H. (1982). *In search of excellence.* New York: Harper & Row.

Pincus, I. D. (1986). Communication satisfaction, job satisfaction and job performance. *Human Communication Research, 12,* 395–419.

Saks, A. M., & Ashforth, B. E. (1996). Proactive socialization and behavioral self-management. *Journal of Vocational Behavior, 48,* 301–323.

Sass, J. S., & Canary, D. J. (1991). Organizational commitment and identification: An examination of conceptual and operational convergence. *Western Journal of Speech Communication, 55,* 275–293.

Schein, E. H. (1968). Organizational socialization and the profession of management. *Industrial Management Review, 9,* 1–16.

Teboul, J. C. B. (1997). "Scripting" the organization: New hire learning during organizational encounter. *Communication Research Reports, 14,* 33–47.

Tompkins, P. K., & Cheney, G. (1983). Account analysis of organization: Decision making and identification. *Communication and organizations: An interpretive approach* (pp. 123–146). Beverly Hills, CA: Sage.

Turner, P. (1999). What if you don't?: A response to Kramer and Miller. *Communication Monographs, 66,* 382–389.

Van Maanen, J., & Schein, E. H. (1979). Toward a theory of organizational socialization. *Research in Organizational Behavior, 1,* 209–264.

Wanous, J. P. (1992). *Organizational entry: Recruitment, selection, and socialization of newcomers.* Reading, MA: Addison-Wesley.

Waung, M. (1995). The effects of self-regulatory coping orientation on newcomer adjustment and job survival. *Personnel Psychology, 48,* 633–650.

Wilkens, A. L. (1983, Autumn). The cultural audit: A tool for understanding organizations. *Organizational Dynamics,* 24–38.

APPENDIX A
Interview Protocol

1. Do you feel more like a part of the company than you did on your first day here?

2. What do you think changed?

3. Do you remember any situation or particular time when you felt that you were becoming a part of the company?

4. What did others (coworkers, managers or your subordinates) say that might have helped you to feel this way?

5. Do you believe you felt differently toward the company and your fellow employees after that time?

6. How so?

7. Do you think you can tell when a new employee has assimilated into the company? What might that person or their coworkers do that might clue you into whether or not that person has become a part of (name of the company)?

8. Can you think of someone who hasn't really assimilated into the company?

9. What might that person or their coworkers do that might clue you into whether or not that person has become a part of (name of the company)?

10. How does a new employee know when they have begun to "fit in" here at (name of the company)? What might their co-workers say that indicates their acceptance?

11. What would someone who had 'fit in' do or say that would be different from someone who had not?

12. What strategies did you use to integrate into the company?

Demographics:

How long have you worked for (name of the organization)?

What is your position? How long have you been in that position?

Do you supervise other employees? How many?

Your age?

What is your ethnicity?

(Information was also recorded about the participant's sex.)

APPENDIX B

Original Items for Organizational Assimilation Index

Familiarity with Others

I consider my coworkers friends.

I feel comfortable talking to my coworkers.

I must work up the courage to talk to my supervisor about a problem.

I can tell when my supervisor would prefer not to talk.

I have shared my problems at work with some of my coworkers.

I spend time away from work with some of my coworkers.

I avoid conversations with my coworkers whenever possible.

I feel like I know my supervisor pretty well.

My supervisor sometimes discusses problems with me.

My supervisor and I talk together often.

I feel like I know my coworkers pretty well.

Acculturation

I know what is expected to succeed in this organization.

I know whom I should talk to about a work-related problem.

I understand the standards of the company.

I think I know "how things happen around here."

I feel more stressed than I should at work.

I think I have a good idea about how this organization operates.

I feel very comfortable in my work environment.

I am tense in my work environment.

I can see how my work benefits our customers.

I know the values of my organization.

I usually feel stressed at the end of my shift.

Recognition

My work is appreciated by the organization.

My supervisor recognizes when I do a good job.

My coworkers tell me I do good work.

My boss listens to my ideas.

I think my supervisor values my opinions.

My supervisor does not recognize the good work I do.

I think my superior recognizes my value to the organization.

My supervisor has told me that he/she trusts my judgment.

I do not think I can perform my work as well as others.

I think the work I do would be missed if I quit.

Involvement

I talk to my coworkers about how much I like it here.

I question why we do things the way we do at this organization.

I volunteer for duties that benefit the organization.

I do not prefer to take on more job responsibility.

I talk about how much I enjoy my work.

I feel involved in the organization.

I tell others that I am only working in this job temporarily.

I would do my best work even if I were not being supervised.

I often start work early or leave work late if they need me.

I am happy to do the work I do for the organization.

Job Competency

I can do others' jobs, if I am needed.

I sometimes feel overwhelmed trying to figure out how to do my work.

I often feel as though I need someone to tell me how to do my job.

I know how to work to accomplish all my duties.

I think I could train someone to do my work.

I have figured out efficient ways to do my work.

I do not feel very competent in my work.

I think I'm an expert at what I do.

I feel unsure of my work when my supervisor watches me.

I often show others how to perform our work.

Adaptation and Role Negotiation

I think I have adapted to my organization's expectations.

I question why we do things the way we do at this organization.

I feel like I have too many responsibilities for my job.

I feel I have to adapt to too many company policies.

Adapting to the organization's ways has helped me in my work.

I do not mind being asked to perform my work according to the organization's standards.

I have offered suggestions for how to improve productivity.

I have helped to change the duties of my position.

I would like to change some of the organization's standards.

Appendix D—An Example of the Embedded Design

Meeting the Challenge of Doing an RCT Evaluation of Youth Mentoring in Ireland

A Journey in Mixed Methods

Bernadine Brady and Connie O'Regan
National University of Ireland, Galway

The youth mentoring program Big Brothers Big Sisters is one of the first social interventions involving youth in Ireland to be evaluated using a randomized controlled trial methodology. This article sets out the design process undertaken, describing how the research team came to adopt a concurrent embedded mixed methods design as a means of balancing ethical, feasibility, and scientific issues associated with the randomized controlled trial method, establishing an epistemological position and integrating data from various methods and multiple sources.

KEYWORDS: *RCT; mixed methods; youth services; youth mentoring*

The move toward evidence-based practice throughout the Western world has led to a renewed focus on randomized controlled trials (RCTs) as a means of establishing impact. RCT studies are difficult, costly, and challenged on philosophical, methodological, and ethical grounds but, for policy makers, remain the method of choice for demonstrating cause and effect. To date, evaluations of

AUTHORS' NOTE: Please address correspondence to Bernadine Brady, Child & Family Research Centre, National University of Ireland, Galway, Ireland; e-mail: Bernadine.brady@nuigalway.ie.

SOURCE: Brady, B., & O'Regan, C. (2009). Meeting the challenge of doing an RCT evaluation of youth mentoring in Ireland: A journey in mixed methods. *Journal of Mixed Methods Research, 3*(3), 265-280. Reprinted with permission of SAGE Publications, Inc.

social interventions in Ireland have not used RCTs (with the notable exception of an evaluation of the community mothers program by Johnson, Howell, & Molloy, 1993), but in recent years there has been a strong drive toward the use of this design as a means of helping to establish an evidence base for children's and young people's services. The authors are part of a research team in the west of Ireland that is currently undertaking an RCT study on the youth mentoring program, Big Brothers Big Sisters (BBBS). This article outlines the design process that the team went through, starting with an intention to develop a standard RCT with a process study built in. However, the challenges faced in the process of designing and implementing the study led the authors to move toward a mixed methods design. This article argues that researchers can respond to the epistemological and practical limitations of the RCT method in the context of social interventions through creative use of mixed methods models and approaches.

The initial part of this article describes the context within which the study takes place. Attention then moves to the RCT model, outlining the tensions inherent in both the paradigmic constraints of the experimental design chosen by the program funders for the evaluation and the practical limitations of the research setting. The final part of the article describes the three phases of the research process and the methodological and paradigmic choices made at each stage.

● CONTEXT FOR THE STUDY

Although there is a long history of gathering statistical data and undertaking survey research, going back into the 19th century (Tovey & Share, 2003), the modern development of the field of evaluation in Ireland has been largely influenced by EU funding programs that rely mostly on process-and indicator-driven methodologies (EU Commission, 1999). However, in recent times, a new driver of evaluation practice has emerged as a consequence of a surge in philanthropic activity in Ireland. One such organization, the Atlantic Philanthropies (AP), has been a significant contributor in Ireland since the late 1990s. In this work, it is guided by a firm belief in the role of education and knowledge creation as a key driver of programs that can change people's lives. As part of its strategic vision for the Children and Youth Services in Ireland, it recognized the fact that funding in this area has been ad hoc and fragmented. The youth sector itself is primarily staffed by volunteers and reliant on a myriad funding sources and streams, often resulting in duplication and lack of coordination across agencies (Lalor, de Roiste, & Devlin, 2007). To address this patchwork approach to service provision and to develop the foundations of an evidence-based practice in social care in Ireland, AP resolved to make the funding of its Children and Youth program strongly linked to the requirement that service providers undertake rigorous, randomized controlled trials of the intervention, where possible. They believe that this will aid the longer term development of better program infrastructure and effective evidence-based policy (The Atlantic Philanthropies, 2007). In addition to investment in service development, they made unparalleled investment in universities, including fourth-level education programs (F. H. T. Rhodes & Healy, 2006), capital and revenue support, including the development of children's research centers, at Trinity College Dublin and NUI, Galway.

The BBBS program is one of the oldest and most established models of youth mentoring in the world, operating since 1905 in the United States and now in more than 30 countries worldwide. The program oversees the creation of supportive relationship between adult mentors and young people. What sets the BBBS approach to mentoring apart from others is its highly structured nature, with each match organized by a case manager who works to agreed standards for the screening, training, and ongoing support and supervision of matches. The program focus is not on specific outcomes but on developing a relationship that will foster positive youth development (Tierney, Grossman, & Resch, 1995).

BBBS was established in Ireland in 2002, with a leading national youth organization, Foroige, as the host agency. An initial pilot project provided community-based mentoring in the west of Ireland, to young people between 10 and 18 years. Foroige received a grant from the AP in 2005 toward the further development of the BBBS mentoring program, the stated aim being "to produce better outcomes for children by demonstrating and testing a proven model of youth mentoring" (The Atlantic Philanthropies, 2007). The support package agreed between Foroige and the AP specified that the program would be supported to expand and that it would undergo a rigorous evaluation.

Although RCTs had not been undertaken in youth services in Ireland before, the BBBS program was an attractive place to start. A number of critical factors created a positive climate for the "trying out" of such a methodology. To begin with, stakeholders were positively predisposed to the study as it would build on studies in the United States, which had shown the BBBS program to be effective. In one of the most high-profile and large-scale RCT studies in the United States, Public/Private Ventures, an independent social research agency, assessed whether the BBBS mentoring program made a tangible difference to young people's lives. They found that youth with a mentor were less likely to start using drugs or alcohol, were less likely to hit someone, had improved school attendance and performance, had improved attitudes toward completing schoolwork, and had improved peer and family relationships (Tierney et al., 1995). Further evidence in relation to mentoring was provided by a meta-analysis of more than 55 studies of mentoring programs. It found that there is a small (.13), but significant, positive effect for mentees in the areas of enhanced psychological, social, academic, and job/employment functioning, as well as reductions in problem behaviors (DuBois, Holloway, Valentine, & Cooper, 2002). DuBois et al. (2002) emphasize that to facilitate attainment of desired outcomes, programs must provide an organized program structure and support.

The BBBS program is considered an exemplar in terms of such programs operated under strict criteria that are associated with good practice in youth mentoring. The meta-analysis confirmed the finding that intensive supervision and support of the mentors by paid staff, a requisite of the BBBS approach, was especially critical to successful outcomes (Furona, Roaf, Styles, & Branch, 1993).

Thus, the Irish BBBS program, which is operated to the same standards as the U.S. model, could be very hopeful that positive effects would be found. Just as the Tierney evaluation in the United States spurred a huge impact on the growth of youth mentoring, Foroige management envisaged that a similar study would provide the evidence they needed to garner policy and financial support for the development of the program in Ireland. Furthermore, aspects of the methodology of the Tierney et al. (1995) study could be replicated in an Irish context, meaning that the evaluation did not have to start from a blank page. In addition, there was great openness on the part of ground-level Foroige staff to the research as they believed it could prove their intuitive sense that mentoring "works."

Another key advantage was that there has been a great growth in mentoring theorization and analysis since the Tierney study was published in 1995. Of particular note is the work of Jean Rhodes, who developed a plausible theory of mentoring, using data from the Tierney et al. evaluation (J. E. Rhodes, 2005; see Figure D.1). In undertaking an RCT, Ghate (2001) recommends a specified causal model that explains what effects are expected and why these effects are likely to occur. For the proposed study, J. E. Rhodes's (2005) model of mentoring offered the opportunity to test not just "if" mentoring works, but "how" it works in an Irish context. Furthermore, it has been argued that RCTs are most suited to testing services that are delivered in a systematic way (Ghate, 2001; Oakley et al., 2003). BBBS was ideal in this regard because the program is underpinned by a detailed manual, which clearly sets out the nature of the intervention.

Figure D.1 Rhodes's Model of Mentoring

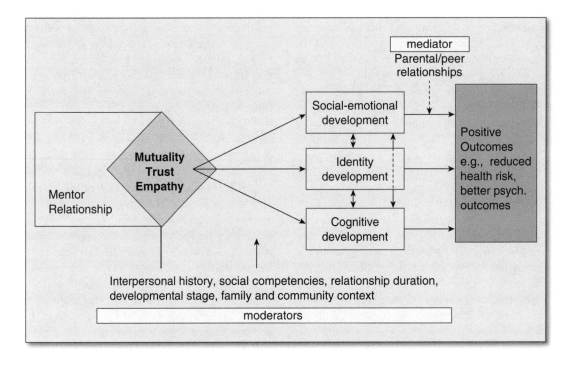

NOTE: Adapted from the *Handbook of Youth Mentoring* (p. 32), edited by David L. DuBois and Michael J. Karcher, 2005. Thousand Oaks, CA: Sage. Copyright © 2005 by Sage. Adapted with permission.

Finally, the fact that a philanthropic organization was willing to fund the study meant that cost was not a prohibitive factor as it often can be in studies of this nature. As part of their targeted initiative to enhance the evidence base of children's services in Ireland, this organization was also to provide capacity building for the research team to facilitate them to learn "how to do an RCT." An expert advisory group (EAG) was formed, composed of leading researchers and academics, whose role was to guide the research team through the overall research project. In summary, therefore, the conditions merged to make this a positive context within which to undertake an RCT.

● CRITICISMS AND CHALLENGES ASSOCIATED WITH THE RCT METHOD

Although RCTs have been described as the "Rolls Royce" and gold standard of evaluation methods (Chelimsky, 1997), the method and its underlying postpositivist paradigm have also been subject to intense criticism and epistemological debate (Greene, 2003). To place these criticisms in context, it is useful to refer to the main philosophical choices that exist in the conduct of evaluations of social programs (Greene, 2000). Apart from the post-positivist paradigm underpinning experimental design, there is also "utilitarian pragmatism," a position that matches research method to the particular research question and avoids consideration of which method is superior. In addition, there is an "interpretative" stance, which privileges the voice and experiences of the stakeholders in a given situation. The approach is not to search for one objective account but instead to seek a representation of a multilayered complex reality. Finally, there is the "critical social sciences" stance, which focuses on the power imbalances inherent in a given evaluand and seeks to promote the equal participation and empowerment of less powerful stakeholders (Greene, 2000).

At a fundamental level, the application of postpositivist laboratory experimental design to the field of social research is criticized on the basis of its incompatibility to the open complex reality that is the social world. The implicit assumptions in the paradigm that it is possible to separate facts from values and that the objective facts about a program can be established using the experimental method have been vigorously contested.

Some have also questioned the external validity of the method on the basis that participants are not selected at random from all members of a given population; instead, participants are randomly assigned from a sample of people already referred to a given program. It is therefore difficult to establish how representative this population is of the wider sample, which in turn limits the degree to which findings can be generalized. In addition, the experimental design relies on the use of a linear understanding of causality, asserting that it can be proved mathematically that any difference between two groups randomly assigned to a treatment or control condition can be said to be because of the treatment. This focus on the input/output model alone is criticized as a reductionist approach to understanding the nature of causality in the social world (Pawson & Tilley, 1997). Another concern attached to this design is the reliance on quantitative, usually survey, techniques to measure the construct of interest. Issues arise regarding the

suitability of measuring outcomes in this way, the construct validity of such items, and the application of standard instruments to populations that are different from the original population the measure was designed for. For some, this issue is addressed by the piloting and redesign of survey instruments, others add qualitative items, although others reject this as a way of "measuring" reality at all. A further challenge to the use of the experimental design in the area of children's research has been the growth in popularity and influence of participatory and inclusive research designs.

The above criticisms have been leveled at the RCT method mainly from those who consider it an inappropriate means of evaluating social interventions. However, even for those who believe in the value of this form of impact evaluation, a range of ethical, technical, and feasibility difficulties associated with the RCT method must be faced. To begin with, it is argued that, as a result of random allocation, the control group may be deprived of something seen as beneficial. These ethical issues mean that the method may not be appropriate in certain situations. A second issue relates to sample size. As highlighted earlier, the effect size found by Tierney et al. (1995) in the U.S. evaluation of BBBS can be considered small. For interventions that are likely to have small or variable effects, both experimental and control groups must be quite large. The larger the number of units studied the more likely the experimental and the control groups are to be statistically equivalent, and the likelihood of Type 2 errors is reduced (Rossi, Lipsey, & Freeman, 2004). With smaller sample sizes there is a risk that the treatment and control groups will not be statistically equivalent, despite being randomly assigned. A third issue relates to the state of development of the intervention under study. It is generally accepted that an RCT is not suitable for programs in the early stages of implementation as, if the program changes during the intervention, there is no easy way to determine what effects are produced by any given form of the intervention. Rossi et al. (2004) suggest that a minimum of 2 years of running the program is necessary. Likewise, Ghate (2001) suggests that the services have time to "bed down" so that teething problems can be overcome.

Fourth, considerable time is needed to ensure buy-in from stakeholder staff. Previous studies have shown that there can be a resistance to random allocation because of practitioners' aspiration to get the best services for the most needy cases (Little, Kogan, Bullock, & van der Laan, 2004). Fifth, long lead times for facilitating and measuring attitudinal and behavioral change are a major methodological problem in measuring the impact of social interventions. Ghate (2001) suggests that the timetable should allow for preevaluation research and for careful detailed planning. Finally, although randomly formed experimental and control groups are statistically equivalent at the start of an evaluation, nonrandom processes may threaten their equivalence as the experiment progresses. Attrition can affect the validity of results because it tends to be more pronounced for members of excluded groups, and differential attrition may produce differences between groups. Oakley (2000) urges that particular consideration be given to how best to avoid the "resentful demoralization" (Shadish, Cook, & Campbell, 2002, p. 80) often experienced by control group members and to encourage control groups to feel that it is worthwhile to make an active contribution to the research. Another concern is that the control group may receive treatment that contaminates the experiment.

● THE DESIGN CHALLENGE

Given the philosophical and practical difficulties with the RCT method just outlined, two key challenges faced the research team in relation to the study design. First, there was a need to find a paradigmatic stance, and second, the design had to be able to address the ethical, feasibility, and technical challenges associated with the RCT method.

With regard to the former, the research team struggled with the epistemological and ontological limitations of the RCT method. A key difficulty in relation to the RCT method is its linear understanding of causality and lack of attention to context. Because the BBBS program was being evaluated in a different cultural context, the research team recognized the need to describe and account for how this context may affect the program. This was especially the case because the program was concerned with developing supportive relationships for young people, and the mentoring research has indicated the need for analytic approaches that are sensitive to detecting how mentoring relationships may be shaped by and shape features of the settings and environments in which they occur (Dubois, Doolittle, Yates, Silverthorn, & Kraemer Tobes, 2006). Furthermore, the research team was cognizant of the strengths of both the U.S.-based mentoring research referred to earlier, which is primarily quantitative in focus, and the insights and critical approach adopted by the more qualitative focus of the U.K. mentoring research (see, e.g., Philip, Shucksmith, & King, 2004). We saw the opportunity to bridge these two traditions in a study that could address questions of impact as well as of process and implementation.

Although RCT studies are primarily quantitative in nature, it is recommended that process designs be incorporated as a means of overcoming some of the perceived difficulties with the use of an RCT in evaluating complex social interventions (Oakley et al., 2003). The integration of a process study into the overall design could allow a focus on program fidelity, compliance, and strength and the collection of data on the experience of stakeholders.

By incorporating a process element, therefore, the opportunity presented itself to move toward a mixed methods approach. Greene and Caracelli (1997) outline three stances that are usually taken on the question of whether it is possible to establish a paradigmic stance from which to combine methods. Those adopting a purist stance believe that postpositivist and interpretivist approaches cannot be combined in a single study due to their differing ontological and epistemological worldviews. This position is also described as the incommensurability thesis (Tashakkori & Teddlie, 2003). However, alternative view points are the pragmatic and dialectical stances. The pragmatic stance holds that there are differences between the worldviews as held by the purists but that these should not prohibit researchers from matching research methods to the research question at hand in order to meet the particular needs of the stakeholders. The dialectical position also holds that there are differences in the postpositivist and interpretivist worldviews but, instead of prohibiting their combination, it encourages the development of designs that actively seek to create deeper and more integrated understandings of complex phenomena through interrogating and comparing the data arising from each worldview (Greene, 2007; Greene, Benjamin, & Goodyear, 2001). Having made the decision to undertake a mixed methods study of the impact of the mentoring, the section that follows describes the experience of the research team in finding a framework within which to mix paradigms.

The second challenge was related to the "nuts and bolts" of the evaluation design. To begin with, the design would have to meet the *ethical* standards of both the researchers and service providers and answer potential criticisms regarding withholding valuable services from young people. Another critical issue was that of *sample* size. As recruitment of participants would be undertaken by Foroige youth projects, the design process would involve negotiating with Foroige to assess whether it would be possible for them to recruit a minimum of 200 participants. Furthermore, as outlined earlier, it is recommended that *programs* undergoing RCT are well established. In the case of BBBS, although the program had been established for 5 years in the west of Ireland, it was in the process of being "rolled out" nationally. The logical solution was to limit the study to the western area but this in turn would have implications in terms of the ability to recruit an adequate sample. Another challenge would be to ensure *stakeholder commitment* to a lengthy study and complete the study within the timeframes set down by funders. In addition, the design had to include

strategies to avoid "resentful demoralization" on the part of the control group and ensure that control group participants were sufficiently motivated to continue with the study over the proposed 1.5-year timeframe. They also had to ensure that they did not receive *alternative treatment* that would threaten the integrity of the experiment.

The next part of this article reviews our journey in trying to resolve these tensions and develop an integrated study design. The design process was very much a journey of three stages for the research team, reflecting a move from impact/quantitative dominant to a more rounded mixed methods design. It shows that the attempts to resolve the "nuts and bolts" issues influenced the paradigmic stance and vice versa.

Resolving the Tensions: Phase 1 of the Design Journey

Because the task of designing and implementing an RCT was such a challenging one, the research team initially applied themselves to the nuts and bolts of the impact study. The practical challenges outlined above had to be resolved through consultation with stakeholders and the EAG. In developing the design, a balance had to be struck between ethical practice, scientific validity, and feasibility in terms of what the BBBS program could take.

In relation to sample size, we were supported in our work by members of our EAG, who had particular experience in experimental design. This group advised that a minimum sample size of 200 would be required in order to potentially identify the expected effect size of a Cohen's *d* of just under .2. However, the recruitment of 200 study participants would represent a challenge for the program. At the time, Foroige, the service provider, was supporting 60 mentoring pairs in the western region and had just received funding to roll out the program nationally. Given, as mentioned earlier, that programs undergoing RCT should be well established, the decision was made to restrict the study to the western region where the BBBS program was in operation for 5 years. This meant the program had to grow exponentially from supporting 60 matches to supporting an additional 100 to conduct the study.

The ethical issue of denying young people a service was addressed in a number of ways. Both intervention and control groups would be offered a basic youth service and mentoring would represent an "add-on" service for the intervention group. Thus, all research participants would be offered a service. This meant that mentoring would be evaluated as an additional element of youth service provision rather than as a stand-alone program.[1] Furthermore, the youth in the control group would be placed on a waiting list for support. However, as a result, the target sample age group would have to be reduced from 10–18 years to 10–14 years, so that the young people on the waiting list would have a chance to be matched and benefit from a mentor's support before being ineligible for the program when they reached the age of 18 years. In addition, we agreed on a "free pass" system with the staff, whereby any vulnerable young person deemed to be in need of mentoring support and who the staff were not comfortable with possibly being randomly allocated to the control condition, could be forwarded for the intervention and not included in the study. Detailed information materials were developed in conjunction with program staff to ensure that the research study was communicated clearly to potential participants and full written consent was required from all participants.

Like the Tierney et al. (1995) study in the United States, it was planned that the Irish study would take measures at baseline, 12 months, and 18 months from young people, parents, teachers, and mentors. We were supported in our work by Dr. Jean Rhodes, a member of our international advisory group, who agreed to provide the research team with a set of quantitative research instruments to be used in the study, which would enable us to explore whether the implementation of the BBBS program in Ireland could be understood in terms of her theory of mentoring (J. E. Rhodes, 2005).

In terms of reducing attrition and avoiding "resentful demoralization" of the control group, the fact that control group participants would be engaged in Foroige services meant that they would be less likely to "drop out" and more accessible to the research team than if they were not receiving any intervention. The research team worked extensively with the program staff, developing communication and data management protocols with them. To avoid threats to the integrity of the experiment, it was critical that the program staff were aware of the need to offer similar activities to both control and intervention groups and not favor those not receiving a mentor in any way. Data systems were established to record the precise dosage of "intervention as normal" activities received by both intervention and control groups.

A design document was drafted that described the impact study in detail. A flow chart summarizing the initial design choices is outlined in Figure D.2.

In practice, as illustrated in Figure D.3, the need to address the feasibility and ethical issues associated with RCT impinged on the recruitment of the sample and, consequently, the power of the study. By restricting the study to the west of Ireland and operating a waiting list control that meant that the age range of recruits had to be lowered, the pool of potential recruits was reduced. In addition, the breadth of the study was reduced in that it would focus just on the 10-14 year age group and would evaluate mentoring as an add-on service rather than a stand-alone intervention. However, the ethical and feasibility demands on the study could not be ignored and compromises had to be made.

From the outset, the research team had identified that the evaluation would need to answer the following three research questions:

1. What is the impact of the BBBS program on the participating youth?

2. How is the program experienced by stakeholders?

3. How is the program implemented?

Figure D.2 Overview of Impact Study Deisgn

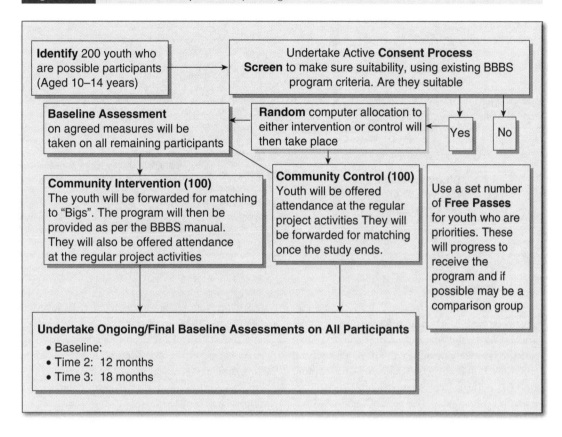

It would be fair to say that during Phase 1 of the design development the research team was consumed by a focus on the impact measurement because of the range of issues to be addressed to make this methodology applicable to the local context.

We had yet to develop a design that would not only incorporate these research questions but also provide a framework for integration of the various data sources. Our position in this phase could be summarized as maintaining a *pragmatic stance* in our intention to use both quantitative and qualitative approaches to answer the different research questions. Our progress in selecting an appropriate mixed methods design is set out below.

Impact and Process: Phase 2 of the Design Journey

As described earlier, the intention at the outset had been to have some type of process study incorporated into the RCT, as a result of the research team's own methodological orientation, a recognition of the importance of understanding process in mentoring studies as illustrated through U.K. research, and in compliance with good practice in RCT studies, which highlights the importance of process studies to describe implementation. At this stage of the design process, some additional forces emerged to place further emphasis on the need for a stronger process element.

Figure D.3 How Ethics and Feasibility Issues Affected Ability to Recruit Sample

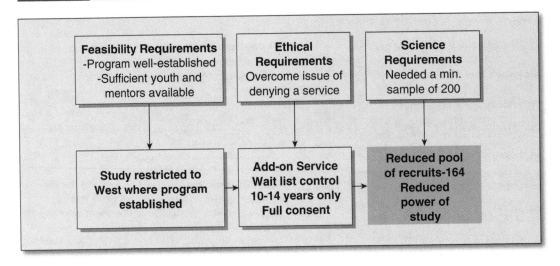

First, as mentioned earlier, the research culture in Ireland is very much focused on process studies. While the research team had to grapple with this new form of inquiry, the program staff and other stakeholders also had difficulties in accepting the RCT methodology. When "selling" the impact study to program staff, the research team was frequently asked if there would be an opportunity for them to provide feedback on the program as part of the study. Thus, there was a demand from stakeholders for a mixing of methods. From the perspective of the research team, the promise of a process element was a means of "softening the blow" in terms of the rigidity of the RCT methodology and providing stakeholders with a little bit of what they were familiar with in terms of research to lessen their anxiety or resistance in the face of the RCT. This was of critical importance given the central role of program staff in liaising between the research team and the study participants. Another factor of relevance at this stage was the difficulty associated with recruitment of the sample. The search for sufficient numbers of participants took longer and was more difficult than anticipated. The data collection time points had to be extended,

and the eventual final sample size was reduced to 164. The fact that the projected sample size would limit the statistical impact of the study gave us renewed focus on considering how we could strengthen the study through a strong combination of both quantitative and qualitative approaches.

Returning to our research questions, our plan was to use a survey-based methodology to collect outcome data to answer the first question in relation to the impact of the program, as just described. The research team now had to agree on the appropriate means of answering the second and third research questions, in relation to stakeholder experiences and program implementation, respectively. A design proposal was circulated to the EAG that placed the RCT study as primary, with a process study taking a secondary role, examining issues of implementation, process, and meaning.

The data for the second research question regarding the experiences of stakeholders were be answered through interviews with key program participants, including youth, mentors, parents, and staff. A purposive sample of 12 mentoring pairs was to be selected from across the study area reflecting differences in age, gender, and location, whether urban or rural. Interviews were to be undertaken on two occasions, once when the relationship was established and the next following an interval of 6 months or more. This process would enable us to collect data on stakeholder perspectives and also allow an exploration with each pair of how the relationship develops over time. In relation to the third research question regarding program implementation, it was planned that a review of the case files of mentored youth would be undertaken to establish whether the program was implemented according to the manual. Focus groups with the program staff were also included in the design to collect data regarding their experience of implementing the program.

In this phase, we had moved our stance from a *pragmatic* stance to a *dialectical* position in that we now intended to use the data from both the impact and case study streams to inform each other in the analysis. This stance is facilitated by contrasting the data findings from the deductive framework of the quantitative impact study with the inductive framework of the qualitative case study (Greig, Taylor, & Mackay, 2007). The measurement of the impact of youth mentoring on the participant youth outcomes was heavily influenced by the developmental focus on much of the North American literature on youth mentoring (Philip, 2003). Our design incorporated this focus both in terms of approach through the impact study and through the use of U.S. data instruments to measure the impact of mentoring. However the focus on the U.K.-based youth-mentoring literature has been more influenced by a sociological approach that has recognized youth agency and the effects of structural limitations in trying to develop an understanding of youth-mentoring programs (Colley, 2003; Liabo, Lucas, & Roberts, 2005; Philip & Spratt, 2007). The conduct of the case studies allows for a more inductive exploration of the mentoring in context from the perspective of those involved. Taking a dialectical stance would provide the opportunity to compare and contrast both these approaches to the exploration of youth

mentoring. By using NVivo software to analyze the qualitative case studies and match files, we would be able to link each case study narrative to the quantitative survey scores for that participant. In so doing we would be able to create an analysis of the mentoring relationship that used both qualitative and quantitative data.

However, feedback from the EAG challenged the research team to give more consideration to how the impact and process studies would be integrated. Their feedback highlighted risk that the qualitative aspect may go off on a tangent and that the findings of the two studies would not "speak to each other." J. E. Rhodes's (1995) model of mentoring, as described earlier, was suggested as offering a unifying framework for which the qualitative and quantitative could offer different types of evidence.

Integration at Last? Phase 3 of the Design Journey

In the final stage of our design journey, Rhodes's theory of mentoring (see Figure D.1) was placed as central to the design to achieve coherence across the research questions and integrate qualitative and quantitative data sources. As had been planned at all stages of the study design, this theory would guide the analysis of the impact data. However, our breakthrough in Phase 3 of the design process was to conceptualize the program implementation or process element of the study as providing evidence to enable us to test an essential part of the program theory, namely whether the strength of program implementation is a "moderator" of the program impact as predicted by the Rhodes model. Furthermore, we also established that the model could be used to guide the analysis of the qualitative case study data in a number of ways. First, as described above, the qualitative data could be linked to the quantitative data for the case study participants and used to develop an integrated analysis at the individual level. Second, from the case study and the program staff interview data we could seek qualitative evidence in support of

Figure D.4 Evaluation of Big Brothers Big Sisters Ireland: A Concurrent Embedded Model

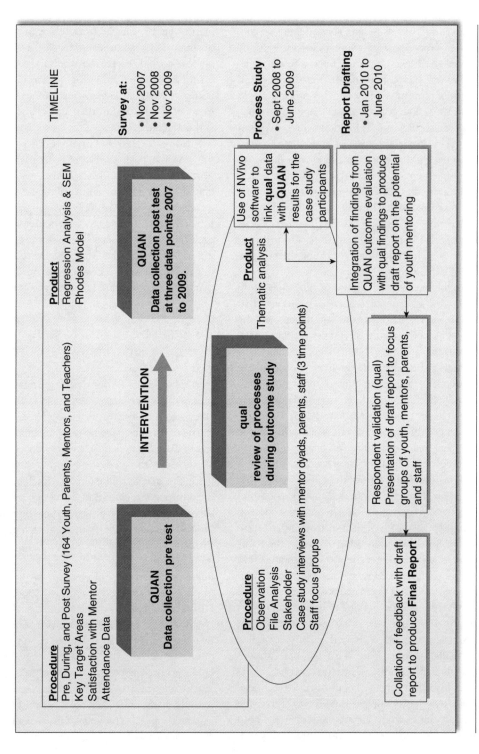

NOTE: QUAN = quantitative; qual = qualitative.

Rhodes's theory of mentoring, and thus explore its goodness of fit to understand the developing of mentoring relationships in the Irish setting.

We resolved to address the concern that the data streams were not integrated by developing a mixed methods research question that would illustrate how the data sources could be interlinked. Therefore, we added a fourth research question to our evaluation to enable us to complete an "integrated" design. Our final and fourth research question was "What results emerge regarding the potential of this youth mentoring program from comparing the outcome data from the impact study with the case study data from the mentoring pairs?" Placing the Rhodes model as the core framework for our analysis meant that this overarching question could be answered through a comparative examination of both the qualitative and quantitative data sources.

At this final stage of the planning the design of our primarily experimental study, we believe that the qualitative stream is both independent and interrelated. By maintaining our efforts to use the various sources of data to inform each other, we have maintained a dialectical stance. By placing the underlying theory as the guiding framework for considering each data source in isolation, transformation, and in comparison, we have developed a concurrent embedded mixed methods design (Creswell & Plano Clark, 2007). The design map to illustrate the various components of our design and their interrelationships is set out in Figure D.4.

● CONCLUSION

This article has described how a research team in the west of Ireland responded to the task of designing an RCT study. Although the context was supportive in terms of stakeholder buy-in, funding, established evidence base, and a strong program theory and infrastructure, there was a range of challenges to be faced. First, the study would have to accommodate the feasibility, ethical, and

scientific difficulties associated with RCT studies. Second, some means would have to be found to accommodate pressures to incorporate implementation and process data in a meaningful way and find a paradigmatic fit for the study. As described, this was one of the first RCT studies of its nature to be undertaken in Ireland and thus the research team was on a "learning curve."

In cases such as the one described wherein the practical constraints of program size and stage of development prohibit the undertaking of generalizable impact studies, we believe it is pragmatic to look toward the mixing of methods as a way of benefiting from the strengths of the RCT method but compensating for its weaknesses. It is also worth looking creatively at how the data and methods available can work dialectically to inform each other and enable a consideration of both causality and meaning as interconnected and contingent concepts. The final design is particularly suitable to mentoring, which not only has a tradition and literature to be mined both on the qualitative and quantitative sides but also requires its evaluation to capture both the general impacts and the specific case-by-case interpersonal magic that makes mentoring work (or not). Furthermore, the opportunity to incorporate stakeholder feedback throughout the study is an aspect of the design that references the more participatory values that traditional experimental design can overlook. However, the endeavor is not problem free. Chen (1997) makes the point that rigor may be sacrificed in mixed method designs as the evaluator may not have the time or resources to pursue standards of dual rigor. However, Chen argues that, under theory-driven evaluations, the strength of inferences comes from both methodological rigor and theoretical reasoning, which means that the impact of a reduced rigor is less than in a method-driven evaluation alone.

To conclude, our experience suggests that when assigned a task to undertake a certain type of study within worldview, rather than engaging in an argument about the incompatibility between concepts such as objective and subjective accounts, it is preferable to see mixed methods theory and practice as a resource to conceptualize how learning from the research opportunity can be maximized.

● NOTE

1. In Ireland, BBBS is offered as part of youth service provision but in the United States it is a stand-alone program.

● REFERENCES

The Atlantic Philanthropies. (2007). 2006 *Annual report— Giving while living.* Retrieved May 30, 2008, from http://atlanticphilanthropies.org/content/download/4518/69535/file/2006APAR.pdf

Chelimsky, E. (1997). Thoughts for a new evaluation society. *Evaluation, 3,* 97-118.

Chen, H. (1997). Applying mixed methods under the framework of theory driven evaluations. In J. C. Greene & V. J. Caracelli (Eds.), *Advances in mixed method evaluation: The challenges and benefits of integrating diverse paradigms* (pp. 61–72). San Francisco: Jossey-Bass.

Colley, H. (2003). *Mentoring for social inclusion: A critical approach to nurturing mentoring relationships.* London: Routledge.

Creswell, J. W., & Plano Clark, V. L. (2007). *Designing and conducting mixed methods research.* Thousand Oaks, CA: Sage.

Dubois, D. L., Doolittle, F., Yates, B. I., Silverthorn, N., & Kraemer Tobes, J. (2006). Research methodology and youth mentoring. *Journal of Community Psychology, 34,* 657–676.

Dubois, D. L., Holloway, B. E., Valentine, J. C., & Cooper, H. (2002). Effectiveness of mentoring programs for youth: A meta-analytic review. *American Journal of Community Psychology, 30,* 157–197.

EU Commission. (1999). Council regulation (EC) No 1260/1999 [Electronic Version]. *Official Journal of the European Communities.* Retrieved April 7, 2009, from www.esf.gov.sk/documents/1999-1260.pdf

Furona, K., Roaf, P. A., Styles, M. B., & Branch, A. Y. (1993). *Big Brothers/Big Sisters: A study of program practices.* Philadelphia: Public/Private Ventures.

Ghate, D. (2001). Community-based evaluations in the UK: Scientific concerns and practical constraints. *Children & Society,* 15, 23–32.

Greene, J. C. (2000). Understanding social programs through evaluation. In N. K. Denzin & Y. S. Lincoln (Eds.), *Handbook of qualitative research* (pp. 981-1000). Thousand Oaks, CA: Sage.

Greene, J. C. (2003). Understanding social programs through evaluation. In N. K. Denzin & Y. S. Lincoln (Eds.), *Collecting and interpreting qualitative materials* (pp. 590–618). Thousand Oaks, CA: Sage.

Greene, J. C. (2007). *Mixed methods social inquiry.* San Francisco: Jossey-Bass.

Greene, J. C., Benjamin, L., & Goodyear, L. (2001). The merits of mixing methods in evaluation. *Evaluation, 7,* 25–44.

Greene, J. C., & Caracelli, V. J. (1997). Defining and describing the paradigm issue in mixed-method evaluation. In J. C. Greene & V. J. Caracelli (Eds.), *Advances in mixed-method evaluation: The challenges and benefits of integrating diverse paradigms* (pp. 5–17). San Francisco: Jossey-Bass.

Greig, A., Taylor, J., & Mackay, T. (2007). *Doing research with children.* London: Sage.

Johnson, Z., Howell, F., & Molloy, B. (1993). Community mothers' programme: Randomised control trial of non-professional intervention in parenting. *British Medical Journal, 306,* 1449–1452.

Lalor, K., de Roiste, A., & Devlin, M. (2007). *Young people in contemporary Ireland.* Dublin, Ireland: Gill & Macmillan.

Liabo, K., Lucas, P., & Roberts, H. (2005). International: The UK and Europe. In D. Dubois & M. Karcher (Eds.), *Handbook of youth mentoring* (pp. 392–407). Thousand Oaks, CA: Sage.

Little, M., Kogan, J., Bullock, R., & van der Laan, P. (2004). ISSP: An experiment in multi-systemic responses to persistent young offenders known to children's services. *British Journal of Criminology, 44,* 225–240.

Oakley, A. (2000). *Experiments in knowing.* New York: Polity.

Oakley, A., Strange, V., Toroyan, T., Wiggins, M., Roberts, I., & Stephenson, J. (2003). Using random allocation to evaluate social interventions: Three recent UK examples. *Annals of the American Academy of Political and Social Science, 589,* 170–189.

Pawson, R., & Tilley, N. (1997). *Realistic evaluation.* London: Sage.

Philip, K. (2003). Youth mentoring: The American dream comes to the UK? *British Journal of Guidance and Counselling, 31,* 101–112.

Philip, K., Shucksmith, J., & King, C. (2004). *Sharing a laugh? A qualitative study of mentoring interventions with young people.* York, UK: Joseph Rowntree Foundation.

Philip, K., & Spratt, J. (2007). *A synthesis of published research on mentoring and befriending* [Electronic Version]. Retrieved April 7, 2009, from http:// www .mandbf.org.uk/resources/research/

Rhodes, F. H. T., & Healy, J. R. (2006). Investment in knowledge: A case study of a philanthropy's partnership with government. *Administration, 54,* 63–84.

Rhodes, J. E. (2005). A model of youth mentoring. In D. L. Dubois & M. J. Karcher (Eds.), *Handbook of youth mentoring* (pp. 30-43). Thousand Oaks, CA: Sage.

Rossi, P. H., Lipsey, M. W., & Freeman, H. E. (2004). *Evaluation: A systematic approach.* Thousand Oaks, CA: Sage.

Shadish, W. R., Cook, T. D., & Campbell, D. T. (2002). *Experimental design and quasi-experimental designs for generalized causal inferences.* Boston: Houghton Mifflin.

Tashakkori, A., & Teddlie, C. (Eds.). (2003). *Handbook of mixed methods in social and behavioral research.* Thousand Oaks, CA: Sage.

Tierney, J. P., Grossman, J. B., & Resch, N. L. (1995). *Making a difference: An impact study of Big Brothers/Big Sisters.* Philadelphia: Public/Private Ventures.

Tovey, H., & Share, P. (2003). *A Sociology of Ireland.* Dublin, Ireland: Gill & Macmillan.

Appendix D

Appendix E—An Example of the Transformative Design

Telling it All

A Story of Women's Social Capital Using a Mixed Methods Approach

Suzanne Hodgkin

La Trobe University, Albury-Wodonga, Australia

The aim of this article is to demonstrate how quantitative and qualitative methods can be used together in feminist research. Despite an increasing number of texts and journal articles detailing mixed methods research, there are relatively few published reports of its use in feminist study. This article draws on a study conducted in regional Australia, exploring gender and social capital. Through the analysis and interpretation of data derived from a large survey and in-depth interviewing, the author will demonstrate the power of the mixed methods approach to highlight gender inequality. Despite past reluctance of feminists to embrace quantitative methods, the big picture accompanied by the personal story can bring both depth and texture to a study.

KEYWORDS: *mixed methods; feminist methodology; social capital*

Much has been written in the literature concerning the differences in epistemology between positivist and naturalistic forms of inquiry. Indeed, they have been represented as two distinct research paradigms, drawing on different bodies of thought, and using different methods of data collection. Within the paradigm of positivism, established theory drives the empirical focus of inquiry. Alternatively, within the paradigm of naturalism, knowledge is understood to be socially constructed, and theory is generated from getting an "inside" interpretation. The two approaches have been viewed

SOURCE: Hodgkin, S. (2008). Telling it all: A story of women's social capital using a mixed methods approach. *Journal of Mixed Methods Research, 2*(3), 296-316. Reprinted with permission of SAGE Publications, Inc.

as incompatible, because of their essential epistemological differences.

Despite vigorous support for the incompatibility thesis, there is growing support in the literature for mixed methods approaches (Creswell, 2003; Creswell & Plano Clark, 2007; Johnson, Onwuegbuzie, & Turner, 2007; Morgan, 2007; Tashakkori & Teddlie, 2003). Bryman (1988), among others, has argued that differences between the traditions have been exaggerated and the overlaps ignored. Similarly, Epstein, Jayaratne, and Stewart (1991) contended that much of the debate about quantitative and qualitative methods has been "sterile and based on false polarization (p. 89)." Whereas qualitative methods provide for richly textured data, quantitative methods allow for the incorporation of a large number of contextual variables (Epstein et al., 1991). Quantitative data may assist in providing the big picture, but it is the personal story, accompanied by thoughts and feelings, that brings depth and texture to the research study.

The aim of this article is to demonstrate the use of mixed methods in feminist research. Sands (2004) highlighted that although there are many genres of feminist research, feminist research is about women's experiences as gendered subjects and "efforts to understand and meet challenges related to their status as women" (p. 50). Despite an increasing number of texts and journal articles detailing mixed methods research, there are relatively few published studies that report on its use in feminist study. Drawing on a study of gender and social capital, this article will provide an example of such a mixed methods research approach. Through the interpretation and analysis of the findings of research conducted in regional Australia, the power of the mixed methods approach to highlight gender inequality will be demonstrated. It will be argued that past conceptualization of social capital, and the research drawing on it, has shown only limited sensitivity to gender. The author will provide examples of quantitative data to demonstrate the existence of different social capital profiles for men and women. Stories will also be presented to provide a picture of gender inequality and

expectation. The author will conclude by arguing that despite reluctance on the part of feminists to embrace quantitative methods, the big picture accompanied by the personal story can bring both depth and texture to a study.

● THE STUDY OF SOCIAL CAPITAL

A quick search of the concept *social capital* in any database will provide many and varied results. Such a search will provide evidence of the concept's many applications in both scholarly discussion and public debate. In the scholarly literature, the concept has been empirically measured across countries and across disciplines. In Australia, the concept is referred to quite loosely across political viewpoints to advocate policies of government withdrawal, mutual obligation, and community capacity building. It has gained increased prominence in these social policy discussions across all levels of government (Healy & Hampshire, 2002).

All forms of government have operated under the assumption that social capital is eroding. What they differ on is the aspects of contemporary society that undermine social capital. This is hardly surprising given that definitions of social capital tend to be at best vague and often open to interpretation. The literature is dominated by discussions of the different definitions of social capital, different ideological standpoints, and different interpretations. However, most scholars agree that the concept *social capital* describes the norms and networks that enable people to work collectively together to address and resolve problems they face in common (Saunders & Winter, 1999; Stewart-Weeks & Richardson, 1998; Stone & Hughes, 2000). It is concerned with how people trust each other and how they help one another, whether they are acquaintances, friends, family, workmates, fellow committee members, or club members. It is promoted as a social good: People are happier if they ultimately feel they belong to a community and that they are connected in various ways (McMichael & Manderson, 2004).

As a concept, it has been the subject of different empirical measurement. In Australia alone, several major studies have set out to provide quantitative measurement of social capital (Baum et al., 2000; Onyx & Bullen, 2000; Stone & Hughes, 2002). These large quantitative studies using survey methods have looked for evidence of trust, reciprocity, the exercising of social norms, and participation in social, civic, and community life.

Criticism has been leveled at empirical studies that have measured the extent of participation in associational life, with little consideration for the informal networks to which people belong. This debate raised by Cox (1996) has led to a reexamination of the concept and acknowledgment that there may be different types of social capital that include participation in both informal networks and more formal associations. Putnam (2000) made the distinction between bonding and bridging forms of social capital. He defined bonding social capital as those networks that encompass strong informal social networks accompanied by strong in-group loyalty. Bonding types of social capital are crucial in generating a shared sense of identity (Onyx, 2001). Bonding types of social capital assist people to "get by." In contrast, bridging social capital is found in outward-looking networks and includes people across race, gender, and class. This type of social capital brings people together from outside their familial and close networks, to achieve common goals. Bridging forms of social capital are thought to assist in "getting ahead." This form of social capital is what interested Bourdieu (1986), as it can be used by individuals to foster their position in society. Increasing one's connections outside personal networks enhances individuals' opportunities for employment and other forms of social standing.

Although acknowledging that there may be different types of social capital, the research drawing on it has shown limited sensitivity to systematic inequalities associated with gender,

race, and class where being civic or caring for others means different things to different people. As Bryson and Mowbray (2005) noted, a very White, middle-class notion of community dominates discussions of social capital. Conflict and exclusion from participation is rarely considered. This lack of consideration is important. Structural inequities of gender, age, and class are very closely related to distribution of civic resources (Norris & Inglehart, 2003). The question that is rarely examined is whether women in particular have access to associational life that brings the returns associated with bridging forms of social capital (Parks-Yancy, DiTomaso, & Post, 2006).

The gender-neutral examination of social capital is concerning, as it fails to consider long-established structural inequalities. Bezanson (2006) explained this by arguing that the most prominent and enthusiastic supporters of the concept tend to be men who are not specialists in family or feminist theory. Studies conducted in the United States and the United Kingdom found evidence that certain types of organizations remain disproportionably male; political parties, sporting clubs, labor unions, and professional groups are examples here (Lowndes, 2000; Norris & Inglehart, 2003; Sapiro 2003). In contrast, other studies have found women's predominance in unpaid, domestic roles with limited status (Alessandrini, 2003).

These concerns have led academics such as Lowndes (2000) to call for the examination of social capital to focus on different social capital profiles for men and women. The argument made by Lowndes (2000) and Alessandrini (2003) is that women's caring and community-based responsibilities may constrain their civic and political aspirations.

The present study decided to test out this argument. It sought to explore social capital in two different ways: first, to map the different patterns of participation based on gender, and second, to explore how the role of "mother" alters both the activities women become involved in and the reasons for this.

● COMBINING METHODS IN FEMINIST RESEARCH

Choosing a research design that best captured the story of women's social capital required careful consideration. As this study was concerned with drawing attention to the lack of gender focus in studies of social capital and with making more visible women's contributions, the researcher was seeking an approach that would be considered valid. Kohler Riessman (1994) contended that feminists have traditionally placed themselves within the postmodern paradigm of research methodology. However, she also believes that there is a growing school of thought that research about a problem may be strengthened when "various kinds of data are brought to bear" (p. X). Similarly, Finch (2004) argued that although feminist research is linked very closely to qualitative forms of inquiry, this link is tenuous and may ultimately disadvantage women. Research that is population based may illuminate issues for women on a broad scale. Despite previous reluctance to use quantitative methods, quantitative and qualitative methods can be used together to give a more powerful voice to women's experiences (Brannen, 1992; Epstein et al., 1991; Oakley, 1999; Shapiro, Setterlund, & Cragg, 2003). Brannen (1992) reassessed previous viewpoints that reject quantitative approaches on the basis that surveys are "imbued with masculinist assumptions" and looked to the new school of thought that believes there is no one feminist methodology; rather, it is more important that the researcher locates herself as feminist within the research process. Brannen argued,

> Moreover there are grounds for arguing that both qualitative and quantitative approaches need to be applied in combination, especially where investigations are carried out on social groups whose material situations and perspectives have been under- or mis-represented in social research. While the qualitative approach may overcome some of

the problems of giving a voice and language to such groups, through which they may better express their experiences, the quantitative approach would serve to indicate the extent and patterns of their inequality at particular historical junctures. (p. 22)

The present study is located within the transformative research paradigm. Mertens (2007) argued that the transformative paradigm provides a framework for addressing issues of social justice in the research process. The ontological assumption of the transformative paradigm holds that socially constructed realities are influenced by power and privilege. The transformative paradigm recognizes that "voices of those who are disenfranchised on the basis of gender, race/ethnicity, disability or other characteristic" (Mertens, 2007, p. 214) can be excluded in research. Within this paradigm, mixed methods are preferred to highlight issues of need (quantitative data) and to give voice to these issues (qualitative data). The transformative paradigm with its ontological, epistemological, and methodological assumptions provides a logical framework for different types of feminist research. Research about women should explore the research question first and remain open to a range of data collection methods to arrive at a better understanding (Oakley, 1999). In the past and more recently, arguments have been made that feminist methodology should bring together both the subjective and objective ways of knowing the world (Rose, 1982; Shapiro et al., 2003). In addition, feminist research that draws on evidence from a variety of sources is more likely to be seen as valid and reliable and is thus more likely to be heard in the policy arena (Shapiro et al., 2003).

Advantages and disadvantages inherent in both approaches should be recognized. With large quantitative research, women's voices as an oppressed group have remained unheard (Oakley, 1999). With qualitative research, problems with poor representation and a tendency to overgeneralize need to be highlighted. A mixed methods approach can alleviate some of these inherent problems. As the literature concerned with mixed methods continues to grow, there are numerous studies demonstrating the power of both breadth and depth in studies of complex social issues.

The majority of research studies undertaken about social capital have primarily used quantitative techniques, and there was a need to find a different approach that included multiple ways of knowing. Back in 1998, Cox called for an increase in studies that combine research methods. This present study is thus different from many other studies of social capital. In the present study, a mixed methods approach was taken to best understand the research problem, with the intent of capturing the best of both qualitative and quantitative methodology (Creswell, 2003). Capturing women's social capital, and the complexities associated with this, led the researcher to want to examine it from different angles (Darlington & Scott, 2002). Thus, both quantitative and qualitative methods of data collection were used in a sequential study that had two distinct phases.

● PRESENT STUDY

The present study makes a unique contribution to the mixed methods literature, examining its use in feminist research. There are several key texts that provide a solid discussion of the history of mixed methods research and its application in social sciences (Brewer & Hunter, 2006; Bryman, 1988; Creswell, 2003; Creswell & Plano Clark, 2007; Greene & Caracelli, 1997; Mertens, 2007; Tashakkori & Teddlie, 1998, 2003). Journals such as *Quality & Quantity* and *Journal of Mixed Methods Research* are committed to publishing a range of mixed methods studies. Despite these publications and an extensive search in many and varied journals, there are very few published studies that combine a feminist approach with a mixed methods approach. Finch (2004) argued that research that uses an approach other than qualitative to address feminist questions remains underdeveloped (p. 64).

This study was conducted in a regional city in Australia. Prior to commencing the study, the researcher's research experience had been with qualitative forms of inquiry. Despite this, the researcher determined that a large representative sample of data was required to generalize to the population. For instance, the researcher determined that the survey method would be the best method for identifying whether men and women had different social capital profiles. However, only qualitative methods could illuminate the stories behind these profiles.

The study used an explanatory sequential design (Creswell & Plano Clark, 2007). Extensive data collection occurred in both stages of the study. This led to intensive data analysis at two different stages of the research process. Padgett (1998) discussed the temptation to revert to a dominant-less-dominant design, compromising both data collection and data analysis in one of the methods. Equal time was spent on Stage 1 and Stage 2 of the research process.

A survey method was chosen for Stage 1 of the study. This enabled the researcher to describe; explore; and, to some extent, explain aspects of the differences between men and women on social, community, and civic participation within the sample.

In the second stage, the study focused on exploring, from the viewpoint of women, their processes of interacting in their social, community, and civic worlds and how they felt about their lives and the activities in which they became involved. The researcher was interested in describing aspects of their lives, based on their telling of their experiences. Thus, the participant became the expert, and the data generated were qualitative.

Greene, Caracelli, and Graham (1989) identified five reasons for conducting mixed methods studies: triangulation, complementarity, development, initiation, and expansion. In this present study, the aim of complementarity provided the justification for the mixed methods approach. When the aim is to complement findings, the researcher is seeking elaboration, enhancement, illustration, and clarification from the results of one method with those of the other. The findings of the first stage illustrated some complexities in the data. Women demonstrated a different pattern of participation in social and community and to a lesser extent, civic activities. This was particularly the case for women aged between 29 and 49 years. The researcher was seeking an extended and deeper view of this difference. One of the criticisms leveled at research that produces only quantitative data is that the data can become overinterpreted. Here was the advantage of complementarity; the qualitative study elaborated on and enhanced some of the results from the quantitative study. The qualitative study also provided vivid illustrations of some of the results found in Stage 1.

The present study had several limitations. Social capital has been measured and conceptualized in several different ways. The researcher set out to find a measure of social capital that could be easily adapted to the nonmetropolitan context and that made a distinction between informal and formal types of participation. This was particularly important, as the researcher was attempting to highlight the range of activities that women became involved in, not just those that occur within the public eye. The first stage of the study thus used a measure developed by the South Australian Community Research Unit (Baum et al., 2000). In using this, the researcher acknowledges the other valuable tools that have since been developed, which focus on family social capital (Hughes & Stone, 2003) and different types of networks (Onyx & Bullen, 2000).

In the second stage of the study, the researcher chose to follow up on a subsample of women between the ages of 29 and 49 years. In doing so, the researcher also acknowledges that it would have been beneficial to conduct in-depth interviews with a subsample of men. Although motherhood changed women's participation, the quantitative data indicated that this was also true to a lesser extent for men. Similarly, women 29 and younger and 50 and older were excluded from the interviews. Both groups may have provided some useful comparative

data. The scope and size of the research prevented further subsamples to be explored.

● RESEARCH QUESTIONS

There were two clear research questions driving the study:

> *Research Question 1:* Do men and women have different social capital profiles?

> *Research Question 2:* Why do women participate more in social and community activities than in civic activities?

● METHODS

The researcher used sequential mixed methods sampling (Teddlie & Yu, 2007). Participants were selected sequentially through probability and purposive sampling strategies.

Stage 1: Participants

Simple random sampling was the method chosen for determining the sample. To compensate for the poor response rate to self-administered surveys and reduce sampling error, a large sample was chosen. Because of new privacy legislation, the researcher was unable to select names and addresses from the electoral roll. The questionnaire went out to 4,000 households, randomly selected from a database of residential addresses provided by the local government authority. Replies were received from 1,431 residents (a response rate of 35%). Of these, 403 were male (28.8%) and 998 were female (71.2%). Women are therefore disproportionately represented in the survey. The mean age of respondents was 48.7 years. More than 32% of the respondents worked full-time, 21% worked part-time, 17% were engaged in home duties, 4% were students, 3% were permanently unable to work, 18.5% were retired, 2% were unemployed, and the rest were "other." Of those who were working, the mean working hours was 41.5 for men and 31.1 for women. Almost two thirds of the respondents reported that they lived in households as a couple or a couple with children. Lone-parent households made up 7.1% of the sample. Respondents were asked to identify their highest level of education completed. Those with no formal schooling totaled 0.4%, 5.9% had primary school as their highest level of education, 40.2% had secondary school, 14.3% had a Technical and Further Education (TAFE) qualification, 13.5% had trade, 27.8% had a university and/or higher degree, with the rest "other."

It should be noted that the study's focus was squarely on adults and social capital, thus only people aged 18 and older were invited to complete the questionnaire. Members of the sample were sent a cover letter explaining the study's purpose and usefulness, and describing how the respondent was selected. Respondents were also sent an additional form seeking an expression of interest in participating in the second stage of the study.

Stage 2: Subsample

All participants were recruited as a result of the first stage of the research. Those participants who were interested in being interviewed for the second stage signed the agreement form sent with the initial questionnaire. This meant that forms were returned from both men and women. As this form was returned to the researcher separately from their questionnaire to ensure confidentiality, the researcher could not match up the survey results with any one respondent or with any demographic information.

Seventy-five women responded by filling in and signing their form. Those who had expressed an interest in being interviewed were posted an information sheet and an informed consent form. The researcher decided on a cluster random sampling technique. In cluster random sampling, "already formed groups of individuals within the population are selected as sampling units" (Tashakkori & Teddlie, 1998, p. 75). The researcher contacted by phone those who had returned the informed consent form explaining the purpose of the study and its interest in women between the ages of 29 and 49 years. Those who did not fit the criteria for the sample were excluded. The researcher had determined initially on a quota sample of 6 participants, thus the first 6 participants who met the criteria and were willing to be interviewed were chosen. This sample grew until the researcher felt comfortable that saturation had been achieved.

The final sample comprised 12 participants. There were an equal number of participants aged 29 to 39 years and 39 to 49 years. Similar to the broader sample, 4 participants worked in a full-time paid capacity, 6 participants worked in a part-time paid capacity, and 2 participants provided full-time home duties. Three of the 12 women were sole parents, a figure higher than that of the broader sample. The participants' level of education was also higher than that of the broader sample. Two of the participants had completed secondary education, 1 participant held a TAFE diploma, 6 participants had gone on to complete a university degree, and 2 participants had completed the equivalent of a master's degree.

Procedures

Quantitative data. The study's initial theory testing required data from a large representative sample to generalize to the population. By using a survey as a form of data collection, the researcher was able to describe; explore; and, to some extent, explain aspects of social, community, and civic participation within the sample. The survey was cross-sectional, with data collected at one point in time.

In Australia, instruments have been developed to measure social capital in different communities (Baum et al., 2000; Onyx & Bullen, 2000; Stone & Hughes, 2002). It is important to note that the

Table E.1 Items Contained in Each Participation Category

Social participation—informal (3 items)

If the respondent had done any of the following activities: Visited family or had family visit, visited friends or had friends visit, visited neighbors or had neighbors visit.

Social participation—in public spaces (4 items)

If the respondent had done any of the following activities: Been to a cafe or restaurant, been to a social club, been to the cinema or theater, been to a party or dance.

Social participation—group activities (6 items)

If the respondent had done any of the following activities: Played sport, been to the gym or exercise class, been involved in a hobby group, been involved in a self-help or support group, singing/acting/musician in a group, gone to a class.

Civic participation—individual activities (7 items)

If the respondent had done any of the following activities: Signed a petition, contacted a local Member of Parliament, written to the Council, contacted a local councilor, written a letter to the editor of a newspaper, attended a Council meeting, attended a protest meeting.

Civic participation—collective activities (4 items)

If the respondent had been involved in any of the following activities: Resident or community action group; political party, trade union, or political campaign; campaign or action to improve social or environmental conditions; local government.

Community group participation—mix of social and civic (4 items)

If the respondent had been involved in any of the following activities: Volunteer organization or group; school-related group; service club; been involved with a children's group.

measurement of social capital is not uniform, as academics search for different forms of measurement. Approval was granted by the South Australian Community Health Research Unit to draw on a survey instrument developed to measure social capital and health in Adelaide. This instrument was chosen for a number of methodological reasons. The validity of the instrument had already been demonstrated (Baum et al., 2000). This instrument was developed using a combination of preexisting measures and some measures specifically developed for the Adelaide study. The survey instrument was sufficiently sensitive to gender issues, particularly its focus on caring and the amount of hours devoted to caring for children. The researcher particularly favored the distinction made between those activities that were of a social nature and those that were conducted on behalf of the civic or community good.

Respondents were asked how often they had been involved in different activities. As in the Adelaide study, this survey measured levels of participation by the number of activities in which individuals were involved. A key distinction is made here between social, community, and civic participation. Social participation embodies those activities performed in a company such as visiting friends, going to the cinema, and going to a party. Civic activities are performed for a different reason, usually to promote the civic or community good. Community participation embodies those activities that have a mixture of social and civic activities, such as being involved in a service club and involvement in child-related activities. Items that are contained in each participation category are outlined in Table E.1. To ensure reliability, these scales of participation were drawn from the Adelaide study (Baum et al., 2000). These scales were developed "through a process of discussion between the research team that was informed by their knowledge of the literature of participation and the measurement of social capital" (Baum et al., 2000, p. 417).

Data were analyzed using the computer software package Statistical Package for Social Scientists (SPSS). A series of measures was used to provide scope for a thorough testing of the hypotheses that men and women would have different social capital profiles. Initially, frequencies were gathered on each variable contained within levels of social, community, and civic participation (six types), and the data file was divided into male and female.

All items contained within the six participation types were then totaled to compute a total score for each respondent for the following six items: informal social participation, social participation in public places, social participation in groups, individual civic participation, collective civic participation, and community group participation. Using the participation types as dependent variables and gender as an independent variable, a one-way between-groups multivariate analysis of variance was performed. In addition, multivariate analysis of variance (MANOVA) was also conducted to compare the mean differences between men and women on a combined dependent variable of participation.

Qualitative data. The second stage of the study was concerned with understanding participation from the participant's perspective. In this, the researcher wanted to develop an understanding of what motivates women's involvement in social, civic, and community life and the social realities of their experiences. These motivations and experiences are crucial to our understanding of social capital.

In-depth one-to-one interviews were conducted with 12 women who had already been involved in Stage 1 of the research. It was decided that participants would be interviewed on two separate occasions, 1 week apart. Each participant was interviewed for approximately 2 hours in total. Each interview was tape-recorded with the permission of the participant. The interview guide consisted of open-ended questions that explored women's daily lives. In Stage 1, the quantitative data had revealed a difference between men and women in social and community group participation. The intent here was to develop an understanding of why this might be so. It was also the intent to explore fully the range and types of participation they were involved in. For instance, in the first interview, participants were asked to describe a week in their lives and how they viewed their lives as being different from their partner's. The participants were interviewed again 1 week later. A diary the participants had been asked to keep for the week was used as the basis for the second interview. Several of the participants used the week to talk to other women about the research project. Not all kept a formal diary; however, all of the participants produced some written reflections about their week. The diary or the written reflections gave a focus to the first part of the second interview.

Again, participants were asked open-ended questions. They were initially asked to reflect on the first interview. Participants were asked to talk about their week and to describe how they felt about their involvement in certain activities. They were asked to think about their caring responsibilities and how these may have affected their life experiences, goals, or ambitions. In the final question, participants were asked whether there was anything in their life that they would like to do differently. In all, there were very few questions as the researcher wanted participants to provide the direction for the interview by telling their story.

The data were analyzed using a model of narrative analysis (Ezzy, 2002; Sands, 2004). The researcher's intent was to analyze the data in successive stages, looking for plot, characters, metaphors, interpretations, and cultural norms; how the stories compared and contrasted; and how the researcher was viewed by the participant. This required careful reading of the transcripts at each stage of the analysis. It was anticipated that this type of multistage layering of systematic analysis would add rigor to the study (Stevens & Doerr, 1997).

● RESULTS

In a mixed methods study, reporting on findings is complex because of the vast amount of data collected (Gioia, 2004). The findings presented here were selected to demonstrate how a mixed methods study can provide both statistical data and narrative data to increase understanding. Thus, only some of the findings from Stage 1 and Stage 2 will be presented. First, some examples of the quantitative data will be presented. Examples of the qualitative data will then be presented by means of three particular narrative themes.

Quantitative Results

The first stage of the research was driven by the following research question: *Do men and women have different social capital profiles?*

To answer this question, a key distinction was made between social, community, and civic participation (see Table E.1). This decision to differentiate between the different types of participation reflects more accurately the different social worlds that men and women occupy.

Levels of participation by gender are displayed in Table E.2. In general, higher levels of social participation were recorded for women with the exception of group, hobby, and sporting activities. Low levels of civic participation were recorded apart from signing petitions. Generally, higher levels of participation were recorded for community group participation, especially those involving children and school-related activities.

Informal social participation. Each respondent was given a score out of 18 on the first of the new variables, informal social participation. This category contained three items: whether the respondent had visited family or had family visit, visited friends or had friends visit, and had visited neighbors or had neighbors visit.

Social activities in public spaces. This participation category includes social activities that occurred outside the home and consists of the following items: whether the participant had been to a café/restaurant, social club, cinema/theater, and party/dance. Each respondent was given a score out of 24 on social participation in public places.

Social participation in group hobby or sporting activities. This participation category includes social activities in group, hobby, or sport and contains the following items: whether the respondent had played sport, had been to the gym or exercise class, had been involved in a hobby group, had been involved in a self-help group, had been singing/acting or had been musician in a group, and had gone to a class. The six items were grouped together, and respondents were given a total mean score out of 24 on social participation in group hobby or sporting activities.

Civic participation—Individual items. This participation category includes civic activities conducted on an individual basis such as signing a petition, contacting the local Member of Parliament, and writing to council, writing a letter to the editor of a newspaper, attending a council meeting, contacting a local councilor, and attending a protest meeting. These seven items were grouped, and respondents were given a total score out of 7.

Civic participation—Collective activities. This form of participation comprises those civic

	Men	Women
Social participation informal[a]		
Visited family/family visit	69.02	75.02
Visited friends/friends visit	59.06	61.0
Visited neighbors/neighbors visit	53.4	49.6
Social activities in public space[a]		
Went to a café/restaurant	40.09	44.00
Went to a social club	23.00	17.09
Went to cinema/theater	18.59	20.01
Went to a party/dance	8.03	9.08
Group hobby/sporting activities[a]		
Played sport	9.00	6.04
Went to a gym/exercise class	4.00	3.04
Went to class	3.05	7.06
Involved in a hobby group	5.03	6.05
Went to self-help/support group	2.00	4.07
Singing/acting/music group	2.03	1.05
Individual civic participation[b]		
Signed a petition	56.10	61.50
Contacted local member of parliament	14.02	9.08
Written to the council	15.07	13.00
Contacted local councilor	10.02	6.05
Attended a protest meeting	6.03	4.08
Attended a council meeting	7.03	4.01
Written letter to editor of newspaper	6.01	5.04
Collective civic participation[b]		
Resident/community action group	8.05	9.02
Campaign/action to improve social or environmental conditions	8.07	7.08
Political party/trade union/political campaign	4.08	2.04
Local government	5.08	3.09
Community group participation[b]		
Volunteer group or organization	30.01	32.02
School-related group	12.05	23.03
Children's group	7.00	18.02
Service club	13.05	6.06

Table E.2 Levels of Participation Reported by Respondents in Social, Community, and Civic Activities, by Gender (in percentages)

a. Did activity monthly or more often in the past year.

b. Did activity at all in the past year.

activities that are performed with other people such as belonging to a resident or community action group; belonging to a political party, trade union, or political campaign; joining a campaign or action to improve social or environmental conditions; and being involved in local government. Individual items were grouped, and respondents were given a score out of 4 on this item.

Community group participation. The types of activities in this category included involvement in a children's group, school-related group, service club, and volunteer group. Respondents were given a score out of 4 on this item.

A one-way between-groups multivariate analysis of variance was performed. The six participation scales were used as dependent variables. The independent variable was gender. Preliminary assumption testing was conducted to test for normality, linearity, univariate and multivariate outliers, homogeneity of variance-covariance matrices, and multicollinearity, with no serious violations noted. There was a statistically significant difference between men and women on the combined dependent variables, $F(6, 1372) = 6.16$, $p = .000$; Wilks's Lambda $= .97$; partial $\eta^2 = .03$. When the results for the dependent variables were considered separately, significant differences between men and women were found on three of the six participation scales; informal social participation, $F(1, 1378) = 10.63$, $p = .001$, partial $\eta^2 = .01$; social participation in group, hobby, or sporting activities, $F(1, 1378) = 2.81$, $p = .000$, partial $\eta^2 = .01$; community group participation, $F(1, 378) = 11.43$, $p = .001$, partial $\eta^2 = .01$. A Bonferroni-adjusted alpha level of .008 was used. An inspection of mean scores indicated that women reported higher levels of informal social participation, social participation in groups, and community group participation.

Summary of Quantitative Findings

A look at social, civic, and community participation provides evidence of difference between men and women, suggesting a prevalent gendered pattern of participation. These findings lend support to Lowndes's (2000) contention that men's and women's social capital is different, with women more involved in informal sociability. Lowndes (2004) has since argued that women are more likely to draw on informal sociability to help them "get by" by balancing the competing responsibilities of work, home, and children (Lowndes, 2004, p. 61).

This argument seems to be reflected in the findings presented here. The descriptive statistics point to some differences between men and women. On individual items that involved social participation, women reported higher levels than men. They were also involved more in activities that had a group focus or community group focus. Their level of participation was higher for those activities that were focused on children. In contrast, men reported higher participation in activities that were generally more formal in focus. For instance, they were more involved in traditional service clubs, social clubs, sporting clubs, and political parties/trade union groups. Men recorded slightly higher involvement in civic activities.

The quantitative results presented here show a difference between men and women on social and community participation and to a limited extent civic participation. Women scored higher on items contained in the informal social participation scale, with women more likely to make and receive visits with family and friends. Gender did not appear to affect social participation in public places. Women were more involved in social participation in group activities. This look at social participation suggests women predominate in the informal arenas built around family and friends. Turning to civic and community group participation, men reported slightly higher rates of civic participation. Women, however, participate significantly more in activities that have a community group focus. They participate more in school and children's groups. These findings may well reflect women's social worlds, constructed around family responsibilities.

What they are not able to do is expand on why this might be so. The researcher was attempting to understand why men and women had different social capital profiles. Although the quantitative data give a broad snapshot of women's and men's participation in social and community activities, they do not tell the reader what is the underlying motivation for such participation, the experience of this, and the feelings associated with giving up other types of participation. These motivations and experiences are crucial to our understanding of social capital. In the present study, the researcher wanted to delve into what lies behind motivations to participate and the experiences of these. This type of approach had been missing in the social capital literature.

Qualitative Findings

The following research question drove the second stage of the study: Why do women participate more in social and community activities than in civic activities?

The data presented here provide insight into the participants' understandings of participation. This is important, because it extends understandings beyond *what* women become involved in to *why* they become involved. What is striking in the findings is the different ways the women thought about how they participate and their motivations behind this. A pattern started to emerge in their responses.

The women interviewed were all in their thirties and forties, all mothers, and all struggling to find the time to pursue their own interests. When the researcher delved into some motivations behind their participation, it became apparent that the experience of motherhood, more than anything else, influenced these motivations. All the participants felt overburdened by the responsibilities associated with motherhood. They reflected on how they measured up as a "good mother." Here, some were influenced by the ideology of familialism and expressed guilt at not always being there for their children. They felt that they should participate in social and civic life for

the sake of their children. A second and overlapping group became involved as mothers to avoid social isolation. Another group rallied against participating as "mothers" and instead wanted to participate as active citizens. Their experiences in doing so were surprisingly similar; they all had experienced exclusion closely tied to their gender and their role of mother. Each theme will be discussed in turn.

Wanting to be a "good mother." The first group of women focused on the traditional volunteer activities associated with women and, more particularly, mothers: help in the school classroom, mother's club, fruit and milk preparation at preschool, assistance with school canteen, and help with school excursions. They had taken the concept of a "good mother," doing it all for their children, and had located themselves within this image. They often felt good about their involvement, which they saw as contributing to their children's development. Members of this group in this study are motivated by an overwhelming sense that this is what they are supposed to do. They are either actively involved or wish for more time to be further involved in their children's preschool and school life. Motivations behind this vary, from wanting knowledge of their children's development to the pleasure their involvement gives to their children. A strong desire to foster their child's human capital was evident. They were not striving for individual recognition; rather, there was a strong sense that this is what a "good mother" does. An example of this follows:

> But I think being a working mother is definitely hard work. You've got to really juggle your commitments at work and then your responsibilities to your family. That sometimes makes you feel like you're a bad mother because you can't get there all the time. Trevor's job is more flexible so he drops them off and picks them up. Instead of Mum going to do those things, this year it's been swapped around. He's going to any parent/teacher interviews, anything that's in

the hours that I can't get to. That's hard, because sometimes you feel like you should be there, and you can't be there . . . And that's hard because I know, being a working mother, you feel that you're not fulfilling your whole motherly role, if you can't be there.

This focus on being a good mother did not automatically mean that all the participants enjoyed their participation in these child-focused activities. For some it represented hard work, different from what they were used to and often quite tedious. They stuck with it because of their strong sense of doing the right thing by their children. One participant reflected,

So I'll do fruit duty then. Because it's from 9:00 a.m.-11:00/11:30 a.m., I stay the whole session then, until 2:00 p.m. It's such a long session, I don't know how preschool teachers do it; I get such a headache. The kids really like it and Kerry loves showing me, so I stay the whole session.

The above quotes provide some insight into one group's motivations behind participating as mothers in the community. Their construction of this is that they tolerate their participation for the sake of their children. The metaphor "good mother" aptly describes what lies behind such motivations. Their gender and role as mothers strongly influence their degree of social responsibility.

Wanting to avoid social isolation. Although there was one group who demonstrated strong bonding types of social capital, involved with their family and friends, there was another group who were new to the regional area. This group comprised seven nonlocal women, each striving for ways to establish roots and connections in the community, with varying success. They each looked to the community in an attempt to avoid social isolation. Motivations were not always altruistic in nature; sometimes, altruistic motivations combined with motivations of self-preservation.

One participant's story is a powerful example of this. She describes herself as a "fringe dweller"—someone who has tried knocking on several doors, but has ultimately felt excluded from the larger margins of society. She reflected on this:

Int: Do you often feel isolated or lonely?

E: Yes, very much so. I heard something this morning on the radio, on the ABC, and they were talking about how people who are part of community groups and social groups actually live longer than the people who are isolated, and I thought, I can really feel that in the pit of my stomach, that isolation, how it's just not healthy, not good.

Another participant, also new to the area, has had to work hard at establishing social connections. She used emotive language to describe her 1st year in the area. *Horrible* and *ugly* are the words she chose. When she arrived in the community, she immersed herself in children's activities as a way to get out of the house, with varying success. She has reinvented herself as the "mothers' group junkie," desperate to belong to a group and to feel included again.

I'm not by nature an extrovert, but I just hated being stuck at home. I'm not really into babies, I realized. I thought it might be different with my own, but it wasn't. I became a real mothers' group junkie and because I like structure, I'd have Monday, playgroup, then Tuesday I'd go to the library, there was a little craft session. I joined Nursing Mothers because I was really into breast-feeding at the time. I did kids' gym and swimming lessons. Every day we had an activity, and I got to meet people like that. Some groups were better than others. The kids were doing gym, and the group of women there I didn't particularly get on with, so they just stopped doing gym. And I thought, am I doing this for the kids, or am I doing this for me? And it was really for both of us.

There were several stories of social exclusion and social isolation coming through the data.

Several of the participants had tried to participate through their children, and not all had had positive experiences. Although the quantitative data showed gender polarization in participation based on children, the qualitative data help explain why this may be so. The participants often had two different motivations for participation; trying to be a good mother and trying to avoid loneliness and social exclusion.

Wanting to be a good citizen. The third group is primarily focused on being a good civic citizen, and the involvement of members of this group was more civic in nature. They had extended their involvement to political parties, committees of management, and collective civic action. Although they did not reject the stereotype of the "good mother," their sense of community extended beyond their children. The following quotes provide examples of this:

The question was, What motivates me? I think it's a real sense of obligation and responsibility. It's a really good thing for the community if people become more involved. I'm just astounded when people say no. I think, you can do that?

I feel really strongly about our responsibility to make sure that, you know, we [don't] take, take, take all the time. I use the leisure center and use facilities and services that are provided by state, local, and federal governments, so I feel compelled to somehow put back in. Do you know what I mean? I feel like I need to put back in. The conscience is there. I think women also go through the guilts, and we do have a desire to please and to put back in. I think that's a big part of our makeup, if you can generalize.

This group's participation was structured by altruistic motivations. They each shared a developed sense of civic conscience and wished to be involved in more public activities, often the activities that social capital theorists advocate. They

rejected gendered roles associated with volunteering, in favor of more formal roles usually associated with men.

Despite this, they had each experienced the pain of exclusion and believe strongly that this is related to their gender and their role as mother. The following quote, from a participant who had attempted to become involved in a political party, provides an example of this:

I can remember when we moved here, a good friend of mine was quite involved in a political party, and I'm not really political but I thought this would be a good way of meeting people. I thought, I won't become a member; I'll go to a few meetings. Again, it was oh f . . . . The party up north were quite left-wing, whereas here they're much more right-wing and union orientated. There were a lot of men at this meeting. I turn up, and I was wearing one of my hippy tops with Kayla, breast-feeding baby, and the men, honest to God, as soon as I started feeding, they all just stopped. And I thought, have I done something? I felt very uncomfortable, and I never went back.

Another participant described taking on the responsibility for organizing a political campaign. In becoming the manager, she encountered a great deal of opposition coming from women themselves. She reflected on this experience and concluded she had overstepped the boundaries. She was given the message that she had deviated from the script and should leave "the political stuff" to men and get on with the hands-on work. She has since found that women do most of the fund-raising activities for the party, particularly the catering. She describes an experience when the Deputy Leader came up for a major function. The women did all the work before and after the dinner:

Oh, the money-raising area, that's fascinating to see what happens. That is women-dominated work to the extent that we had a function, and we all worked our butt off, but it was predominantly the women, then it was predominantly the women during the dinner, and it was predominantly the women after that dinner. And the men came in and out and grabbed a tea towel and dried some dishes, and that was all very nice, but it was fascinating to see. I thought, no, I haven't got the time or energy for this, so I backed off a bit. I got a very clear impression that there was women's work and men's work here.

There were several examples in the data where exclusion was experienced. The participants in this group were motivated to contribute at a civic level, but all have faced considerable opposition. They attributed this to traditional gender roles being more tightly scripted in rural areas. All had retreated, not knowing how to confront the hostility they encountered. Although their domestic skills were valued, their political skills were not. These stories help to explain why the levels of civic participation for women were so low in the quantitative data.

Summary of Qualitative Findings

The three themes presented in the narratives—wanting to be a "good mother," wanting to avoid social isolation, and wanting to be an active citizen—provide some understanding of how women construct meaning around their involvement in social, civic, and community life. A look at the literature more concerned with women and volunteer work finds evidence of similar themes (Petrzelka & Mannon, 2006). They summarize how women have framed their volunteerism as (a) an expression of their maternal instincts, (b) a means for providing socialization, and (c) a means of involvement in public life.

There were examples of gender strategizing (Hochschild, 1989), with the first group devoting themselves selflessly to the care of their children—as this is what "good mothers" do. For others, there was clear evidence of becoming involved as a

way to "get by" in two different ways: (a) to survive the experience of motherhood and its many competing demands and (b) to avoid loneliness. Self-preservation, as opposed to altruism, appears to drive participation. Those motivated by the spirit of altruism experienced exclusion directly related to their gender.

The qualitative findings provide a deeper story and help enhance the findings from Stage 1. These stories highlight a need to focus not just on what people do but also on why and what their subsequent experiences are. They help explain some of the quantitative findings and provide a more complete story of women's social capital. For instance, in the present study, women are involved more in bonding types of social capital, but this is very closely tied to their socially constructed roles as mothers. The stories also provide insight into why women's civic participation is limited. Issues such as time constraints, role constraints, and exclusion now surface and become part of the whole story of women's social capital.

● INTEGRATION OF FINDINGS

This study sought to capture women's social capital. It has brought together and interpreted both the quantitative data and the qualitative data collected in this mixed methods study. The quantitative data have assisted in providing the big picture, revealing a different pattern of participation for men and women. The qualitative data have assisted in developing and sharpening this picture, assisting to explain why this may be so.

This study has allowed for a much richer understanding of gender and social capital. Whereas the theoretical and empirical study of social capital has largely discounted gender, the findings of this mixed methods study underscore its importance. In the present study there was evidence of gendered patterns of participation. A pattern of difference between men and women was observed around levels of social; community group; and, to a lesser extent, civic participation. Overall, the quantitative findings provide evidence of women's predominance in informal sociability and to a lesser extent men's predominance in associational life.

The quantitative findings also highlighted significant differences between men and women in community participation centered on children. The researcher used the narratives to delve more into the motivations behind the range of participation. Here, the responses reflected more complex motivations. In several cases, the participants measured their own contributions according to socially constructed norms of behavior. Some measured themselves against the myth of the "good mother" embedded in the ideology of familialism. They are driven by duty and guilt and attempt to foster their own children's human capital, as this is what a "good mother" does. Hays (1996, p. 131) argues that all mothers share recognition of the ideology of "intensive mothering." Lareau (2003) argued the existence of a "new standard of child rearing in the middle class" (p. 248) supporting children in a range of creative and sporting activities.

Thurer (1994) claimed that this socially constructed myth of the "good mother" is synonymous with self-sacrifice. It is considered both normal and good to place children's needs above a mother's needs. This group volunteered as mothers and volunteered their time assisting in domestic service as their own mothers did. The problem here is that they also have competing demands on their time and do not always enjoy volunteering as a mother. These unpaid family and community support roles lack status and authority (Alessandrini, 2003).

Another story also came through, that of self-preservation. The participant who described herself as a "mothers' group junkie" was desperately joining various groups through her children to avoid social isolation. An assumption made in the social capital literature is that involvement in community life is driven by altruism. By delving into motivation, this present study has found a more complex set of motivation and experiences.

Women's limited access to, and involvement in, political activity has been extensively studied elsewhere (Burns, Lehman Schlozman, & Verba, 2001). Overall, in the present study, there was limited support for gender difference in civic participation. The descriptive statistics highlighted that men scored higher on individual items such as involvement in politics, trade unions, service clubs, and social clubs. Previous studies report greater gender polarization in participation (Lowndes, 2000; Onyx & Leonard, 2000). Similarly, an Australian Bureau of Statistics (2001) study found polarization in men's and women's volunteering, with men dominant in management, coaching, and maintenance and women dominant in fund-raising, preparation and serving of food, and supportive listening and counseling.

On their own, the quantitative findings do not tell the entire story. They highlight women's increased role in informal social participation, social participation in groups, and community participation. However, the quantitative findings do not explain why their civic levels of participation were so low. Here the benefit of the participants discussing their daily lives was immense. Conflict and exclusion were keenly experienced. This was closely related to gender and gender expectations. They believed they were expected to volunteer as mothers. This kind of volunteering is reminiscent of the unpaid domestic work they do at home. Some enjoyed this, whereas the majority believed their skills were wasted. When they have tried to participate at a civic level, they have felt excluded, as if in attempting to enter the civic world they have attempted to enter some exclusive men's club. There was evidence here of the existence of Connell's (2002) gender order. Here, again, the qualitative data provided the story behind the statistics.

Despite a considerable body of literature devoted to social constructions of gender roles, there is little discussion in the social capital literature on the effect of gender. The power of the mixed methods research approach has been to build a comprehensive picture that challenges this lack of attention in the social capital literature.

CONCLUSION

Feminist research seeks to illuminate women's experiences through the eyes of women. Qualitative research has been favored by feminist researchers to explore these subjective experiences. This article concludes by arguing that, despite this, some feminist research questions may best be answered using a combination of data collection methods. Those seeking to influence the policy and practice agenda around women's issues might consider the types of data that are most highly regarded by the audience they are seeking to persuade. This can be important in convincing nonfeminist decision makers. A feminist mixed methods approach might provide the best and most convincing avenue to answer complex social issues. It can provide a more powerful and vivid story, illuminating issues of gender difference on a broad scale and providing the personal story to accompany this.

To date, the theoretical and empirical work of social capital has shown limited sensitivity to gender. The present study sought to redress this, using a mixed methods design. Despite previous reluctance to use quantitative methods, a mixed methods approach can be used to give a more powerful voice to gender inequality. In the present study, the two sets of data together capture the statistics and the story of women's social capital. The quantitative data have provided detail of genderized patterns of participation, thus providing the big picture of gender inequality. The qualitative data have provided the personal story, accompanied by thoughts and feelings that have brought depth and texture to the research study. The mixed methods approach to data collection has provided a convincing research basis to argue a particular gender order to women's participation.

REFERENCES

Alessandrini, M. (2003, September). *The double shift and policy implementation: A gendered analysis of the supply end of social capital.* Paper presented at the Australasian Political Studies Association Conference, Hobart, Tasmania.

Australian Bureau of Statistics. (2001). Voluntary work Australia report (Catalogue No. 4441.0). Canberra: Australian Bureau of Statistics.

Baum, F. E., Bush, R. A., Modra, C. C., Murray, C. J., Alexander, K., & Potter, R. (2000). Epidemiology of participation: An Australian community study. *Journal of Epidemiology and Community Health, 54*(6), 414.

Bezanson, K. (2006). Gender and the limits of social capital. *Canadian Review of Sociology & Anthropology, 43*(4), 427–443.

Bourdieu, P. (1986). The forms of capital. In D. Richardson (Ed.), *Handbook of theory and research for the sociology of education* (pp. 241–258). New York: Greenwood.

Brannen, J. (1992). *Mixing methods: Qualitative and quantitative research.* Aldershot, UK and Brookfield, VT: Avebury.

Brewer, J., & Hunter, A. (2006). *Foundations of multimethod research: Synthesizing styles.* Thousand Oaks, CA: Sage.

Bryman, A. (1988). *Quantity and quality in social research.* London and Boston: Unwin Hyman.

Bryson, L., & Mowbray, M. (2005). More spray on solution: Community, social capital and evidence based policy. *Australian Journal of Social Issues, 40*(1), 91.

Burns, N., Lehman Schlozman, K., & Verba, S. (2001). *The private roots of public action: Gender, equality, and political participation.* Cambridge: Harvard University Press.

Connell, R. (2002). *Gender.* Malden, MA and Cambridge, UK: Blackwell.

Cox, E. (1996). Beyond economics: The building of social capital. *Refractory Girl, 50,* 52–53.

Cox, E. (1998). Measuring social capital as part of progress and well-being. In R. Eckersley (Ed.),

Measuring progress: Is life getting better? (pp. 157–167). Collingwood, Australia: CSIRO.

Creswell, J. W. (2003). *Research design: Qualitative, quantitative, and mixed methods approaches* (2nd ed.). Thousand Oaks, CA: Sage.

Creswell, J. W., & Plano Clark, V. L. (2007). *Designing and conducting mixed methods research.* Thousand Oaks, CA: Sage.

Darlington, Y., & Scott, D. (2002). *Qualitative research in practice: Stories from the field.* St Leonards, Australia: Allen & Unwin.

Epstein, T., Jayaratne, S., & Stewart, A. (1991). Quantitative and qualitative methods in the social sciences. In M. Fonow & J. Cook (Eds.), *Beyond methodology, feminist scholarship as lived research* (pp. 85–106). Bloomington: Indiana University Press.

Ezzy, D. (2002). *Qualitative analysis: Practice and innovation.* Crows Nest, Australia: Allen & Unwin.

Finch, J. (2004). Feminism and qualitative research. *International Journal of Social Research Methodology, 7*(1), 61–64.

Gioia, D. (2004). Mixed methods in a dissertation study. In D. Padgett (Ed.), *The qualitative research experience* (pp. 119–140). Belmont, CA: Wadsworth.

Greene, J., & Caracelli, V. (Eds.). (1997). *Advances in mixed-method evaluation: The challenges and benefits of integrating diverse paradigms.* San Francisco: Jossey-Bass.

Greene, J., Caracelli, V., & Graham, W. (1989). Toward a conceptual framework for mixed-method evaluation designs. *Educational Evaluation and Policy Analysis, 11*(3), 255–274.

Hays, S. (1996). *The cultural contradictions of motherhood.* New Haven, CT: Yale University Press.

Healy, K., & Hampshire, A. (2002). Social capital: A useful concept for social work? *Australian Social Work, 55*(3), 227–238.

Hochschild, A., with Machung, A. (1989). *The second shift.* New York: Viking.

Hughes, J., & Stone, W. (2003, Winter). Family and community life: Exploring the decline thesis. *Family Matters, 65,* 40–47.

Johnson, R. B., Onwuegbuzie, A. J., & Turner, L. A. (2007). Toward a definition of mixed methods research. *Journal of Mixed Methods Research, 1*(2), 112–133.

Kohler Riessman, C. (1994). *Qualitative studies in social work research.* Thousand Oaks, CA: Sage.

Lareau, A. (2003). *Unequal childhoods.* Berkeley, CA: University of California Press.

Lowndes, V. (2000). Women and social capital: A comment on Hall's "Social capital in Britain." *British Journal of Political Science, 30*(3), 533.

Lowndes, V. (2004). Getting on or getting by? Women, social capital and political participation. *British Journal of Politics and International Relations, 6,* 45–64.

McMichael, C., & Manderson, L. (2004). Somali women and well-being: Social networks and social capital among immigrant women in Australia. *Human Organization, 63*(1), 88–99.

Mertens, D. (2007). Transformative paradigm: Mixed methods and social justice. *Journal of Mixed Methods Research, 1*(3), 212–225.

Morgan, D. L. (2007). Paradigms lost and pragmatism regained: Methodological implications of combining qualitative and quantitative methods. *Journal of Mixed Methods Research, 1*(1), 48–76.

Norris, P., & Inglehart, R. (2003, May). *Gendering social capital: Bowling in women's leagues?* Paper presented at the Gender and Social Capital Conference, Manitoba, Winnipeg, Canada.

Oakley, A. (1999). People's way of knowing: Gender and methodology. In S. Hood, B. Mayall, & S. Oliver (Eds.), *Critical issues in social research: Power and prejudice* (pp. 154–170). Philadelphia: Open University Press.

Onyx, J. (2001). Third sector as a voice: The importance of social capital. *Third Sector Review, 7*(2), 73–88.

Onyx, J., & Bullen, P. (2000). Sources of social capital. In I. Winter (Ed.), *Social capital and public policy in Australia* (pp. 105–135). Melbourne: Australian Institute of Family Studies.

Onyx, J., & Leonard, R. (2000). Women, volunteering and social capital. In J. Warburton & M. Oppenheimer (Eds.), *Volunteers and volunteering* (pp. 113–124). Leichhardt, Australia: Federation Press.

Padgett, D. (1998). *Qualitative methods in social work research: Challenges and rewards.* Thousand Oaks, CA: Sage.

Parks-Yancy, R., DiTomaso, N., & Post, C. (2006). The social capital resources of gender and class groups. *Sociological Spectrum, 26*(1), 85–113.

Petrzelka, P., & Mannon, S. E. (2006). Keepin' this little town going: Gender and volunteerism in rural America. *Gender & Society, 20*(2), 236.

Putnam, R. D. (2000). *Bowling alone: The collapse and revival of American community.* New York: Simon & Schuster.

Rose, H. (1982). Making science feminist. In E. Whitelegg (Ed.), *The changing experience of women* (pp. 352–372). Oxford, UK: Martin Robinson in association with the Open University.

Sands, R. (2004). Narrative analysis: A feminist approach. In D. Padgett (Ed.), *The qualitative research experience* (pp. 48–75). Belmont, CA: Wadsworth.

Sapiro, V. (2003, May). *Gender, social capital, and politics.* Paper presented at the Gender and Social Capital Conference, Manitoba, Winnipeg, Canada.

Saunders, P., & Winter, I. (1999). Social capital and social policy conference. *Family Matters, 52,* 49–50.

Shapiro, M., Setterlund, D., & Cragg, C. (2003). Capturing the complexity of women's experiences: A mixed-method approach to studying incontinence in older women. *Affilia, 18*(1), 21–33.

Stevens, P. E., & Doerr, B. T. (1997). Trauma of discovery—Women's narratives of being informed they are HIV-infected. *AIDS Care, 9*(5), 523–538.

Stewart-Weeks, M., & Richardson, C. (1998). *Social capital stories: How 12 Australian households live their lives.* St Leonards, Australia: Centre for Independent Studies.

Stone, W., & Hughes, J. (2000). What role for social capital in family policy? *Family Matters, 56,* 20–27.

Stone, W., & Hughes, J. (2002). *Social capital: Empirical meaning and measurement validity* (No. 27). Melbourne: Australian Institute of Family Studies.

Tashakkori, A., & Teddlie, C. (1998). *Mixed methodology: Combining qualitative and quantitative approaches.* Thousand Oaks, CA: Sage.

Tashakkori, A., & Teddlie, C. (2003). *Handbook of mixed methods in social & behavioral research.* Thousand Oaks, CA: Sage.

Teddlie, C., & Yu, F. (2007). Mixed methods sampling: A typology with examples. *Journal of Mixed Methods Research, 1*(1), 77–100.

Thurer, S. (1994). *The myths of motherhood: How culture reinvents the good mother.* Boston: Houghton Mifflin.

Appendix F—An Example of the Multiphase Design

Mixed Methods in Intervention Research

Theory to Adaptation

Bonnie K. Nastasi

Walden University, Minneapolis, Minnesota

John Hitchcock

Caliber, an ICF International Company, Fairfax, VA
Walden University, Minneapolis, Minnesota

Sreeroopa Sarkar
Gary Burkholder

Walden University, Minneapolis, Minnesota

Kristen Varjas

Georgia State University, Atlanta

Asoka Jayasena

Peradeniya University, Sri Lanka

AUTHORS' NOTE: The initial phases of this work were funded by grants to the first author from the Society for the Study of School Psychology and the State University of New York at Albany. An earlier version of this article was presented at the annual meeting of the American Educational Research Association, San Francisco, April 2006. Address correspondence to Bonnie K. Nastasi, PhD, Director, School Psychology Program, Walden University, 155 Fifth Avenue, Suite 100, Minneapolis, MN 55401; e-mail: bnastasi@waldenu.edu or bonnastasi@yahoo.com.

SOURCE: Nastasi, B. K., Hitchcock, J., Sarkar, S., Burkholder, G., Varjas, K., & Jayasena, A. (2007). Mixed methods in intervention research: Theory to adaptation. *Journal of Mixed Methods Research, 1*(2), 164-182. Reprinted with permission of SAGE Publications, Inc.

The purpose of this article is to demonstrate the application of mixed methods research designs to multiyear programmatic research and development projects whose goals include integration of cultural specificity when generating or translating evidence-based practices. The authors propose a set of five mixed methods designs related to different phases of program development research: (a) formative research, Qual →/+ Quan; (b) theory development or modification and testing, Qual → Quan →/+ Qual → Quan . . . Qual → Quan; (c) instrument development and validation, Qual → Quan; (d) program development and evaluation, Qual →/+ Quan →/+ Qual →/+ Quan . . . Qual →/+ Quan, or Qual →← Quan; and (e) evaluation research, Qual + Quan. We illustrate the application of these designs to creating and validating ethnographically informed psychological assessment measures and developing and evaluating culturally specific intervention programs within a multiyear research program conducted in the country of Sri Lanka.

KEYWORDS: *mixed methods; intervention research; evaluation research; culture specificity*

Given the current emphasis on both evidence-based practice and culturally competent practice, it is critical for researchers and interventionists to identify models for developing culturally appropriate evidence-based practice (e.g., Ingraham & Oka, 2006; Nastasi & Schensul, 2005). Mixed methods designs applicable to intervention research can take a number of forms depending on the specific purpose or stage of the project (for an in-depth discussion of mixed methods designs, see Tashakkori & Teddlie, 2003). Most mixed methods discussions (e.g., Creswell, 2003; Tashakkori & Teddlie, 2003) do not cover multiphase evaluation projects in detail, nor do they address the potential role of mixed methods designs for developing culturally appropriate practices in applied fields such as education and psychology. Morse (2003) discussed the application of mixed methods designs across individual studies within a program of research but did not present an integrative multiphase model for conducting programmatic research. Furthermore, although qualitative research designs (e.g., ethnography) are well suited for understanding culture and context, the integration of qualitative and quantitative methods to facilitate development of culture-specific instruments (e.g., psychological assessment tools) and interventions has received minimal attention (see Hitchcock et al., 2005).

We propose that the process of program development research is best characterized by a recurring sequence of qualitative and quantitative data collection culminating in a recursive qualitative-quantitative process depicted as Qual → Quan → Qual → Quan . . . (Qual→←Quan). Qualitative methods (Qual) are used to generate formative data to guide program development, followed by quantitative evaluation (Quan) to test program effectiveness. Application in another setting can be facilitated by subsequent qualitative data collection (Qual) leading to program design adapted to the new context and participants, which is then followed by quantitative data collection (Quan) to test program outcomes. This sequence can occur across multiple settings and participant groups. Following initial adaptations to local context, program implementation and evaluation can be characterized by a recursive process (Qual→←Quan) in which collection of both qualitative and quantitative data inform ongoing modifications as well as implications for future program development and application.

The purpose of this article is to demonstrate the application of mixed methods research designs to multiyear programmatic research and development projects, whose goals include the integration of cultural specificity into development of an evidence base for practice. In particular, we illustrate the application of mixed methods designs to the development and validation of ethnographically informed psychological assessment measures, and the development and evaluation of culturally specific intervention programs.

A HEURISTIC MODEL: THEORY TO ADAPTATION

We propose a general heuristic for depicting multiyear research and development projects as an iterative research↔intervention process (see Figure F.1), based on the Participatory Culture-Specific Intervention Model (PCSIM; Nastasi, Moore, & Varjas, 2004). The research process begins with formative data collection to test the proposed conceptual model based on existing theory and research. At this stage, qualitative research methods are used to identify and define the constructs/variables specific to a particular culture or context (e.g., individual and environmental factors that explain/predict mental health, violent behavior, or academic achievement in a specific cultural group). Findings from the qualitative research are used to construct a modified model and develop assessment and intervention tools to test the model. Quantitative research methods are then used to test the model, for example, using instrument validation techniques and/or experimental or quasi-experimental designs. Evaluation research involves the triangulation of qualitative and quantitative methods to examine acceptability, integrity, and effectiveness of intervention methods as both a formative and summative process. The application of research as an ongoing formative evaluation process can assist in systematic modification of the intervention model and program design to meet context-specific needs (e.g., application of intervention to particular school or community). Summative research provides evidence of program effectiveness and informs application and translation to other settings. As interventions are applied to multiple populations and settings, the iterative use of mixed methods can help to inform adaptations and development of a general intervention model.

APPLYING MIXED METHODS DESIGNS TO MULTIYEAR RESEARCH AND DEVELOPMENT PROJECTS: AN ILLUSTRATION

As depicted in Figure F.1, the multiple purposes for research within any given multiyear project (e.g., formative research, instrument development, evaluation research) necessitate the use of mixed methods designs. Drawing on the general model (Figure F.1), we propose a set of five designs applicable across various phases of the theory→adaptation process (see Table F.1). The remainder of this article is focused on description and illustration of these five designs, based on our own intervention research experiences across an ongoing multiyear project, the Sri Lanka Mental Health Promotion Project (SLMHPP). (Although Figure F.1 provides the heuristic for depicting the theory to adaptation process of program development, the remainder of this article is focused on representing the five designs depicted in Table F.1. (For other examples of the application of mixed methods to multiyear research and development projects, see Nastasi et al., 1998-1999; Nastasi, Schensul, Balkcom, & Cintrón-Moscoso, 2004; Schensul, Mekki-Berrada, Nastasi, & Saggurti, in press; Schensul, Nastasi, & Verma, 2006; Schensul, Verma, & Nastasi, 2004.)

In the SLMHPP, conducted in the Central Province of Sri Lanka, we applied various mixed methods designs to (a) conduct formative research, (b) develop and test culture-specific theory, (c) develop and validate culture-specific instruments, and (d) develop and evaluate a culture-specific intervention program. Attempts to further test and modify culture-specific theory and mental health programming in India and other Sri Lankan contexts are ongoing. Although we attempt to represent the use of mixed methods for specific purposes or phases in the theory→adaptation

Figure F.1 Mixed Methods in Intervention Research Process: Theory to Adaptation

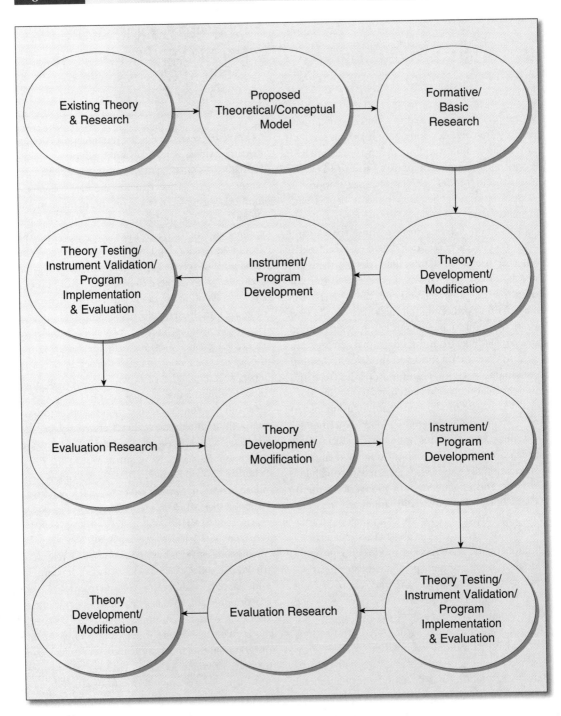

process, the distinctions across phases are artificial (as reflected in Figure F.1). Thus, for example, formative research and theory development phases overlap as do theory testing and instrument development. Furthermore, the phases are not always sequential but may occur concurrently or recursively. (As noted throughout, some of the findings from various phases of the project have been published or presented elsewhere. This article, however, reflects an integration of the work within a multiphase mixed methods framework.)

Formative/Basic Research Phase: Qual →/+ Quan

The application of mixed methods to the formative phase of intervention research is characterized by sequential or concurrent collection of qualitative and quantitative data (see Table F.1). In SLMHPP, we used a sequential process in which initial qualitative data collection informed theory development and design of psychological measures. These measures were then used to collect quantitative data on a larger and more representative sample and, thus, extend and confirm formative research findings.

As an outgrowth of a project focused on sexual risk among Sri Lankan youth, researchers from the United States developed knowledge of the Sri Lankan youth and educational cultures, identified the need for mental health services, and formed partnerships with professionals and community members. A formative research study was conducted in Sri Lanka in 1995 to examine individual and cultural constructs related to mental health of the school-aged population in the country and to assess the need for mental health services in the schools (Nastasi, Varjas, Sarkar, & Jayasena, 1998).

Underlying the work was a conceptual model of mental health based in ecological-developmental theory (Bronfenbrenner, 1989). A major assumption of the model is that critical individual and cultural factors influence mental health. That is, mental health status of an individual is influenced by (a) *personal vulnerabilities* due to personal and

family history (e.g., early school failure, family alcoholism), (b) *social-cultural stressors* (e.g., community violence), (c) the extent to which the individual possesses *culturally valued competencies* (e.g., academic competence, social skills), (d) culture-specific *socialization practices* (e.g., school discipline practices) and *cultural agents* (e.g., family, teacher, media) responsible for promoting the development of competencies, (e) *personal resources* (e.g., problem-solving skills) for coping with daily stresses and major life changes, and (f) *social-cultural resources* available to youth (e.g., peers, family, mental health facilities) to facilitate coping. This conceptual framework has been applied to the development of mental health programs in schools within the United States (Cowen et al., 1996; Nastasi et al., 1998; Nastasi, Moore, & Varjas, 2004; Roberts, 1996).

Formative research data, collected in 18 schools in the Central Province of Sri Lanka, were used to develop an understanding of the individual and cultural factors (described above) that influenced mental health of youth in Sri Lanka. Qualitative data collection methods included 51 focus group interviews with students (33) and teachers (18), individual interviews with school principals and teachers, participant observation in schools, archival materials such as school discipline reports, historical and cultural literature, popular mental health literature, and popular media. In addition, secondary analysis of qualitative (in-depth interviews) and quantitative (ethnographically informed psychological measures) data from the previous sexual risk project (Nastasi et al., 1998-1999) focused on older adolescents and young adults from the same community.

Findings

The primary qualitative data provided culture-specific definitions of the major mental health constructs (e.g., stressors, competencies) and the basis for elaboration of the proposed conceptual framework (i.e., identification and definition of

Table F.1 Mixed Methods Designs Applied to Multiyear Research and Development Projects

Project Phase	Design	Types of Data Collected in SLMHPP
Formative/basic research	Qual →/+ Quan	Focus group interviews Individual in-depth interviews Key-informant interviews Participant observation Archival materials (e.g., school records) Cultural and historical literature Popular mental health literature and popular media Secondary data analysis (qualitative and quantitative data from previous project on sexual risk among older adolescents and young adults from same community)
Theory development or modification and testing	Qual → Quan →/+ Qual → Quan . . . Qual → Quan	Development of culture-specific theory and quantitative psychological measures (self- and teacher report) based on formative research data
Instrument development and validation	Qual → Quan	Administration of psychological measures to 600 students and 100 teachers Instrument validation and theory testing through combined factor analysis of quantitative (psychological measure) data and reanalysis of qualitative formative data Further theory development through parallel formative research in India (qualitative interviews)

Program development and evaluation	(a) Qual →/+ Quan →/+ Qual →/+ Quan . . . Qual →/+ Quan; or (b) Qual→←Quan	Program development based on formative research data
		Formative program evaluation (program monitoring):
		    Participant observations
		    Teacher interviews/meetings
		    Session logs (teachers and observers)
		    Teacher session evaluations
		    Student session evaluations
		    Staff field notes
		    Student products (from session activities)

Evaluation research	Qual + Quan	Experimental pre-post control group design (summative program evaluation):
		    Pre-post student and teacher psychological measures
		    Postintervention teacher interviews
		    Final session student evaluation activity
		    Reanalysis of formative evaluation data

NOTE: Qual = Qualitative methods; Quan = Quantitative methods; → = followed by [sequential design]; + = concurrent with [concurrent design]; →/+ = sequential or concurrent; →← = recursive, interactive; SLMHPP = Sri Lanka Mental Health Promotion Project.

factors specific to Sri Lanka; Nastasi et al., 1998). Findings from this formative stage also suggested gender differences and similarities in definition of mental health as described by the adolescent students (Sarkar, 2003).

Competencies. Both male and female adolescent students argued that a socially competent individual is respectful to others, loyal, trustworthy, helpful, and caring. They also suggested that such a person advises or guides others, and is socially responsible (e.g., loves her or his country, works for its development). However, friendliness was viewed as an important quality only by the female students. (Unless otherwise indicated, the qualitative findings presented in this section are drawn from Nastasi et al., 1998, and Sarkar, 2003).

Student definitions of academic competencies were directly associated with high academic achievement and striving for maximum performance in academics. An academically competent person is one who is "good at studies as well as at extracurricular activities." In defining behavioral competence, students identified good behavior, humility, and obedience as the most important qualities of a behaviorally well-adjusted person. Students argued that such an individual follows rules, obeys laws of the land, and does not harm the country. Showing respect to the elders was another critical feature of behavioral competence as indicated by the students irrespective of their gender.

Adjustment difficulties. Students recognized several adjustment difficulties among Sri Lankan adolescents. For example, smoking, substance abuse, and suicide were viewed as major adjustment difficulties. Suicide rate was reportedly high among the adolescents in Sri Lanka (Nastasi et al., 1998), and the concern for adolescent suicide was reflected in the interviews with students as well. Female respondents indicated that they suffered from anxiety, whereas male respondents described that they felt restless due to the uncertainty of their future. Students from both genders argued that academic adjustment difficulties were primarily related to poor academic achievement or concerns about

performance. These included neglecting studies, academic failure (e.g., failure in the examination), and performance anxiety (e.g., "worry about results in the examination").

Social adjustment difficulties among the Sri Lankan students included aggression, neglecting responsibilities or duties, and being untrustworthy and not helpful to others. Sri Lankan boys also described engagement in criminal activities such as stealing, robbing, and joining gangs as forms of social adjustment problems. Sri Lankan girls suggested that interfering in others' personal affairs and slandering or stigmatizing others were indicative of social adjustment difficulties.

Stressors. Academic stressors identified by students included academic failure, rigorous examination processes, high level of academic pressure with limited opportunities for recreation or leisurely activities, parental or societal pressure for high academic achievement, high level of competition in academics, and uncertainty about the future due to limited access to higher education and high rate of unemployment. Only about 2% of students are allowed access to university study and economic prospects are limited for the rest.

Major family stressors included alcoholism of parents (mainly fathers), poverty or financial difficulties, domestic violence, parental fights, parental divorce or separation, and separation from parents. Students also considered lack of care and attention from parents and abandonment by parents as stressors. Adolescents discussed parentification of children (e.g., children assume household responsibilities in absence of their parents). This was particularly visible in the families where mothers were working in the Middle East. Adolescents also described the physical and sexual abuse of the children in absence of their mothers. Both Sri Lankan males and females spoke of restrictions on male-female interaction in their culture as problematic.

Students also identified financial problems and poverty as major social stressors. In addition, male students spoke of war, terrorism, and injustices in the society as other social problems. Unemployment was another problem that was cited frequently by male students as social stressors. On the contrary, girls did not mention unemployment as a problem. This may be linked to the societal emphasis on the role of men as the primary providers of the family. Furthermore, girls identified gender inequity (such as lack of freedom for women and differential expectations for men and women) and sexual harassment as social stressors specific to their gender. The male students indicated that the ethnic conflicts within Sri Lanka were a major concern for them. In addition, male students exhibited concern about the political violence and the widespread corruption in the country.

Students suggested death of loved ones, loss of relationships, betrayal, and misunderstandings as relationship stressors. According to them, peer ridicule, fighting with friends, and being ignored or neglected by friends hurt their feelings. In addition, fighting with parents, being pushed by the parents to study without recreation, or the controlling behavior of the parents strained the adolescents' relationships with their parents. Despite these relationship concerns, adolescents viewed peers and parents as potential sources of support in the event of relationship stressors.

Vulnerability. Students also indicated poverty, lack of family support, alcohol and drug abuse, and academic failure as major personal history factors that made them vulnerable to mental health problems. A striking gender difference was noted in reports of anxiety, nervousness, and health problems, which were restricted to female respondents.

Socialization. When describing socialization processes and agents, adolescents from Sri Lanka argued that the educational system played an important role in the process of socializing youth. Students discussed the cultural emphasis on high academic achievement and the resultant pressure on adolescents for academic study with limited time for recreation. Students also indicated heavy reliance on tuition classes (private tutoring) for additional academic support that could be related to the prime importance of academic performance.

In addition, adolescents indicated the Sri Lankan society valued and underscored the importance of professional jobs (e.g., doctors, engineers). Performance on standardized examinations at Grades 10 (O/L, ordinary level) and 12 (A/L, advanced level) determined admission to government-funded colleges and one's major area of study (those with highest scores were admitted to medicine, then engineering, etc.).

Cultural norms. With regard to cultural norms, students suggested that society reinforces high levels of respect for elders in Sri Lanka. This norm influences parent-child relationships and may explain reported social and emotional distancing between adults and children. Students also indicated a restriction on male-female interaction. They talked about parental and societal disapproval of relationships between boys and girls. Respondents, particularly females, described the lack of freedom or independence of girls, in contrast to the boys, who were considerably more independent. Among other prominent cultural norms, Sri Lankan adolescents spoke about arranged marriage (i.e., parents arrange and/or approve marriage) and the practice of dowry.

Data collected at this formative stage not only contributed to development of culture-specific theory but also contributed to development of culture-specific assessment, intervention, and teacher training materials that continued over a period of 5 years. The subsequent steps also reflect mixed methods designs.

Theory Development/Modification and Testing Phase: Qual → Quan →/+ Qual → Quan . . . Qual → Quan

The process of theory development and testing can be depicted as a sequence of qualitative data collection to inform theory development, followed by testing theory quantitatively and modifying theory through qualitative data collection conducted sequentially or concurrently, followed by quantitative methods to test modified theory, and so on. The repeated application of mixed methods across cultures, contexts, and populations can be used to develop theory that reflects both universal and culturally specific constructs. In the SLMHPP project, we developed a culture-specific framework for conceptualizing the individual and social-cultural factors related to mental health (see previous formative stage). The combined use of qualitative and quantitative data analysis informed theory development. This work is reflected in the next section on instrument development. In addition, subsequent qualitative data collection in Calcutta, India, helped to extend theory development to another Asian country (Sarkar, 2003). Ongoing work will examine the application within posttsunami contexts within Sri Lanka.

Instrument Development and Validation Phase: Qual → Quan

A number of sources suggest that assessment of abstract psychological phenomena will differ by culture (see Hitchcock et al., 2005); and this is the case when assessing self-concept (Harter, 1999). Instrument development in the SLMHPP project was predicated on the application of mixed methods to instrument development, using a sequential qualitative-quantitative design to develop culturally relevant measures. Qualitative research methods were used to gather data to inform instrument development. Quantitative methods were subsequently employed to conduct instrument validation. As discussed later, this approach has the potential to yield findings that quantitative or qualitative approaches, by themselves, cannot yield. In the SLMHPP, we employed a sequential qualitative-quantitative design to develop culture-specific instruments designed to assess psychological constructs related to mental health. The process of instrument development and validation illustrated in this section overlaps with the process of theory development and testing, which involves a repeated Qual → Quan design as described in the preceding section.

Psychological instruments were developed based on the aforementioned findings on culturally valued competencies, generated via formative research, and self-concept theory of Harter (1999). Harter suggested that positive adjustment requires congruency between culturally valued expectations and self-rated competencies. For example, a male United States–based researcher would typically be in a culture that values skills with statistical analysis over, say, cross-stitching. If this researcher believed he had adequate skills with statistics, there would be congruency between his perceived competencies and what is valued. Meanwhile, his competency with cross-stitching would likely have no impact on the valence of his self-beliefs.

The investigators entered the context with this general theory of self-concept, believing that Sri Lankan adolescent mental health concerns might be tied to disparities between their perception of their competencies and what is valued in the culture. To clarify, we made limited a priori guesses as to what competencies might be valued but did assume that congruence between values and self-beliefs would indicate positive adjustment and vice versa. An example of an a priori expectation we did make was that an adolescent would be experiencing distress if she did not consider herself to be a strong student. Recall that Sri Lankan society places great expectations on educational achievement; indeed, it was believed that a student can shame family members by not performing well on exams, which are high-stakes in nature because they are a gateway to postsecondary education. Other a priori expectations were that Sri Lankan adolescents would have culturally specific stressors, coping mechanisms, support structures, and ways of expressing emotions related to stress. It also was believed that some of these phenomena would be gender-specific. Qualitative data collected during the formative research phase (see previous section) provided the basis for testing these assumptions and developing culture-specific understanding of key constructs.

Two types of scales were developed via a series of individual studies combining ethnographic and factor analytic techniques. The first scale type (a total of five scales were developed) assesses the relationship between culturally specific competencies and values (Nastasi, Jayasena, et al., 1999a). The second scale type (a total of seven scales) includes culturally specific scenarios that adolescents should find stressful (based on formative data), and follow-up items to assess how adolescents might respond to such stress (i.e., emotionally and via active coping, seeking support, or maladjusted behavior; Nastasi, Jayasena, et al., 1999b). Scales in the latter type were used as outcome measures for an exploratory evaluation of an intervention tailored to the needs of Sri Lankan youth (in the forthcoming evaluation design).

The work presented in the *Journal of School Psychology* (Hitchcock et al., 2005) illustrated a mixed method approach for this sort of Qual → Quan instrument development and validation. The article offered a detailed illustration of the approach using the responses of 611 Sri Lankan adolescents provided to five ethnographically informed psychological measures. Such instruments offer a key connection between the primary methodologies used (i.e., ethnographic and factor analytic approaches) as they are predicated on qualitative inquiry, can translate these ideas into quantitative data and allow for the application of factor analysis. If the qualitatively derived constructs are comparable to factor analytic results, then triangulation across methods is achieved and a standardized measure can be developed that is sensitive to culturally specific phenomena. The illustration of this approach used data from an ethnographically informed psychological measure of self-concept, which, again, was predicated on Harter's (1999) work.

The scales were back-translated (e.g., English → Sinhala → English to ensure accuracy of meaning), piloted, and refined after obtaining input from local experts with knowledge of the target culture. They were then administered to students ($n = 611$; 315 males, 296 females), Grades 7 to 12, ages 12 to 19, across six schools that represented the range of the student population in terms of ethnicity, religion, and socioeconomic status. A reanalysis of data from

focus groups and individual interviews (i.e., with students, parents, and school personnel) and archival information from the culture (e.g., newspapers, school documents, etc.) resulted in the identification of the range of responses to various target questions/issues. Examples of these might be as follows: describe a stressful school scenario, or describe a stressful home scenario, and so on. Qualitative analyses inform the generation of psychological constructs to explain the variation of responses and in turn the development of psychological measures that are highly targeted toward the context of interest.

Secondary analyses of quantitative data were conducted because prior analyses of qualitative data (Sarkar, 2003) indicated that the constructs identified via the factor analyses might be gender specific (Hitchcock et al., 2006). Factors from the first self-concept measure (self-rating of competencies and behaviors) were used to develop subscale scores. MANOVA analyses were performed to test for gender differences. Statistically significant differences were found, as expected, on the Suitable Behavior subscale. Furthermore, structured means analyses demonstrated that the Unsuitable Behavior scale was different for boys and girls. That is, boys and girls appeared to recognize the Unsuitable Behavior construct but report on it in different ways. To summarize, no gender differences were evident on the Personal/Interpersonal Needs construct. Girls reported higher values on the Suitable Behaviors construct, suggesting they engage in suitable behaviors more often than boys. Boys and girls differed on how they answered Unsuitable Behavior items. This last finding is probably due to the fact that the (quantitative) construct/factor is formed by items that deal with joining gangs, carrying weapons, and substance abuse; and the qualitative data indicate that these behaviors are only relevant to males. Meanwhile, female behavior is more rigid and less permissive. Of course, cross-method data triangulation was needed to reach the conclusions and to develop a scale that is sensitive to both gender and culture.

As noted previously, this mixed method approach to scale development yielded insights to Sri Lankan youth culture that could not have been obtained with singular approaches. The formative ethnographic work provided the initial identification of culturally relevant constructs. These constructs in turn generated items that could be administered to hundreds of students (of course, it is generally inefficient to apply qualitative methods when working with larger samples). Analyses of responses provided additional insights into the culturally relevant constructs via cross-method triangulation, clarification of how the constructs appear in quantitative factors, and the opportunity to apply statistical tests of null hypotheses to verify presumed gender differences. One result of SLMHPP is an assessment battery that can be used for future work, and as noted below, this general method also yielded culturally specific outcome measures that can be employed in randomized controlled trials testing culturally specific interventions.

Recall that the second set of scales we developed assessed how Sri Lankan adolescents might respond to culturally specific stressors. Hypothetical stressors were identified via a series of group and individual interviews with stakeholders in the culture, specifically, students, administrators, teachers, and parents (see formative research phase). Three types of stressors emerged from the data: academic, family, and social. Respondents also noted that stressors might be dealt with via emotion-focused coping (or lack thereof), problem-focused strategies, and seeking support from others.

To assess how students might respond to hypothetical culturally specific scenarios, seven scenarios were generated from prior qualitative analyses conducted in the formative research phase (see Table F.2) and presented as vignettes in the ethnographically informed psychological measures (Nastasi et al., in press). A series of follow-up items were generated, also from prior qualitative analyses of the formative data, to assess how students might respond to these scenarios and the resulting scales were used as outcome measures to evaluate the effects of a culturally specific intervention (more on this below). Each scenario (and follow-up item) was

translated into the primary language of the group, using a back translation method (e.g., English → Sinhala → English) to ensure accuracy of meaning. The instruments were then administered to 120 Sri Lankan students coming from urban and suburban areas, a range of socioeconomic status (SES) levels, and different ethnic groups.

With the exception of the demographic questions, each item utilized a 3-point response format (i.e., *a lot, some, not at all*), and adolescents were asked to rate themselves on a set of culturally defined items capturing perceptions of stress and coping. To assess reactions to each scenario, students were asked to respond to items that assessed their emotional responses; coping strategies; social support (i.e., emotional or instrumental help from others); and behavioral, emotional, or health-related difficulties resulting from stressful experiences such as alcohol abuse, suicidal ideation/attempts, aggression toward peers, and physical symptoms such as headaches or stomachaches.

Qualitative analyses generated the a priori expectation that students would, if faced with the hypothetical stressors, identify with the indicators of adjustment difficulties, coping strategies, and social supports listed in the measure. Note, however, that it was anticipated that factors would likely include a mix of feelings, coping, support, and adjustment difficulty items. To verify these expectations, principal component analyses (PCAs) were conducted (Nastasi et al., in press). Across all scenarios, the analyses yielded the following factors: Adjustment Difficulties—Externalizing, engaging in acting-out behaviors labeled "undesirable/unsuitable" in the culture; Social Support, perceived effectiveness of social resources (family, peer, school/mental health personnel); and Feelings of Distress, affective reactions (e.g., sad, angry, confused) without active coping. The analyses yielded scales that were consistent with qualitative expectations. Furthermore, the factor analyses indicated variation in reactions to stressors as a function of stressful situation and raised questions about the cultural meaning of suicide.

Table F.2   Hypothetical Scenarios for Assessing Coping With Stressors

Academic Scenarios

Scenario #1

You are currently studying for O/L exams. Your mother is a doctor and your father is an engineer. Your parents want you to be a doctor, so it is important you do well on your O/Ls. You attend tuition 7 days a week and spend all of your free time studying. You have no time to visit with friends or for recreation.

Scenario #2

You have failed A/L exams by a few points and are concerned about your future. You want to be an engineer. Your family cannot afford to send you to private school or to study abroad. You are not sure what you should do.

Scenario #3

You are in a mathematics class with 50 other students and the teacher is explaining a new topic in math. You don't understand but don't ask the teacher because the other students will get at you for using class time.

Relationship Scenario

Scenario #4

You have been having a secret love affair. You and your boy/girlfriend just broke up. You cannot talk to your family or your teacher about it. You have trouble sleeping. Your parents and teachers have asked you what is wrong but you cannot talk to them. You do not know what to do. Meanwhile one of the prefects who searched your school bag found a love letter and gave the letter to the class teacher. The class teacher called your parents. The parents and teacher forbid you to communicate with your lover.

Family Scenarios

Scenario #5

You are living on the street with your family. You have a school uniform but no shoes. You usually feel hungry and sleepy at school, but are a very good student. You like to do handwriting and ask the teacher for her lunch bag to practice writing. After school, you and your sisters and brothers beg on the street.

Scenario #6

Your mother has been working in the Middle East for about a year. She sends money home regularly for the family, but there is little direct communication with the children. You are the eldest child and have been taking care of the four younger children. Your father has brought a stepmother from the village to live with you to help with household tasks. When you object to the stepmother living in the house, your father beats you severely. Because of the severe abuse, you are considering leaving home. Some of your friends have already left home and have formed a gang and invited you to become a member.

Scenario #7

It [is] the day before a big exam in school. You [come] home from school and, when you [enter] your home, your father is yelling at your mother. You father has been drinking arrack. He asks your mother for dinner. She says that dinner is not ready because she had to find money to buy rice. Your parents start arguing about money. When your mother serves dinner, the rice is overcooked. Your father starts yelling and throws the rice on the floor. Your mother says, "I'll cook more," and begins to cry. Your father tells you to clean up the mess he has made. Your mother says that you should study, not to clean up the mess that your father has made. Your father then starts beating your mother.

SOURCE: Reprinted with permission of Sage Publications from Nastasi et al. (in press).

NOTE: O/L = ordinary level; A/L = advanced level.

Overall, these factors are largely consistent with qualitative findings, providing additional evidence that the three constructs for responding to the scenarios presented above are valid in Sri Lankan youth culture. To assess the reliability of these scales, alpha coefficients computed separately by scenario indicated good to excellent internal consistency (alphas ranging from .70 to .95).

Program Development and Evaluation Phase: Qual →/+ Quan →/+Qual →/+ Quan . . . Qual →/+ Quan; Alternatively, Qual→←Quan

Mixed methods applied to program development and evaluation (see Table F.1) is characterized by repeated sequential or concurrent use of qualitative and quantitative methods, to design, modify, and evaluate the program. For example, formative qualitative and quantitative data inform program design, and formative evaluation through concurrent or sequential qualitative and quantitative data collection during program implementation informs program modification or adaptation to meet local needs. Alternatively, this process might be characterized as an interactive or recursive process, in which qualitative and quantitative data collected on an ongoing basis inform program design, formative evaluation, and modification/adaptation.

The formative research phase of the SLMHPP provided the basis for designing a mental health promotion program (Nastasi, Varjas, et al., 1999), which was pilot tested in one school in the Central Province of Sri Lanka. The researchers employed a randomized-controlled trial to test the effectiveness of the program, and concurrent and sequential qualitative-quantitative data collection for the purposes of formative evaluation, program monitoring and adaptation, and outcome evaluation. The program consisted of 18 sessions conducted each weekday over a 4-week period with 60 students in Grades 7 through 12. Sessions were cofacilitated by teachers (from participating school) and teacher educators (from participating local university). Students engaged in individual, small group, and large group activities designed to facilitate identification of cultural expectations, stressors, coping mechanisms, and social supports in key ecological contexts (community, family, school, peer group); development and practice of culturally appropriate coping strategies; and participation in peer support activities. An example of the cultural specificity of the program was the sequence of ecological contexts in which students were encouraged to identify stressors and social supports. In contrast to typical social-emotional learning curricula designed for U.S. population, the SLMHPP curriculum focused on the self only in relationship to others (with minimal focus on the self in isolation) and began with an exploration of self within community/societal context and progressing to increasingly more intimate contexts such as school, peer group, and family. Typical programs in the United States begin with focus on self-identity (and self-care), progress to self within interpersonal relationships (caring for others), and conclude with self within society/community (community service).

During program implementation, researchers collected formative evaluation data for each session that focused on examining program acceptability, cultural relevance and social validity, integrity, and immediate impact. The data collection tools included participation observation of curriculum sessions and weekly teacher training meetings; key informant interviews with teachers, students, and school administrators; session evaluation forms completed by students, teachers, and observers; and session products (e.g., student narratives, visual depictions of stressors and supports within ecological contexts; more detailed information about evaluation methods and tools can be obtained from the first author). These data were reviewed after each session and used to inform curricular adaptations and ongoing teacher training and support. Subsequent data collection provided feedback about the success of adaptations and teacher training and support. Thus, an iterative process was reflected in the ongoing integration and application of qualitative and quantitative data to inform decision making during program implementation.

Evaluation Research Phase: Qual + Quan

Application of mixed methods to evaluation research can be characterized by concurrent use of multiple qualitative and quantitative data collection methods to facilitate data triangulation and evaluate programs in a comprehensive manner. Comprehensive approaches to program evaluation extend beyond traditional notions of evaluating effectiveness to assessment of program acceptability, social validity (application to daily life) and cultural specificity (relevance and appropriateness to cultural background and experiences of participants), integrity or quality of program implementation, immediate and long-term outcomes, and sustainability and institutionalization of program efforts (see Nastasi, Moore, & Varjas, 2004). Furthermore, comprehensive evaluation includes data collection from multiple informants and interpretation from multiple perspectives.

A concurrent qualitative-quantitative design was reflected in the evaluation of the SLMHPP pilot program. As described above, formative evaluation (reflecting an iterative mixed method design) addressed issues of acceptability, social validity and culture specificity, integrity, and immediate program impact. In addition, outcome evaluation was conducted using a pre-post control group design ($N = 120$; 60 experimental, 60 control) with concurrent qualitative and quantitative data collection. Outcome measures included student pre-post self-report measures (culture-specific psychological measures designed from formative data; described in an earlier section), student feedback reflected in final session products (resulting from

structured session activity designed for evaluative purposes), and postintervention group interviews with program implementers (teachers and teacher educators).

We used a series of null-hypothesis significance tests and estimates of effects to analyze program impacts, supplemented by analysis of qualitative data collected during program implementation. A 2 × 2 multivariate analysis of covariance (MANCOVA; controlling for pretest scores) was performed for each of the stressful scenarios (depicted in Table F.2) to test for intervention effects and gender by intervention group. Tests of the overall MANCOVA were significant for Scenarios 4 (romantic relationship), 5 and 6 (family scenarios); follow-up tests indicated a significant Group × Gender interaction for those scenarios (Nastasi et al., 2006). (The full presentation of outcome data is beyond the scope of this article. Please contact the first author for more information.)

The quantitative outcomes indicated that the SLMHPP may have heightened the awareness of girls, but not boys, to the potential feelings of distress and limited helpfulness of social support, particularly with regard to situations in which they may have limited control. In addition, exploratory analyses of anticipated responses to complex family stressors (parental alcohol abuse and domestic violence) suggests that the intervention may have heightened girls' awareness of the potential negative impact of such stressors for them personally, that is, internalizing adjustment difficulties. However, the intervention may also have heightened girls' sense of responsibility for resolving complex family problems. The quantitative results were consistent with qualitative data collected during the intervention sessions and during the formative research phase. For example, the heightened sense of responsibility resulting from complex family problems such as absent mother or family alcoholism was evident also in qualitative depictions of stressful situations. These findings have important implications regarding the need for gender specificity in mental health promotion and social-emotional learning programming, and the need for addressing context specificity (e.g., family vs. peer contexts) of coping.

Program acceptability data indicated that students responded positively to activities and opportunities to discuss common stressors and ways of coping. Observations and student reports indicated enjoyment of opportunities to be creative; curriculum activities provided opportunities to express themselves through drawing, writing, role-playing, and discussion. Teachers responded well to on-site support and ongoing skills training. They generally responded favorably to the curriculum; these responses seemed to be influenced by student responses and participation (Bernstein, 2000). For example, teachers reported satisfaction with the program when students showed interest and enjoyment and seemed to benefit from activities. Teachers reported gaining a better understanding of the lives of their students and perceived themselves in a new role as facilitator of students' social-emotional development. These perceptions were consistent with students' favorable reports of emotional support from teachers during the program. Furthermore, in follow-up interviews after program completion, teachers reported that students (both those who had participated in the program and those who were non-participants) sought them out for emotional support in the larger school context.

Keep in mind that the initial piloting of the intervention was a small, exploratory study designed to obtain preliminary findings on the effects of a culturally specific intervention. Hence, multiple analyses were conducted using promising outcome measures but nevertheless are still in a development phase. The number of analyses elevate the possibility of making a Type I error, and in all cases the tests were underpowered. In addition, the program was implemented in one school in one community of Sri Lanka and thus the results may not be generalizable to all students and schools within the country. Despite these limitations, the data yield important findings that can be used to guide future intervention work and larger experimental investigations.

As an extension of this work, Nastasi and Jayasena are currently engaged in developing long-term recovery programs for students and parents living in tsunami-affected coastal communities of Sri Lanka. The ongoing data collection using mixed methods designs as described herein is providing information about the applicability of the intervention program to address context-specific stressors such as natural disasters and to extend the program by involving parents as agents for promoting children's mental health. For example, the adapted intervention program included focus on coping with environmental stressors such as natural disasters (Nastasi & Jayasena, 2006). (For more information on this work, contact the first author.)

● IMPLICATIONS: MIXED METHODS DESIGNS IN INTERVENTION RESEARCH

The work presented in this article illustrates the application of mixed methods designs to the development and evaluation of culturally specific psychological assessment measures and interventions. In this work, formative qualitative data collection was used to identify culturally relevant constructs and develop a culturally specific model of mental health. This model and the qualitative data were then used to develop assessment measures and an intervention program. Mixed methods were used to validate the assessment measure and evaluate the acceptability, integrity, social validity, and outcomes of a pilot intervention. For example, the combination of qualitative analysis of ethnographic data and factor analysis of quantitative data was used to validate scales to measure constructs related to self-concept and coping with stress, which in turn could serve as outcome measures for interventions. Similarly, the evaluation of intervention outcomes was informed by both quantitative indices and qualitative data collected during program implementation. Furthermore, mixed methods were used to monitor and adapt the program to meet context-specific and individual needs of students and teachers. Finally, a new cycle of mixed methods research was instituted to adapt the program model to a new population and context (i.e., students and parents living in tsunami-affected communities).

The repeated application of a recursive research↔intervention process using mixed methods can facilitate the development of culture-specific interventions and translation of evidence-based practices to diverse populations and settings. Using a mixed methods approach, researchers can engage local stakeholders (e.g., community members, educators, school administrators) in developing intervention programs that address local cultural, contextual, and population needs (e.g., community violence, drug abuse among middle school students, poor academic performance within a school district); adapting programs across multiple settings (e.g., adapting a sexual risk education program across grade levels and diverse student populations); and translating evidence-based practices to new contexts and populations. The successful application (or translation) of evidence-based interventions developed through randomized-controlled trials to naturalistic settings requires research to identify the conditions necessary for ensuring established program outcomes (see National Institute of Mental Health, 2001). Mixed method designs, as described in this article, are particularly relevant to the comprehensive evaluation of conditions necessary for effective intervention and can thus help to facilitate translational research (e.g., extension of the worked portrayed herein to tsunami-affected areas as described above).

The illustration presented here reflects a multi-year effort to develop and test theory, instruments, and interventions that are specific to culture and context, with the purpose of demonstrating the application of mixed methods designs across the multiple phases of research and development projects. The designs can of course be applied to shorter term and more focused efforts to develop culturally

and contextually appropriate interventions. Moreover, as the illustration suggests, the process of ensuring cultural specificity is ongoing through the multiple stages of program design, implementation, evaluation, and translation. Mixed methods designs provide an important mechanism for facilitating development of culturally sensitive interventions and evidence-based practices.

Finally, this article also contributes to the development of multistage program evaluation models. Bamberger, Rugh, and Mabry (2006) and Stufflebeam (2001) noted that mixed methods evaluations are complex and can take the form of multistage projects. However, there appears to be a dearth of examples of such projects in the literature. We have attempted to address this shortcoming here, while advancing mixed methods conceptual frameworks to help others think through how to plan multiphase evaluation projects that use mixed methods.

● REFERENCES

Bamberger, M., Rugh, J., & Mabry, L. (2006). *Realworld evaluation working under budget, time, data, and political constraints*. Thousand Oaks, CA: Sage.

Bernstein, R. (2000). *A demonstration of the acceptability of a mental health project through a participatory culture-specific model of consultation in the country of Sri Lanka*. Unpublished doctoral dissertation, Department of Educational and Counseling Psychology, State University of New York at Albany.

Bronfenbrenner, U. (1989). Ecological systems theory. In R. Vasta (Ed.), *Annals of child development* (Vol. 6, pp. 187–249). Greenwich, CT: JAI.

Cowen, E. L., Hightower, A. D., Pedro-Carroll, J. L., Work, W. C., Wyman, P. A., & Haffey, W. G. (1996). *School-based prevention for children at risk: The Primary Mental Health Project*. Washington, DC: American Psychological Association.

Creswell, J. W. (2003). *Research design: Qualitative, quantitative, and mixed methods approaches*. Thousand Oaks, CA: Sage.

Harter, S. (1999). *The construction of the self: A developmental perspective*. New York: Guilford.

Hitchcock, J. H., Nastasi, B. K., Dai, D. C., Newman, J., Jayasena, A., Bernstein-Moore, R., et al. (2005).

Illustrating a mixed-method approach for identifying and validating culturally specific constructs. *Journal of School Psychology, 43*(3), 259–278.

Hitchcock, J. H., Sarkar, S., Nastasi, B. K., Burkholder, G., Varjas, K., & Jayasena, A. (2006). Validating culture- and gender-specific constructs: A mixed-method approach to advance assessment procedures in cross-cultural settings. *Journal of Applied School Psychology, 22*(2), 13–33.

Ingraham, C. L., & Oka, E. R. (2006). Multicultural issues in evidence-based interventions. *Journal of Applied School Psychology, 22*(2), 127–149.

Morse, J. M. (2003). Principles of mixed methods and multimethod research design. In A. Tashakkori & C. Teddlie (Eds.), *Handbook of mixed methods in social & behavioral research* (pp. 189–208). Thousand Oaks, CA: Sage.

Nastasi, B. K., Hitchcock, J. H., Burkholder, G., Varjas, K., Sarkar, S., & Jayasena, A. (in press). Assessing adolescents' understanding of and reactions to stress in different cultures: Results of a mixed-methods approach. *School Psychology International*.

Nastasi, B. K., & Jayasena, A. (2006). *Mental health promotion post-tsunami curriculum*. Minneapolis, MN: Walden University.

Nastasi, B. K., Jayasena, A., Hitchcock, J., Burkholder, G., Varjas, K., & Sarkar, S. (2006, April). *Reactions of female adolescents to a school-based mental health program*. Paper presented at the 10th National Convention on Women's Studies, sponsored by Centre for Women's Research, Columbo, Sri Lanka.

Nastasi, B. K., Jayasena, A., Varjas, K., Bernstein, R., Hitchcock, J. H., & Sarkar, S. (1999a). *Student questionnaire: Perceived competencies measure for Sri Lanka Mental Health Promotion Program*. Albany: School Psychology Program, State University of New York at Albany.

Nastasi, B. K., Jayasena, A., Varjas, K., Bernstein, R., Hitchcock, J. H., & Sarkar, S. (1999b). *Student questionnaire: Stress & coping measure for Sri Lanka Mental Health Promotion Program*. Albany: SUNY Press.

Nastasi, B. K., Moore, R. B., & Varjas, K. M. (2004). *School-based mental health services: Creating comprehensive and culturally specific programs*. Washington, DC: American Psychological Association.

Nastasi, B. K., & Schensul, S. L. (2005). Contributions of qualitative research to the validity of intervention research. *Journal of School Psychology, 43*(3), 177–195.

Nastasi, B. K., Schensul, J. J., Balkcom, C. T., & Cintrón-Moscoso, F. (2004). Integrating research and practice to facilitate implementation across multiple contexts: Illustration from an urban middle school drug and sexual risk prevention program. In K. E. Robinson (Ed.), *Advances in school-based mental health: Best practices and program models* (chap. 13). Kingston, NJ: Civic Research Institute.

Nastasi, B. K., Schensul, J. J., deSilva, M. W. A., Varjas, K., Silva, K. T., Ratnayake, P., et al. (1998–1999). Community-based sexual risk prevention program for Sri Lankan youth: Influencing sexual-risk decision making. *International Quarterly of Community Health Education, 18*(1), 139–155.

Nastasi, B. K., Varjas, K., Bernstein, R., Hellendoorn, C., Brewster, M., Hitchcock, J., et al. (1999). *Program for Mental Health Promotion in Sri Lankan Schools: Curriculum manual & instructional guide*. (Developed for implementation in the Central Province Schools, Kandy, Sri Lanka). Albany: School Psychology Program, State University of New York at Albany.

Nastasi, B. K., Varjas, K., Sarkar, S., & Jayasena, A. (1998). Participatory model of mental health programming: Lessons learned from work in a developing country. *School Psychology Review, 27*(2), 260–276.

National Institute of Mental Health. (2001). *Blueprint for change: Research on child and adolescent mental health* (Report of the National Advisory Mental Health Council Workgroup on Child and Adolescent Mental Health Intervention Development and Deployment). Washington, DC: Author.

Roberts, M. C. (Ed.). (1996). *Model programs in child and family mental health*. Mahwah, NJ: Lawrence Erlbaum.

Sarkar, S. (2003). *Gender as a cultural factor influencing mental health among the adolescent students in India and Sri Lanka: A cross-cultural study*. Unpublished doctoral dissertation, State University of New York at Albany.

Schensul, S. L., Mekki-Berrada, A., Nastasi, B. K., & Saggurti, N. (in press). Healing traditions and men's sexual health in Mumbai, India: The realities of practiced medicine in urban poor communities. *Social Sciences and Medicine*.

Schensul, S. L., Nastasi, B. K., & Verma, R. K. (2006). Community-based research in India: A case example of international and interdisciplinary collaboration [Electronic version]. *American Journal of Community Psychology*. Available from http://dx.doi .org/10.1007/s10464-006-9066-z

Schensul, S. L., Verma, R. K., & Nastasi, B. K. (2004). Responding to men's sexual concerns: Research and intervention in slum communities in Mumbai, India. *International Journal of Men's Health, 3,* 197–220.

Stufflebeam, D. L. (2001). Evaluation models. *New Directions for Evaluation, 89,* 7–99.

Tashakkori, A., & Teddlie, C. (2003). *Handbook of mixed methods in social & behavioral research*. Thousand Oaks, CA: Sage.

GLOSSARY

Advocacy and expansion period in the history of mixed methods involved authors advocating for mixed methods research as a separate methodology, method, or approach to research, and interest in mixed methods being extended to many disciplines and many countries.

Case-oriented merged analysis display is a display for merged data analysis that positions cases on a quantitative scale along with qualitative text data about the individual cases.

Category/theme display in merged data analysis is a display that arrays the qualitative themes derived from the qualitative analysis with quantitative categorical or continuous data from items or variables from the quantitative statistical results.

Closed-ended questions are used in quantitative research to collect data. These questions are based on predetermined response scales, or categories.

Combination mixed methods questions are research questions about mixing the quantitative and qualitative data in a mixed methods study in which the researcher makes explicit both the methods and the content of the study.

Concurrent timing occurs when the researcher implements both the quantitative and qualitative strands during a single phase of a research study.

Connecting is a mixing strategy in which the results from one strand of data shape the collection of data in the second strand.

Connecting mixed methods data analysis involves the analysis of the first data set and its connection to data collection for the second data set.

Constructivism, which is typically associated with qualitative approaches, is based on understanding or meaning of phenomena, formed through participants and their subjective views.

Content-focused mixed methods research questions are research questions about mixing the quantitative and qualitative data in a mixed methods study in which the researcher makes explicit the content of the study and implies the research methods.

Convergent and divergent findings in a merged data analysis display is a table displaying congruent or incongruent (or discrepant) findings along the horizontal dimension. Along the vertical dimension, the researcher may indicate different topics and/or participant types as indicated by their numeric scores. Within the cells of this display could be quotes, numbers, or both.

Convergent parallel design is a mixed methods design in which the researcher uses concurrent timing to implement the quantitative and qualitative strands during the same phase of the research process, prioritizes the methods equally, keeps the strands independent during analysis, and mixes the results during the researcher's overall interpretation of the data.

Critical realism is a theoretical or philosophical position that integrates a realist ontology (there is a real world that exists independently of our perceptions, theories, and constructions) with a constructivist epistemology (our understanding of this world is inevitably a construction from our own perspectives and standpoint).

Data collection decisions for the convergent design include who will be selected for the two samples, the size of the two samples, the design of the data collection questions, and the format and order of the different forms of data collection.

Data collection decisions for the embedded design include the rationale for embedding one form of data, the timing of the embedded data, and how to address problems that may arise from the embedding.

Data collection decisions for the explanatory design include who should be the participants in the second phase, what sample sizes to use for both strands, what data to collect from one phase to the other and from whom, and how to secure institutional review board (IRB) permissions for the two data collections.

Data collection decisions for the exploratory design include the determination of samples for each phase, the decisions about what results to use from the first phase, and, if a middle phase is used, how to design

a rigorous instrument with good psychometric properties.

Data collection decisions for the multiphase design include sampling, using longitudinal designs, and developing a programmatic objective that binds the multiple projects together.

Data collection decisions for the transformative design are related to sampling, benefits to those participating in the study, and collaboration during the data collection process.

Data transformation merged analysis consists of transformation of one type of data into the other type so that both databases can be easily compared and further analyzed.

Data-transformation variant is a variant of the convergent design in which the researcher implements the quantitative and qualitative strands during the same phase of the research process but prioritizes them unequally, placing greater emphasis on the quantitative strand, and uses a merging process of data transformation.

Data-validation variant is a variant of the convergent design in which the researcher includes both open- and closed-ended questions on a questionnaire and uses the results from the open-ended questions to confirm or validate the results from the closed-ended questions.

Decisions in mixed methods data analysis refer to those critical points in data analysis when the researcher needs to decide what options to select for analysis.

Definition of core characteristics of mixed methods research is the collection and analysis of both qualitative and quantitative data (based on research questions), mixing (or integrating or linking) the two forms of data, giving priority to one or to both (in terms of what the research emphasizes), using procedures in a single study or in multiple phases of a program of study, framing these procedures within philosophical worldviews and theoretical lenses, and combining the procedures into specific research designs that direct the plan for conducting the study.

Disability lens transformative variant is a variant of the transformative design in which the researcher frames the study using a disability theoretical lens.

Dynamic approach is an approach to mixed methods design that considers and interrelates multiple components of research design rather than placing emphasis on selecting an appropriate design from an existing typology.

Emancipatory theory in mixed methods involves taking a theoretical stance in favor of underrepresented or marginalized groups, such as a feminist theory, a racial or ethnic theory, a sexual orientation theory, or a disability theory.

Embedded-correlational variant is a variant of the embedded design in which the researcher embeds qualitative data within a correlational design.

Embedded design is a mixed methods design in which the researcher collects and analyzes both quantitative and qualitative data within a traditional quantitative or qualitative design to enhance the overall design in some way.

Embedded-experiment variant is a variant of the embedded design in which the researcher embeds qualitative data within an experimental trial design.

Embedded instrument development and validation variant is a variant of the embedded design in which the researcher embeds qualitative data within a traditional instrument development and validation design.

Emergent mixed methods designs are found in mixed methods studies where the use of mixed methods arises due to issues that develop during the process of conducting the research.

Equal priority is a weighting option that occurs when the quantitative and qualitative methods play equally important roles in addressing the research problem in a mixed methods study.

Evaluate a mixed methods study based on the following criteria: the collection of both quantitative and qualitative data, the use of persuasive and rigorous methods procedures, the mixing of the two sources of data, the use of a mixed methods design, incorporation of philosophical assumptions, and the use of mixed methods research terms.

Explanatory sequential design is a two-phase mixed methods design in which the researcher starts with the collection and analysis of quantitative data, followed by the collection and analysis of qualitative data to help explain the initial quantitative results.

Exploratory sequential design is a two-phase mixed methods design in which the researcher starts with the collection and analysis of qualitative data, followed by the collection and analysis of quantitative data to test or generalize the initial qualitative findings.

Exploring the data in qualitative data analysis involves reading through all of the data to develop a general understanding of the database.

Exploring the data in quantitative data analysis involves visually inspecting the data and conducting a descriptive analysis (the mean, standard deviation [SD], and variance of responses to each item on instruments or checklists) to determine the general trends in the data.

Feminist lens transformative variant is a variant of the transformative design in which the researcher frames the study using a feminist theoretical lens.

Fixed mixed methods designs are found in mixed methods studies in which the use of quantitative and qualitative methods is predetermined at the start of the research process, and the procedures are implemented as planned.

Follow-up explanations variant is a variant of the explanatory design in which the researcher places priority on the initial, quantitative phase and uses the subsequent qualitative phase to help explain the quantitative results.

Formative period in the history of mixed methods began in the 1950s and continued up until the 1980s. This period saw the initial interest in using more than one method in a study.

Independent is a level of interaction that indicates the quantitative and qualitative strands of a mixed methods study are implemented so that they are independent from each other, mixing only at the point of interpretation.

Inferences in mixed methods research are conclusions or interpretations drawn from the separate quantitative and qualitative strands of a study as well as across the quantitative and qualitative strands, called "meta-inferences."

Instrument-development variant is a variant of the exploratory design in which the initial

qualitative phase plays a secondary role, often for the purpose of gathering information to build a quantitative instrument that is needed for the prioritized quantitative phase.

Interactive is a level of interaction that indicates the quantitative and qualitative strands of a mixed methods study directly interact with each other during the design, data collection, or data analysis points of the study.

Intercoder agreement in qualitative research involves having several individuals code (and develop themes) for a transcript and then compare their analysis to determine whether they arrived at the same codes and themes or different ones.

Interpretation of results involves stepping back from the detailed results and advancing their larger meaning in view of the research problems, questions in a study, the existing literature, and perhaps personal experiences.

Interview protocol is a form used in qualitative research to collect qualitative data. On this form are stated questions to be asked during an interview and space for recording information gathered during the interview. This protocol also provides space to record essential data about the time, day, and place of the interview.

Joint display is a figure or table in which the researcher arrays both quantitative and qualitative data so that the two sources of data can be directly compared. In effect, the display merges the two forms of data.

Large-scale program development and evaluation projects are a variant of the multiphase design, which are often federally funded research programs in areas such as education and health services research where investigators conduct projects that require exploration, program development, program testing, and feasibility studies.

Level of interaction is the extent to which the quantitative and qualitative strands of a mixed methods study are kept independent or interact with each other.

Merged data analysis comparisons options are side-by-side comparisons in a results or discussion section or summary table, joint display comparisons in the results or interpretation, or data transformation in the results.

Merged data analysis strategies means using analytic techniques for merging the results, analyzing whether the results from the two databases are congruent or divergent, and, if they are divergent, then analyzing the data further to reconcile the divergent findings.

Merging is a mixing strategy in which quantitative results and qualitative results are brought together through a combined analysis.

Method-focused mixed methods research question is a research question about mixing the quantitative and qualitative data in a mixed methods study in which the researcher writes to focus on the methods of the mixed methods design.

Mixed methods case study is a variant of the embedded design in which the researcher collects both qualitative and quantitative data within a case study.

Mixed methods data analysis consists of analytic techniques applied to both the quantitative and the qualitative data as well as to the mixing of the two forms of data concurrently and sequentially in a single project or a multiphase project.

Mixed methods ethnography is a variant of the embedded design in which the researcher collects both qualitative and quantitative data within an ethnographic design.

Mixed methods interpretation involves looking across the quantitative results and the qualitative findings and making an assessment of how the information addresses the mixed methods question in a study.

Mixed methods narrative research is a variant of the embedded design in which the researcher collects both qualitative and quantitative data within a narrative research design.

Mixed methods purpose statements convey the overall purpose of the mixed methods study, and they include the intent of the study, the type of mixed methods design, quantitative and qualitative purpose statements, and the reasons for collecting both quantitative and qualitative data.

Mixed methods research questions are questions in a mixed methods study that address the mixing or integration of the quantitative and qualitative data.

Mixed methods titles include the topic, the participants, and the research site. They foreshadow the use of mixed methods and the type of mixed methods design that the researcher will use.

Mixing is the explicit interrelating of a mixed methods study's quantitative and qualitative strands.

Mixing at the level of design occurs when the quantitative and qualitative strands are mixed during the larger design stage of the research process before the researcher collects data within the study's strands.

Mixing during data analysis occurs when the quantitative and qualitative strands are mixed during the stage of the research process when the researcher is analyzing the two sets of data.

Mixing during data collection occurs when the quantitative and qualitative strands are mixed during the stage of the research process when the researcher collects a second set of data.

Mixing during interpretation occurs when the quantitative and qualitative strands are mixed during the final step of the research process after the researcher has collected and analyzed both sets of data.

Mixing within a program-objective framework is a mixing strategy that occurs when the researcher mixes quantitative and qualitative strands within an overall program objective that guides the joining of multiple projects or studies in a multiphase mixed methods project.

Mixing within a theoretical framework is a mixing strategy that occurs when the researcher mixes quantitative and qualitative strands within a transformative framework (e.g., feminism) or a substantive framework (e.g., a social science theory) that guides the overall mixed methods design.

Multilevel statewide study is a variant of the multiphase design in which different methods and phases are used to examine different levels within a system.

Multiphase combination timing occurs when the researcher implements the quantitative and qualitative methods in multiple phases that include sequential and/or concurrent timing during a research study.

Multiphase design is a mixed methods design that combines both sequential and concurrent strands, collected over a period of time, and the implementation of distinct projects or phases within an overall program of study.

Observational protocol is a form used in qualitative research to collect observational data. On this form, the researcher records a description of events and processes observed, as well as reflective notes about emerging codes, themes, and concerns that arise during the observation.

Open-ended questions are used in qualitative research to collect data. These are questions in which the researcher does not use predetermined categories or scales to collect the data.

Paradigm debate period, in the history of mixed methods, developed during the 1970s and 1980s when qualitative researchers were adamant that different assumptions provided the foundations for quantitative and qualitative research.

Parallel-databases variant is a variant of the convergent design in which two parallel strands are conducted independently and are only brought together during the interpretation phase of the study.

Participant-selection variant is a variant of the explanatory design in which the researcher places priority on the second qualitative phase but uses initial quantitative results to identify and purposefully select the best participants for qualitative study.

Participatory worldviews are influenced by political concerns, and this approach is more often associated with qualitative approaches than quantitative approaches. It includes the need to improve our society and those in it. Researchers using this worldview address issues such as empowerment, marginalization, hegemony, patriarchy, and other issues affecting marginalized groups, and they collaborate with individuals experiencing these injustices. In the end, the participatory researcher plans for the social world to be changed for the better, so that individuals will feel less marginalized.

Philosophical assumptions in mixed methods research consist of basic beliefs or assumptions that guide a research study.

Point of interface is a point in the research process of a mixed methods study where the quantitative and qualitative strands are mixed.

Postpositivism is often associated with quantitative approaches. Researchers make claims for knowledge based on (1) determinism or cause-and-effect thinking; (2) reductionism, by narrowing and focusing on select variables to interrelate; (3) detailed observations and measures of variables; and (4) the testing of theories that are continually refined.

Pragmatism, typically associated with mixed methods research, focuses on the consequences of research, on the primary importance of the question asked rather than the methods, and on the use of multiple methods of data collection to inform the problems under study.

Priority is the relative importance or weighting of the quantitative and qualitative methods in addressing the research problem in a mixed methods study.

Probabilistic sampling in quantitative research means that the researcher selects a large number of individuals who are representative of the population or who represent a segment of the population.

Procedural development period, in the history of mixed methods, is the period in which writers focused on methods of data collection, data analysis, research designs, and the purposes for conducting a mixed methods study

Purposeful sampling in qualitative research means that researchers intentionally select (or recruit) participants who have experienced the central phenomenon or the key concept being explored in the study.

Qualitative computer software programs can store text documents for analysis; enable the researcher to block and label text segments with codes so that they can be easily retrieved; organize codes into a visual, making it possible to diagram and see the relationship among them; and search for segments of text that contain multiple codes.

Qualitative data analysis involves coding the data, dividing the text into small units (phrases, sentences, or paragraphs), assigning a label to each unit, and then grouping the codes into themes.

Qualitative priority is a weighting option that occurs when a greater emphasis is placed on the qualitative methods and the quantitative methods are used in a secondary role in addressing the research problem in a mixed methods study.

Qualitative purpose statement conveys the overall qualitative purpose of the study and includes a central phenomenon, the participants, the research site for the study, and the type of qualitative design in the study.

Qualitative research questions focus or narrow the qualitative purpose statement and are stated as a central question and several subquestions. The central question and subquestions are concise, open-ended questions that begin with words such as *what* or *how* to suggest an exploration of the central phenomenon.

Qualitative study titles state a question or use literary words, such as metaphors or analogies. Qualitative titles include several components: the central phenomenon (or concept) being examined, the participants, and the site at which the study will occur. In addition, a qualitative title might include the type of qualitative research being used, such as ethnography or grounded theory.

Qualitative validity means assessing whether the information obtained through the qualitative data collection is accurate through such strategies as member-checking, triangulation of evidence, searching for disconfirming evidence, and asking others to examine the data.

Quantitative data analysis consists of analyzing the data based on the type of questions or hypotheses and using the appropriate statistical test to address the questions or hypotheses.

Quantitative priority is a weighting option that occurs when a greater emphasis is placed on the quantitative methods than the qualitative methods in addressing the research problem in a mixed methods study.

Quantitative purpose statement conveys the overall quantitative purpose of the study and includes the variables in the study, the participants, and the site for the research.

Quantitative reliability means that scores received from participants are consistent and stable over time.

Quantitative research questions and hypotheses narrow the quantitative purpose statement through research questions (that relate variables) or through hypotheses (that make predictions about the results of relating variables).

Quantitative study titles convey how investigators compare groups or relate variables. Primary variables are evident in the title, as well as the participants and possibly the site for the research study.

Quantitative validity is validity in quantitative research addressed at two levels: the quality of the scores from the instruments used and the quality of the conclusions that can be drawn from the results of the quantitative analyses.

Reflective period, in the history of mixed methods, is characterized by two intersecting themes: a current assessment of the field and a look into the future and constructive criticisms challenging the emergence of mixed methods and what it has become.

Research problems suited for mixed methods are those in which one data source may be insufficient, results need to be explained, exploratory findings need to be generalized, a second method enhances a primary method, a theoretical stance needs to be described, and an overall research objective can be best addressed with multiple phases or projects.

Sequential timing occurs when the researcher implements the quantitative and qualitative methods in two distinct phases, with the collection and analysis of one type of data occurring after the collection and analysis of the other type.

Side-by-side comparison for merged data analysis involves presenting the quantitative results and the qualitative findings together in a discussion or in a summary table so that they can be easily compared.

Single mixed methods study that combines both concurrent and sequential phases is a variant of the multiphase design in which the researcher conducts a mixed methods study in two sequential phases, with at least one of the phases including a concurrent component.

Social science theory is positioned at the beginning of mixed methods studies, and it provides a framework or theory from the social sciences that guides the nature of the questions asked and answered in a study.

Socioeconomic class lens transformative variant is a variant of the transformative design in which the researcher frames the study using a socioeconomic class theoretical lens.

Statement of the problem conveys a specific problem, or issue, that needs to be addressed in a mixed methods study and the reasons why the problem is important to study.

Steps in mixed methods data analysis refers to the procedures taken in a logical order by the researcher when conducting data analysis for a mixed methods design.

Strand is a component of a mixed methods study that encompasses the basic process of conducting quantitative or qualitative research: posing a question, collecting data, analyzing data, and interpreting results based on that data.

Standards for evaluating a qualitative study depend on the stance taken by the researcher. Qualitative researchers differ in the criteria they use, such as philosophical criteria, participatory and advocacy criteria, or procedural, methodological criteria.

Standards for evaluating a quantitative study often reflect the type of quantitative research design and the methods of data collection and analysis.

Theoretical foundation in mixed methods is a stance (or lens or standpoint) taken by the researcher that provides direction for many phases of a mixed methods project. Two types of theory might inform a mixed methods study: a social science theory or an emancipatory theory.

Theory-development variant is a variant of the exploratory design in which the researcher places priority on the initial qualitative phase and uses the ensuing quantitative phase in a secondary role to expand on the initial results.

Timing is the temporal relationship between the quantitative and qualitative strands within a mixed methods study.

Transformative design is a mixed methods design that the researcher shapes within a transformative theoretical framework seeking to address needs of a specific population and calling for change.

Transforming qualitative data into quantitative data involves reducing themes or codes to numeric information, such as dichotomous categories.

Typology and statistics merged data analysis display combines in merged analysis qualitative theme data and quantitative data based on a typology or classification.

Typology-based approach is an approach to mixed methods design that emphasizes the classification of useful mixed methods designs and the selection and adaptation of a particular design to a study's purpose and questions.

Validity in mixed methods research involves employing strategies that address potential issues in data collection, data analysis, and the interpretations that might compromise the merging or connecting of the quantitative and qualitative strands of the study.

Worldview in mixed methods research is composed of beliefs and assumptions about knowledge that informs a study.

Worldviews differ in the nature of reality (ontology), how we gain knowledge of what we know (epistemology), the role values play in research (axiology), the process of research (methodology), and the language of research (rhetoric).

REFERENCES

American Educational Research Association, American Psychological Association, National Council on Measurement in Education, and Joint Committee on Standards for Educational and Psychological Testing (United States). (1999). *Standards for educational and psychological testing*. Washington, DC: American Educational Research Association.

Ames, G. M., Duke, M. R., Moore, R. S., & Cunradi, C. B. (2009). The impact of occupational culture on drinking behavior of young adults in the U.S. Navy. *Journal of Mixed Methods Research, 3*(2), 129–150.

Andrew, S., & Halcomb, E. J. (Eds.). (2009). *Mixed methods research for nursing and the health sciences*. Chichester, West Sussex, UK: Wiley-Blackwell.

Arnon, S., & Reichel, N. (2009). Closed and open-ended question tools in a telephone survey about "the good teacher": An example of a mixed methods study. *Journal of Mixed Methods Research, 3*(2), 172–196.

Asmussen, K. J., & Creswell, J. W. (1995). Campus response to a student gunman. *Journal of Higher Education, 66*, 575–591.

Axinn, W. G., & Pearce, L. D. (2006). *Mixed method data collection strategies*. Cambridge, UK: Cambridge University Press.

Bailey, T. (2000). Character, plot, setting and time, metaphor, and voice. In T. Bailey (Ed.), *On writing short stories* (pp. 28–79). Oxford, UK: Oxford University Press.

Bamberger, M. (Ed.). (2000). *Integrating quantitative and qualitative research in development projects*. Washington, DC: World Bank.

Baumann, C. (1999). Adoptive fathers and birthfathers: A study of attitudes. *Child and Adolescent Social Work Journal, 16*(5), 373–391.

Bazeley, P. (2009). Integrating data analyses in mixed methods research [Editorial]. *Journal of Mixed Methods Research, 3*(3), 203–207.

Berger, A. A. (2000). *Media and communication research: An introduction to qualitative and quantitative approaches*. Thousand Oaks, CA: Sage.

Bernardi, L., Keim, S., & von der Lippe, H. (2007). Social influences on fertility: A comparative mixed methods study in Eastern and Western Germany. *Journal of Mixed Methods Research, 1*(1), 223–247.

Biddix, J. P. (2009, April). *Women's career pathways to the community college senior student affairs officer*. Paper presented at the meeting of the American Educational Research Association, San Diego, CA.

Bikos, L. H., Çiftçi, A., Güneri, O. Y., Demir, C. E., Sümer, Z. H., Danielson, S., et al. (2007a). A longitudinal, naturalistic inquiry of the adaptation experiences of the female expatriate spouse living in Turkey. *Journal of Career Development, 34*, 28–58.

Bikos, L. H., Çiftçi, A., Güneri, O. Y., Demir, C. E., Sümer, Z. H., Danielson, S., et al. (2007b). A repeated measures investigation of the first-year adaptation experiences of the female expatriate spouse living in Turkey. *Journal of Career Development, 34*, 5–27.

Boland, M., Daly, L., & Staines, A. (2008). Methodological issues in inclusive intellectual disability research: A health promotion needs assessment of people attending Irish disability services. *Journal of Applied Research in Intellectual Disabilities, 21*(3), 199–209.

Bradley, E. H., Curry, L. A., Ramanadhan, S., Rowe, L., Nembhard, I. M., & Krumholz, H. M. (2009). Research in action: Using positive deviance to improve quality of health care. *Implementation Science, 4*(25). doi:10.1186/1748–5908–4-25

Brady, B., & O'Regan, C. (2009). Meeting the challenge of doing an RCT evaluation of youth mentoring in Ireland: A journey in mixed methods. *Journal of Mixed Methods Research, 3*(3), 265–280.

Brett, J. A., Heimendinger, J., Boender, C., Morin, C., & Marshall, J. A. (2002). Using ethnography to improve intervention design. *American Journal of Health Promotion, 16*(6), 331–340.

Brewer, J., & Hunter, A. (1989). *Multimethod research: A synthesis of styles*. Newbury Park, CA: Sage.

Brown, J., Sorrell, J. H., McClaren, J., & Creswell, J. W. (2006). Waiting for a liver transplant. *Qualitative Health Research, 16*(1), 119–136.

Bryman, A. (1988). *Quantity and quality in social research*. London: Routledge.

Bryman, A. (2006). Integrating quantitative and qualitative research: How is it done? *Qualitative Research, 6*(1), 97–113.

Bryman, A. (2007). Barriers to integrating quantitative and qualitative research. *Journal of Mixed Methods Research, 1*(1), 8–22.

Bryman, A., Becker, S., & Sempik, J. (2008). Quality criteria for quantitative, qualitative and mixed methods research: A view from social policy. *International Journal of Social Research Methodology, 11*(4), 261–276.

Buck, G., Cook, K., Quigley, C., Eastwood, J., & Lucas, Y. (2009). Profiles of urban, low SES, African American girls' attitudes toward science: A sequential explanatory mixed methods study. *Journal of Mixed Methods Research, 3*(1), 386–410.

Bulling, D. (2005). *Development of an instrument to gauge preparedness of clergy for disaster response work: A mixed methods study*. Unpublished manuscript, University of Nebraska–Lincoln.

Campbell, D. T. (1974). *Qualitative knowing in action research*. Paper presented at the annual meeting of the American Psychological Association, New Orleans, LA.

Campbell, D. T., & Fiske, D. W. (1959). Convergent and discriminant validation by the multitrait-multimethod matrix. *Psychological Bulletin, 56*, 81–105.

Campbell, M., Fitzpatrick, R., Haines, A., Kinmonth, A. L., Sandercock, P., Spiegelhalter, D., et al. (2000). Framework for design and evaluation of complex interventions to improve health. *British Medical Journal, 321*, 694–696.

Capella-Santana, N. (2003). Voices of teacher candidates: Positive changes in multicultural attitudes and knowledge. *Journal of Educational Research, 96*(3), 182–190.

Caracelli, V. J., & Greene, J. C. (1993). Data analysis strategies for mixed-method evaluation designs. *Educational Evaluation and Policy Analysis, 15*(2), 195–207.

Caracelli, V. J., & Greene, J. C. (1997). Crafting mixed-method evaluation designs. In J. C. Greene & V. J. Caracelli (Eds.), *Advances in mixed-method evaluation: The challenges and benefits of integrating diverse paradigms* (pp. 19–32). San Francisco: Jossey-Bass.

Cartwright, E., Schow, D., & Herrera, S. (2006). Using participatory research to build an effective type 2 diabetes intervention: The process of advocacy among female Hispanic farmworkers and their families in southeast Idaho. *Women & Health, 43*(4), 89–109.

Cerda, P. R. (2005). *Family conflict and acculturation among Latino adolescents: A mixed methods study*. Unpublished manuscript, University of Nebraska–Lincoln.

Cherryholmes, C. H. (1992, August–September). Notes on pragmatism and scientific realism. *Educational Researcher, 14,* 13–17.

Christ, T. W. (2007). A recursive approach to mixed methods research in a longitudinal study of postsecondary education disability support services. *Journal of Mixed Methods Research, 1*(3), 226–241.

Christ, T. W. (2009). Designing, teaching, and evaluating two complementary mixed methods research courses. *Journal of Mixed Methods Research, 3*(4), 292–325.

Churchill, S. L., Plano Clark, V. L., Prochaska-Cue, M. K., Creswell, J W., & Ontai-Grzebik, L (2007). How rural low-income families have fun: A grounded theory study. *Journal of Leisure Research, 39*(2), 271–294.

Classen, S., Lopez, D. D. S., Winter, S., Awadz, K. D., Ferree, N., & Garvan, C. W. (2007). Population-based health promotion perspective for older driver safety: Conceptual framework to intervention plan. *Clinical Intervention in Aging, 2*(4), 677–693.

Clifton, D., & Anderson, E. (2002). *StrengthsQuest: Discover and develop your strengths in academics, career, and beyond.* Washington, DC: Gallup Organization.

Cobb, G. W. (1998). *Introduction to design and analysis of experiments.* New York: Springer.

Collins, K. M. T., Onwuegbuzie, A. J., & Sutton, I. L. (2006). A model incorporating the rationale and purpose for conducting mixed methods research in special education and beyond. *Learning Disabilities: A Contemporary Journal, 4,* 67–100.

Cook, T. D., & Reichardt, C. S. (Eds.). (1979). *Qualitative and quantitative methods in evaluation research.* Beverly Hills, CA: Sage.

Corrigan, M. W., Pennington, B., & McCroskey, J. C. (2006). Are we making a difference?: A mixed methods assessment of the impact of intercultural communication instruction on American students. *Ohio Communication Journal, 44,* 1–32.

Creswell, J. D., Welch, W. T., Taylor, S. E., Sherman, D. K., Greunewald, T. L., & Mann, T. (2005). Affirmation of personal values buffers neuroendocrine and psychological stress responses. *Psychological Science, 16,* 846–851.

Creswell, J. W. (1994). *Research design: Qualitative and quantitative approaches.* Thousand Oaks, CA: Sage.

Creswell, J. W. (1999). Mixed-method research: Introduction and application. In G. J. Cizek (Ed.), *Handbook of educational policy* (pp. 455–472). San Diego, CA: Academic Press.

Creswell, J. W. (2003). *Research design: Qualitative, quantitative, and mixed methods approaches* (2nd ed.). Thousand Oaks, CA: Sage.

Creswell, J. W. (Facilitator). (2005, May). *Mixed methods.* Workshop hosted by the Veterans Affairs Ann Arbor Health Care System, Center for Practice Management and Outcomes Research, Ann Arbor, MI.

Creswell, J. W. (2007). *Qualitative inquiry and research design: Choosing among five approaches* (2nd ed.). Thousand Oaks, CA: Sage.

Creswell, J. W. (2008a, July 21). *How mixed methods has developed.* Keynote address for the 4th Annual Mixed Methods Conference, Fitzwilliam College, Cambridge University, UK.

Creswell, J. W. (2008b). *Educational research: Planning, conducting, and evaluating quantitative and qualitative research* (3rd ed.). Upper Saddle River, NJ: Pearson Education.

Creswell, J. W. (2009a). *How SAGE has shaped research methods.* London: Sage.

Creswell, J. W. (2009b). Mapping the field of mixed methods research [Editorial]. *Journal of Mixed Methods Research, 3*(2), 95–108.

Creswell, J. W. (2009c). *Research design: Qualitative, quantitative, and mixed methods approaches* (3rd ed.). Thousand Oaks, CA: Sage.

Creswell, J. W. (in press-a). Controversies in mixed methods research. In N. K. Denzin & Y. S. Lincoln (Eds.), *The SAGE handbook of qualitative research* (4th ed.). Thousand Oaks, CA: Sage.

Creswell, J. W. (in press-b). Mapping the developing landscape of mixed methods research. In

A. Tashakkori & C. Teddlie (Eds.), *SAGE handbook of mixed methods research in social & behavioral research* (2nd ed.). Thousand Oaks, CA: Sage.

Creswell, J. W., Fetters, M. D., & Ivankova, N. V. (2004). Designing a mixed methods study in primary care. *Annals of Family Medicine, 2*(1), 7–12.

Creswell, J. W., Fetters, M. D., Plano Clark, V. L., & Morales, A. (2009). Mixed methods intervention trials. In S. Andrew & L. Halcomb (Eds.), *Mixed methods research for nursing and the health sciences.* Oxford, UK: Blackwell.

Creswell, J. W., Goodchild, L. F., & Turner, P. (1996). Integrated qualitative and quantitative research: Epistemology, history, and designs. In J. C. Smart (Ed.), *Higher education: Handbook of theory and research* (Vol. 11, pp. 90–136). New York: Agathon Press.

Creswell, J. W., & Maietta, R. C. (2002). Qualitative research. In D. C. Miller & N. J. Salkind (Eds.), *Handbook of social research* (pp. 143–184). Thousand Oaks, CA: Sage.

Creswell, J. W., & McCoy, B. R. (in press). The use of mixed methods thinking in documentary development. In S. N. Hesse-Biber (Ed.), *The handbook of emergent technologies in social research.* Oxford, UK: Oxford University Press.

Creswell, J. W., & Miller, D. L. (2000). Determining validity in qualitative inquiry. *Theory into Practice, 39*(3), 124–130.

Creswell, J. W., & Plano Clark, V. L. (2007). *Designing and conducting mixed methods research.* Thousand Oaks, CA: Sage.

Creswell, J. W., Plano Clark, V. L., Gutmann, M., & Hanson, W. (2003). Advanced mixed methods research designs. In A. Tashakkori & C. Teddlie (Eds.), *Handbook of mixed methods in social & behavioral research* (pp. 209–240). Thousand Oaks, CA: Sage.

Creswell, J. W., & Tashakkori, A. (2007). Developing publishable mixed methods manuscripts [Editorial]. *Journal of Mixed Methods Research, 1*(2), 107–111.

Creswell, J. W., Tashakkori, A., Jensen, K. D., & Shapley, K. L. (2003). Teaching mixed methods research: Practices, dilemmas, and challenges. In A. Tashakkori & C. Teddlie (Eds.), *Handbook of mixed methods in social & behavioral research* (pp. 619–637). Thousand Oaks, CA: Sage.

Creswell, J. W., & Zhang, W. (2009). The application of mixed methods designs to trauma research. *Journal of Traumatic Stress, 22*(6), 612–621.

Cronbach, L. J. (1975). Beyond the two disciplines of scientific psychology. *American Psychologist, 30,* 116–127.

Crotty, M. (1998). *The foundations of social research: Meaning and perspective in the research process.* London: Sage.

Curry, L. A., Nembhard, I. M., & Bradley, E. H. (2009). Qualitative and mixed methods provide unique contributions to outcomes research. *Circulation, 119,* 1442–1452.

Daley, C. E., & Onwuegbuzie, A. J. (2010). Attributions toward violence of male juvenile delinquents: A concurrent mixed method analysis. *Journal of Psychology, 144*(6), 549–570.

Dellinger, A. B., & Leech, N. L. (2007). Toward a unified validation framework in mixed methods research. *Journal of Mixed Methods Research, 1*(4), 309–332.

Denzin, N. K. (1978). *The research act: A theoretical introduction to sociological methods.* New York: McGraw-Hill.

Denzin, N. K., & Lincoln, Y. S. (Eds.). (2005). *The SAGE handbook of qualitative research* (3rd ed.). Thousand Oaks, CA: Sage.

Denscombe, M. (2008). Communities of practice: A research paradigm for the mixed methods approach. *Journal of Mixed Methods Research, 2,* 270–283.

DeVellis, R. F. (1991). *Scale development: Theory and application.* Newbury Park, CA: Sage.

Donovan, J., Mills, N., Smith, M., Brindle, L., Jacoby, A., Peters, T., et al. (2002). Improving design and conduct of randomised trials by embedding them in qualitative research: ProtecT

(Prostate Testing for Cancer and Treatment) study. *British Medical Journal, 325,* 766–769.

Elliot, J. (2005). *Using narrative in social research: Qualitative and quantitative approaches.* London: Sage.

Engel, R. J., & Schutt, R. K. (2009). *The practice of research in social work* (2nd ed.). Thousand Oaks, CA: Sage.

Evans, L., & Hardy, L. (2002a). Injury rehabilitation: A goal-setting intervention study. *Research Quarterly for Exercise & Sport, 73,* 310–319.

Evans, L., & Hardy, L. (2002b). Injury rehabilitation: A qualitative follow-up study. *Research Quarterly for Exercise & Sport, 73,* 320–329.

Farmer, J., & Knapp, D. (2008). Interpretation programs at a historic preservation site: A mixed methods study of long-term impact. *Journal of Mixed Methods Research, 2*(4), 340–361.

Feldon, D. F., & Kafai, Y. B. (2008). Mixed methods for mixed reality: Understanding users' avatar activities in virtual worlds. *Educational Technology Research and Development, 56*(5–6), 575–593.

Fetters, M. D., Yoshioka, T., Greenberg, G. M., Gorenflo, D. W., & Yeo, S. (2007). Advance consent in Japanese during prenatal care for epidural anesthesia during childbirth. *Journal of Mixed Methods Research, 1*(4), 333–365.

Fielding, N., & Fielding, J. (1986). *Linking data: The articulation of qualitative and quantitative methods in social research.* Beverly Hills, CA: Sage.

Fielding, N. G., & Cisneros-Puebla, C. A. (2009). CAQDAS-GIS convergence: Towards a new integrated mixed methods research practice? *Journal of Mixed Methods Research, 3*(4), 349–370.

Filipas, H. H., & Ullman, S. E. (2001). Social reactions to sexual assault victims from various support sources. *Violence and Victims, 16*(6), 673–692.

Flory, J., & Emanuel, E. (2004). Interventions to improve research participants' understanding of informed consent for research. *Journal of the American Medical Association, 13,* 1593–1601.

Forman, J., & Damschroder, L. (2007, February). *Using mixed methods in evaluating intervention studies.* Presentation at the Mixed Methodology Workshop, VA HSR&D National Meeting, Arlington, VA.

Fowler, F. J., Jr. (2008). *Survey research methods* (4th ed.). Thousand Oaks, CA: Sage.

Freshwater, D. (2007). Reading mixed methods research: Contexts for criticism. *Journal of Mixed Methods Research, 1*(2), 134–145.

Fries, C. J. (2009). Bourdieu's reflexive sociology as a theoretical basis for mixed methods research: An application to complementary and alternative medicine. *Journal of Mixed Methods Research, 3*(2), 326–348.

Giddings, L. S. (2006). Mixed-methods research: Positivism dressed in drag? *Journal of Research in Nursing, 11*(3), 195–203.

Goldenberg, C., Gallimore, R., & Reese, L. (2005). Using mixed methods to explore Latino children's literacy development. In T. S. Weisner (Ed.), *Discovering successful pathways in children's development: Mixed methods in the study of childhood and family life* (pp. 21–46). Chicago: University of Chicago Press.

Greene, J. C. (2007). *Mixed methods in social inquiry.* San Francisco: Jossey-Bass.

Greene, J. C. (2008). Is mixed methods social inquiry a distinctive methodology? *Journal of Mixed Methods Research, 2*(1), 7–22.

Greene, J. C., & Caracelli, V. J. (Eds.). (1997). *Advances in mixed-method evaluation: The challenges and benefits of integrating diverse paradigms: New directions for evaluation, 74.* San Francisco: Jossey-Bass.

Greene, J. C., Caracelli, V. J., & Graham, W. F. (1989). Toward a conceptual framework for mixed-method evaluation designs. *Educational Evaluation and Policy Analysis, 11*(3), 255–274.

Greenstein, T. N. (2006). *Methods of family research* (2nd ed.). Thousand Oaks, CA: Sage.

Guba, E. G., & Lincoln, Y. S. (1988). Do inquiry paradigms imply inquiry methodologies? In

D. M. Fetterman (Ed.), *Qualitative approaches to evaluation in education* (pp. 89–115). New York: Praeger.

Guba, E. G., & Lincoln, Y. S. (2005). Paradigmatic controversies, contradictions, and emerging confluences. In N. K. Denzin & Y. S. Lincoln (Eds.), *The SAGE handbook of qualitative research* (3rd ed., pp. 191–215). Thousand Oaks, CA: Sage.

Haines, C. (2010). *Value added by mixed methods research*. Unpublished manuscript, University of Nebraska–Lincoln.

Hall, B., & Howard, K. (2008). A synergistic approach: Conducting mixed methods research with typological and systemic design considerations. *Journal of Mixed Methods Research*, *2*(3), 248–269.

Hall, B. W., Ward, A. W., & Comer, C. B. (1988). Published educational research: An empirical study of its quality. *Journal of Educational Research*, *81*, 182–189.

Hanson, W. E., Creswell, J. W., Plano Clark, V. L., Petska, K. P., & Creswell, J. D. (2005). Mixed methods research designs in counseling psychology. *Journal of Counseling Psychology*, *52*(2), 224–235.

Harrison, A. (2005). *Correlates of positive relationship-building in a teacher education mentoring program*. Unpublished doctoral dissertation proposal, University of Nebraska–Lincoln.

Harrison, R. L. (2010). *Mixed methods designs in marketing research*. Unpublished manuscript, University of Nebraska–Lincoln.

Hesse-Biber, S. N., & Leavy, P. (2006). *The practice of qualitative research*. Thousand Oaks, CA: Sage.

Hilton, B. A., Budgen, C., Molzahn, A. E., & Attridge, C. B. (2001). Developing and testing instruments to measure client outcomes at the Comox Valley Nursing Center. *Public Health Nursing*, *18*, 327–339.

Hodgkin, S. (2008). Telling it all: A story of women's social capital using a mixed methods approach. *Journal of Mixed Methods Research*, *2*(3), 296–316.

Holmes, C. A. (2006, July). Mixed (up) methods, methodology and interpretive frameworks. Paper presented at the Mixed Methods Conference, Cambridge, UK.

Howe, K. R. (2004). A critique of experimentalism. *Qualitative Inquiry*, *10*, 42–61.

Ibrahim, M. F., & Leng, S. K. (2003). Shoppers' perceptions of retail developments: Suburban shopping centres and night markets in Singapore. *Journal of Retail & Leisure Property*, *3*(2), 176–189.

Idler, E. L., Hudson, S. V., & Leventhal, H. (1999). The meanings of self ratings of health: A qualitative and quantitative approach. *Research on Aging*, *21*(3), 458–476.

Igo, L. B., Kiewra, K. A., & Bruning, R. (2008). Individual differences and intervention flaws: A sequential explanatory study of college students' copy-and-paste note taking. *Journal of Mixed Methods Research*, *2*(2), 149–168.

Igo, L. B., Riccomini, P. J., Bruning, R. H., & Pope, G. G. (2006). How should middle-school students with LD approach online note taking? A mixed-methods study. *Learning Disability Quarterly*, *29*, 89–100.

Ivankova, N. V., Creswell, J. W., & Stick, S. (2006). Using mixed methods sequential explanatory design: From theory to practice. *Field Methods*, *18*(1), 3–20.

Ivankova, N. V., & Stick, S. L. (2007). Students' persistence in a Distributed Doctoral Program in Educational Leadership in Higher Education: A mixed methods study. *Research in Higher Education*, *48*(1), 93–135.

Jick, T. D. (1979). Mixing qualitative and quantitative methods: Triangulation in action. *Administrative Science Quarterly*, *24*, 602–611.

Johnson, R. B., & Onwuegbuzie, A. J. (2004). Mixed methods research: A research paradigm whose time has come. *Educational Researcher*, *33*(7), 14–26.

Johnson, R. B., Onwuegbuzie, A. J., & Turner, L. A. (2007). Toward a definition of mixed methods research. *Journal of Mixed Methods Research*, *1*(2), 112–133.

Johnstone, P. L. (2004). Mixed methods, mixed methodology health services research in practice. *Qualitative Health Research*, *14*, 239–271.

Kelle, U. (2006). Combining qualitative and quantitative methods in research practice: Purposes and advantages. *Qualitative Research in Psychology*, *3*, 293–311.

Kelley-Baker, T., Voas, R. B., Johnson, M. B., Furr-Holden, C. D., & Compton, C. (2007). Multimethod measurement of high-risk drinking locations: Extending the portal survey method with follow-up telephone interviews. *Evaluation Review*, *31*(5), 490–507.

Kennett, D. J., O'Hagan, F. T., & Cezer, D. (2008). Learned resourcefulness and the long-term benefits of a chronic pain management program. *Journal of Mixed Methods Research*, *2*(4), 317–339.

Knodel, J., & Saengtienchai, C. (2005). Older-aged parents: The final safety net for adult sons and daughters with AIDS in Thailand. *Journal of Family Issues*, *26*(5), 665–698.

Kruger, B. (2006). Family-nurse care coordination partnership [Grant No. 1R21NR009781–01]. Abstract obtained from RePORTER database: http://projectreporter.nih.gov/reporter.cfm

Kuckartz, U. (2009). Realizing mixed-methods approaches with MAXQDA. Unpublished manuscript, Philipps-Universitaet Marburg, Marburg, Germany.

Kuhn, T. S. (1970). *The structure of scientific revolutions* (2nd ed.). Chicago: University of Chicago Press.

Kumar, M. S., Mudaliar, S. M., Thyagarajan, S. P., Kumar, S., Selvanayagam, A., & Daniels, D. (2000). Rapid assessment and response to injecting drug use in Madras, south India. *International Journal of Drug Policy*, *11*, 83–98.

Kutner, J. S., Steiner, J. F., Corbett, K. K., Jahnigen, D. W., & Barton, P. L. (1999). Information needs in terminal illness. *Social Science and Medicine*, *48*, 1341–1352.

Lee, Y. J., & Greene, J. (2007). The predictive validity of an ESL placement test: A mixed methods approach. *Journal of Mixed Methods Research*, *1*(4), 366–389.

Leech, N. L., Dellinger, A. B., Brannagan, K. B., & Tanaka, H. (2010). Evaluating mixed research studies: A mixed methods approach. *Journal of Mixed Methods Research*, *4*(1), 17–31.

Lehan-Mackin, M. (2007). The social context of unintended pregnancy in college-aged women [Grant No. 5F31NR010287–02]. Abstract obtained from RePORTER database: http://projectreporter.nih.gov/reporter.cfm

Li, S., Marquart, J. M., & Zercher, C. (2000). Conceptual issues and analytic strategies in mixed-methods studies of preschool inclusion. *Journal of Early Intervention*, *23*(2), 116–132.

Lincoln, Y. S., & Guba, E. G. (1985). *Naturalistic inquiry*. Beverly Hills, CA: Sage.

Lincoln, Y. S., & Guba, E. G. (2000). Paradigmatic controversies, contradictions, and emerging confluences. In N. K. Denzin & Y. S. Lincoln (Eds.), *Handbook of qualitative research* (2nd ed.) (pp. 163–188). Thousand Oaks, CA: Sage.

Lipsey, M. W. (1990). *Design sensitivity: Statistical power for experimental research*. Newbury Park, CA: Sage.

Luck, L., Jackson, D., & Usher, K. (2006). Case study: A bridge across the paradigms. *Nursing Inquiry*, *13*(2), 103–109.

Luzzo, D. A. (1995). Gender differences in college students' career maturity and perceived barriers in career development. *Journal of Counseling and Development*, *73*, 319–322.

Mak, L., & Marshall, S. K. (2004). Perceived mattering in young adults' romantic relationships. *Journal of Social and Personal Relationships*, *24*(4), 469–486.

Malterud, K. (2001). The art and science of clinical knowledge: Evidence beyond measures and numbers. *Lancet, 358*, 397–400.

Maresh, M. M. (2009). *Exploring hurtful communication from college teachers to students: A mixed methods study*. Doctoral dissertation, University of Nebraska–Lincoln.

Maxwell, J. A., & Loomis, D. M. (2003). Mixed methods design: An alternative approach. In A. Tashakkori & C. Teddlie (Eds.), *Handbook of mixed methods in social & behavioral research* (pp. 241–271). Thousand Oaks, CA: Sage.

Maxwell, J. A., & Mittapalli, K. (in press). Realism as a stance for mixed methods research. In A. Tashakkori & C. Teddlie (Eds.), *SAGE handbook of mixed methods in social & behavioral research* (2nd ed.). Thousand Oaks, CA: Sage.

May, D. B., & Etkina, E. (2002). College physics students' epistemological self-reflection and its relationship to conceptual learning. *American Journal of Physics, 70*(12), 1249–1258.

Mayring, P. (2007). Introduction: Arguments for mixed methodology. In P. Mayring, G. L. Huber, L. Gurtler, & M. Kiegelmann (Eds.), *Mixed methodology in psychological research* (pp. 1–4). Rotterdam/Taipei: Sense Publishers.

McAuley, C., McCurry, N., Knapp, M., Beecham, J., & Sleed, M. (2006). Young families under stress: Assessing maternal and child well-being using a mixed-methods approach. *Child and Family Social Work, 11*(1), 43–54.

McEntarffer, R. (2003). *Strengths-based mentoring in teacher education: A mixed methods study*. Unpublished master's thesis, University of Nebraska–Lincoln.

McMahon, S. (2007). Understanding community-specific rape myths: Exploring student athlete culture. *Affilia, 22*, 357–370.

McVea, K., Crabtree, B. F., Medder, J. D., Susman, J. L., Lukas, L., McIlvain, H. E., et al. (1996). An ounce of prevention? Evaluation of the "Put Prevention into Practice" program. *Journal of Family Practice, 43*(4), 361–369.

Meijer, P. C., Verloop, N., & Beijaard, D. (2001). Similarities and differences in teachers' practical knowledge about teaching reading comprehension. *Journal of Educational Research, 94*(3), 171–184.

Mendlinger, S., & Cwikel, J. (2008). Spiraling between qualitative and quantitative data on women's health behaviors: A double helix model for mixed methods. *Qualitative Health Research, 18*(2), 280–293.

Mertens, D. M. (2003). Mixed methods and the politics of human research: The transformative-emancipatory perspective. In A. Tashakkori & C. Teddlie (Eds.), *Handbook of mixed methods in social & behavioral research* (pp. 135–164). Thousand Oaks, CA: Sage.

Mertens, D. M. (2005). *Research and evaluation in education and psychology: Integrating diversity with quantitative, qualitative, and mixed methods* (2nd ed.). Thousand Oaks, CA: Sage.

Mertens, D. M. (2007). Transformative paradigm: Mixed methods and social justice. *Journal of Mixed Methods Research, 1*(1), 212–225.

Mertens, D. M. (2009). *Transformative research and evaluation*. New York: Guilford Press.

Miles, M. B., & Huberman, A. M. (1994). *Qualitative data analysis: An expanded sourcebook* (2nd ed.). Thousand Oaks, CA: Sage.

Milton, J., Watkins, K. E., Studdard, S. S., & Burch, M. (2003). The ever widening gyre: Factors affecting change in adult education graduate programs in the United States. *Adult Education Quarterly, 54*(1), 23–41.

Mirza, M., Anandan, N., Madnick, F., & Hammel, J. (2006). A participatory program evaluation of a systems change program to improve access to information technology by people with disabilities. *Disability and Rehabilitation, 28*(19), 1185–1199.

Mizrahi, T., & Rosenthal, B. B. (2001). Complexities of coalition building: Leaders' successes, strategies, struggles, and solutions. *Social Work, 46*(1), 63–78.

Morales, A. (2005). *Family dynamics of Latino language brokers: A mixed methods study*. Unpublished manuscript, University of Nebraska–Lincoln.

Morell, L., & Tan, R. J. B. (2009). Validating for use and interpretation: A mixed methods contribution illustrated. *Journal of Mixed Methods Research, 3*(3), 242–264.

Morgan, D. L. (1998). Practical strategies for combining qualitative and quantitative methods: Applications to health research. *Qualitative Health Research, 8*(3), 362–376.

Morgan, D. L. (2007). Paradigms lost and pragmatism regained: Methodological implications of combining qualitative and quantitative methods. *Journal of Mixed Methods Research, 1*(1), 48–76.

Morse, J. M. (1991). Approaches to qualitative-quantitative methodological triangulation. *Nursing Research, 40,* 120–123.

Morse, J. M. (2003). Principles of mixed methods and multimethod research design. In A. Tashakkori & C. Teddlie (Eds.), *Handbook of mixed methods in social & behavioral research* (pp. 189–208). Thousand Oaks, CA: Sage.

Morse, J. M., & Niehaus, L. (2009). *Mixed methods design: Principles and procedures.* Walnut Creek, CA: Left Coast Press.

Morse, J., & Richards, L. (2002). *Readme first: For a user's guide to qualitative methods.* Thousand Oaks, CA: Sage.

Muñoz, M. (2010). In their own words and by the numbers: A mixed-methods study of Latina community college presidents. *Community College Journal of Research and Practice, 34*(1), 153–174.

Murphy, J. P. (1990). *Pragmatism: From Peirce to Davidson.* Boulder, CO: Westview.

Myers, K. K., & Oetzel, J. G. (2003). Exploring the dimensions of organizational assimilation: Creating and validating a measure. *Communication Quarterly, 51*(4), 438–457.

Nastasi, B. K., Hitchcock, J., Sarkar, S., Burkholder, G., Varjas, K., & Jayasena, A. (2007). Mixed methods in intervention research: Theory to adaptation. *Journal of Mixed Methods Research, 1*(2), 164–182.

National Institutes of Health (NIH). (1999). *Qualitative methods in health research: Opportunities and considerations in application and review.* Washington, DC: Author.

National Research Council. (2002). *Scientific research in education.* Washington, DC: National Academy Press.

Newman, I., & Benz, C. R. (1998). *Qualitative-quantitative research methodology: Exploring the interactive continuum.* Carbondale: Southern Illinois University Press.

Newman, K., & Wyly, E. K. (2006). The right to stay put, revisited: Gentrification and resistance to displacement in New York City. *Urban Studies, 43*(1), 23–57.

O'Cathain, A. (in press). Assessing the quality of mixed methods research: Towards a comprehensive framework. In A. Tashakkori & C. Teddlie (Eds.), *SAGE Handbook of mixed methods in social & behavioral research* (2nd ed.). Thousand Oaks, CA: Sage.

O'Cathain, A., Murphy, E., & Nicholl, J. (2007). Integration and publications as indicators of "yield" from mixed methods studies. *Journal of Mixed Methods Research, 1*(2), 147–163.

O'Cathain, A., Murphy, E., & Nicholl, J. (2008). The quality of mixed methods studies in health services research. *Journal of Health Services Research and Policy, 13*(2), 92–98.

Olivier, T., de Lange, N., Creswell, J. W., & Wood, L. (2010). *Linking visual methodology and mixed methods research in a video production of educational change.* Unpublished manuscript, Nelson Mandela Metropolitan University, Port Elizabeth, South Africa.

Onwuegbuzie, A. J., & Johnson, R. B. (2006). The validity issue in mixed research. *Research in the Schools, 13*(1), 48–63.

Onwuegbuzie, A. J., & Leech, N. L. (2006). Linking research questions to mixed methods data analysis procedures. *The Qualitative Report, 11*(3), 474–498. Retrieved from http://www.nova.edu/ssss/QR/QR11–3/onwue gbuzie.pdf.

Onwuegbuzie, A. J., & Leech, N. L. (2009). Lessons learned for teaching mixed research: A framework for novice researchers. *International*

Journal of Multiple Research Approaches, 3, 105–107.

Onwuegbuzie, A. J., & Teddlie, C. (2003). A framework for analyzing data in mixed methods research. In A. Tashakkori & C. Teddlie (Eds.), *Handbook of mixed methods in social & behavioral research* (pp. 351–383). Thousand Oaks, CA: Sage.

Oshima, T. C., & Domaleski, C. S. (2006). Academic performance gap between summer-birthday and fall-birthday children in grades K–8. *Journal of Educational Research, 99*(4), 212–217.

Padgett, D. K. (2004). Mixed methods, serendipity, and concatenation. In D. K. Padgett (Ed.), *The qualitative research experience* (pp. 273–288). Belmont, CA: Wadsworth/Thomson Learning.

Pagano, M. E., Hirsch, B. J., Deutsch, N. L., & McAdams, D. P. (2002). The transmission of values to school-age and young adult offspring: Race and gender differences in parenting. *Journal of Feminist Family Therapy, 14*(3/4), 13–36.

Parmelee, J. H., Perkins, S. C., & Sayre, J. J. (2007). "What about people our age?" Applying qualitative and quantitative methods to uncover how political ads alienate college students. *Journal of Mixed Methods Research, 1*(2), 183–199.

Patton, M. Q. (1980). *Qualitative evaluation and research methods.* Newbury Park, CA: Sage.

Patton, M. Q. (1990). *Qualitative evaluation and research methods* (2nd ed.). Newbury Park, CA: Sage.

Paul, J. L. (2005). *Introduction to the philosophies of research and criticism in education and the social sciences.* Upper Saddle River, NJ: Pearson Education.

Payne, Y. A. (2008). "Street life" as a site of resiliency: How street life oriented Black men frame opportunity in the United States. *Journal of Black Psychology, 34*(1), 3–31.

Phillips, D. C., & Burbules, N. C. (2000). *Postpositivism and educational research.* Lanham, MD: Rowman & Littlefield.

Plano Clark, V. L. (2005). Cross-disciplinary analysis of the use of mixed methods in physics education research, counseling psychology, and primary care. Doctoral dissertation, University of Nebraska–Lincoln, 2005. *Dissertation Abstracts International, 66,* 02A.

Plano Clark, V. L. (2010). The adoption and practice of mixed methods: U.S. trends in federally funded health-related research. *Qualitative Inquiry.* Prepublished April 15, 2010, DOI: 10.1177/1077800410364609

Plano Clark, V. L., & Badiee, M. (in press). Research questions in mixed methods research. In A. Tashakkori & C. Teddlie (Eds.), *SAGE Handbook of mixed methods in social & behavioral research* (2nd ed.). Thousand Oaks, CA: Sage.

Plano Clark, V. L., & Creswell, J. W. (2010). *Understanding research: A consumer's guide.* Upper Saddle River, NJ: Pearson Education.

Plano Clark, V. L., & Galt, K. (2009, April). *Using a mixed methods approach to strengthen instrument development and validation.* Paper presented at the annual meeting of the American Pharmacists Association, San Antonio, TX.

Plano Clark, V. L., Huddleston-Casas, C. A., Churchill, S. L., Green, D. O., & Garrett, A. L. (2008). Mixed methods approaches in family science research. *Journal of Family Issues, 29*(11), 1543–1566.

Plano Clark, V. L., & Wang, S. C. (2010). Adapting mixed methods research to multicultural counseling. In J. G. Ponterotto, J. M. Casas, L. A. Suzuki, & C. M. Alexander (Eds.), *Handbook of multicultural counseling* (3rd ed., pp. 427–438). Thousand Oaks, CA: Sage.

Powell, H., Mihalas, S., Onwuegbuzie, A. J., Suldo, S., & Daley, C. E. (2008). Mixed methods research in school psychology: A mixed methods investigation of trends in the literature. *Psychology in the Schools, 45*(4), 291–308.

Punch, K. F. (1998). *Introduction to social research: Quantitative and qualitative approaches.* London: Sage.

Quinlan, E., & Quinlan, A. (2010). Representations of rape: Transcending methodological divides. *Journal of Mixed Methods Research, 4*(2), 127–143.

Ragin, C. C., Nagel, J., & White, P. (2004). *Workshop on scientific foundations of qualitative research* [Report]. Retrieved from the National Science Foundation Web site: http://www.nsf.gov/pubs/2004/nsf04219/nsf04219.pdf.

Ras, N. L. (2009, April). *Multidimensional theory and data interrogation in educational change research: A mixed methods case study.* Paper presented at the meeting of the American Educational Research Association, San Diego, CA.

Reichardt, C. S., & Rallis, S. F. (Eds.). (1994). *The qualitative-quantitative debate: New perspectives.* San Francisco: Jossey-Bass.

Rogers, A., Day, J., Randall, F., & Bentall, R. P. (2003). Patients' understanding and participation in a trial designed to improve the management of anti-psychotic medication: A qualitative study. *Social Psychiatry and Psychiatric Epidemiology, 38,* 720–727.

Rossman, G. B., & Wilson, B. L. (1985). Numbers and words: Combining quantitative and qualitative methods in a single large-scale evaluation study. *Evaluation Review, 9*(5), 627–643.

Saewyc, E. M. (2003). *Enacted stigma, gender & risk behaviors of school youth* [Grant No. 1R01DA017979–01]. Abstract obtained from RePORTER database: http://projectreporter.nih.gov/reporter.cfm

Sandelowski, M. (1996). Using qualitative methods in intervention studies. *Research in Nursing & Health, 19*(4), 359–364.

Sandelowski, M. (2000). Combining qualitative and quantitative sampling, data collection, and analysis techniques in mixed-method studies. *Research in Nursing & Health, 23,* 246–255.

Sandelowski, M. (2003). Tables or tableaux? The challenges of writing and reading mixed methods studies. In A. Tashakkori & C. Teddlie (Eds.), *Handbook of mixed methods in social & behavioral research* (pp. 321–350). Thousand Oaks, CA: Sage.

Sandelowski, M., Voils, C. I., & Knafl, G. (2009). On quantitizing. *Journal of Mixed Methods Research, 3*(3), 208–222.

Schillaci, M. A., Waitzkin, H., Carson, E. A., Lopez, C. M., Boehm, D. A., Lopez, L. A., et al. (2004). Immunization coverage and Medicaid managed care in New Mexico: A multimethod assessment. *Annals of Family Medicine, 2*(1), 13–21.

Shapiro, M., Setterlund, D., & Cragg, C. (2003). Capturing the complexity of women's experiences: A mixed-method approach to studying incontinence in older women. *Affilia, 18,* 21–33.

Sieber, S. D. (1973). The integration of fieldwork and survey methods. *American Journal of Sociology, 78,* 1335–1359.

Skinner, D., Matthews, S., & Burton, L. (2005). Combining ethnography and GIS technology to examine constructions of developmental opportunities in contexts of poverty and disability. In T. S. Weisner (Ed.), *Discovering successful pathways in children's development: Mixed methods in the study of childhood and family life* (pp. 223–239). Chicago: University of Chicago Press.

Slife, B. D., & Williams, R. N. (1995). *What's behind the research? Discovering hidden assumptions in the behavioral sciences.* Thousand Oaks, CA: Sage.

Slonim-Nevo, V., & Nevo, I. (2009). Conflicting findings in mixed methods research. *Journal of Mixed Methods Research, 3*(2), 109–128.

Smith, J. K. (1983). Quantitative versus qualitative research: An attempt to clarify the issue. *Educational Researcher, 12*(3), 6–13.

Snowdon, C., Garcia, J., & Elbourne, D. (1998). Reactions of participants to the results of a randomized controlled trial: Exploratory study. *British Medical Journal, 317,* 21–26.

Stake, R. (1995). *The art of case study research.* Thousand Oaks, CA: Sage.

Stange, K. C., Crabtree, B. F., & Miller, W. L. (2006). Publishing multimethod research. *Annals of Family Medicine, 4,* 292–294.

Steckler, A., McLeroy, K. R., Goodman, R. M., Bird, S. T., & McCormick, L. (1992). Toward integrating qualitative and quantitative methods: An introduction. *Health Education Quarterly, 19*(1), 1–8.

Stenner, P., & Rogers, R. S. (2004). Q methodology and qualiquantology. In Z. Tood, B. Nerlich, S. McKeown, & D. D. Clarke (Eds.), *Mixing methods in psychology: The integration of qualitative and quantitative methods in theory and practice* (pp. 101–120). Hove, The Netherlands, and New York: Psychology Press.

Sweetman, D., Badiee, M., & Creswell, J. W. (2010). Use of the transformative framework in mixed methods studies. *Qualitative Inquiry*. Prepublished April 15, 2010, DOI: 10.1177/1077800410364610

Tashakkori, A. (2009). Are we there yet?: The state of the mixed methods community [Editorial]. *Journal of Mixed Methods Research, 3*(4), 287–291.

Tashakkori, A., & Creswell, J. W. (2007a). Exploring the nature of research questions in mixed methods research [Editorial]. *Journal of Mixed Methods Research, 1*(3), 207–211.

Tashakkori, A., & Creswell, J. W. (2007b). The new era of mixed methods [Editorial]. *Journal of Mixed Methods Research, 1*(1), 3–7.

Tashakkori, A., & Teddlie, C. (1998). *Mixed methodology: Combining qualitative and quantitative approaches*. Thousand Oaks, CA: Sage.

Tashakkori, A., & Teddlie, C. (Eds.). (2003a). *Handbook of mixed methods in social & behavioral research*. Thousand Oaks, CA: Sage.

Tashakkori, A., & Teddlie, C. (2003b). The past and future of mixed methods research: From data triangulation to mixed model designs. In A. Tashakkori & C. Teddlie (Eds.), *Handbook of mixed methods in social & behavioral research* (pp. 671–701). Thousand Oaks, CA: Sage.

Tashakkori, A., & Teddlie, C. (Eds.). (in press). *SAGE handbook of mixed methods in social & behavioral research* (2nd ed.) Thousand Oaks, CA: Sage.

Tashiro, J. (2002). Exploring health promoting lifestyle behaviors of Japanese college women: Perceptions, practices, and issues. *Health Care for Women International, 23*, 59–70.

Teddlie, C., & Stringfield, S. (1993). *Schools make a difference: Lessons learned from a 10-year study of school effects*. New York: Teachers College Press.

Teddlie, C., & Tashakkori, A. (2009). *Foundations of mixed methods research: Integrating quantitative and qualitative approaches in the social and behavioral sciences*. Thousand Oaks, CA: Sage.

Teddlie, C., & Yu, F. (2007). Mixed methods sampling: A typology with examples. *Journal of Mixed Methods Research, 1*(1), 77–100.

Teno, J. M., Stevens, M., Spernak, S., & Lynn, J. (1998). Role of written advance directives in decision making. *Journal of General Internal Medicine, 13*, 439–446.

Thøgersen-Ntoumani, C., & Fox, K. R. (2005). Physical activity and mental well-being typologies in corporate employees: A mixed methods approach. *Work & Stress, 19*(1), 50–67.

Victor, C. R., Ross, F., & Axford, J. (2004). Capturing lay perspectives in a randomized control trial of a health promotion intervention for people with osteoarthritis of the knee. *Journal of Evaluation in Clinical Practice, 10*(1), 63–70.

Vidich, A. J., & Shapiro, G. (1955). A comparison of participant observation and survey data. *American Sociological Review, 20*, 28–33.

Vrkljan, B. H. (2009). Constructing a mixed methods design to explore the older drive-copilot relationship. *Journal of Mixed Methods Research, 3*(4), 371–385.

Way, N., Stauber, H. Y., Nakkula, M. J., & London, P. (1994). Depression and substance use in two divergent high school cultures: A quantitative and qualitative analysis. *Journal of Youth and Adolescence, 23*(3), 331–357.

Webb, D. A., Sweet, D., & Pretty, I. A. (2002). The emotional and psychological impact of mass casualty incidents on forensic odontologists. *Journal of Forensic Sciences, 47*(3), 539–541.

Webster, D. (2009). *Creative reflective experience: Promoting empathy in psychiatric nursing*. Retrieved from ProQuest Dissertations & Theses (AAT 3312911).

Weine, S., Knafl, K., Feetham, S., Kulauzovic, Y., Klebic, A., Sclove, S., et al. (2005). A mixed methods study of refugee families engaging in multiple-family groups. *Family Relations, 54,* 558–568.

Wittink, M. N., Barg, F. K., & Gallo, J. J. (2006). Unwritten rules of talking to doctors about depression: Integrating qualitative and quantitative methods. *Annals of Family Medicine, 4*(4), 302–309.

Woolley, C. M. (2009). Meeting the mixed methods challenge of integration in a sociological study of structure and agency. *Journal of Mixed Methods Research, 3*(1), 7–25.

AUTHOR INDEX

SUBJECT INDEX